Lecture Notes in Control and Information Sciences

Edited by A. V. Balakrishnan and M. Thoma

For information about Vols. 1–21 please contact your bookseller or Springer-Verlag.

Lecture Notes in
Control and
Information Sciences

Edited by M. Thoma and A. Wyner

78

Stochastic Differential Systems

Proceedings of the 3rd Bad Honnef Conference June 3–7, 1985

Edited by
N. Christopeit, K. Helmes, M. Kohlmann

Springer-Verlag Berlin Heidelberg GmbH

Editors

Norbert Christopeit
Institut für Ökonometrie und Operations
Research der Universität Bonn
Ökonometrische Abteilung
Adenauerallee 24–42
D-5300 Bonn 1

Kurt Helmes
Institut für Angewandte Mathematik
Universität Bonn
Wegelerstraße 6
D-5300 Bonn 1

Michael Kohlmann
Fakultät für Wirtschaftswissenschaften
und Statistik
Universität Konstanz
Postfach 5560
D-7750 Konstanz 1

Library of Congress Cataloging in Publication Data

Main entry under title:
Stochastic differential systems.
(Lecture notes in control and information sciences; 78)
1. Stochastic systems -- Congresses.
2. Differentiable dynamical systems -- Congresses.
I. Christopeit, N. II. Helmes, K. (Kurt) III. Kohlmann, M. (Michael) IV. Series.
QA402.S846 1986 003 85-32548

ISBN 978-3-540-16228-5 ISBN 978-3-540-39767-0 (eBook)
DOI 10.1007/978-3-540-39767-0

Originally published by Springer-Verlag Berlin Heidelberg New York in 1986

PREFACE

This volume contains the scripts of most of the lectures given at the 3rd Bad Honnef Conference on "Stochastic Differential Systems" held at Bad Honnef, West Germany, from June 3 - 7, 1985. In addition, we have included the contributions of some invited speakers who were unable to present their talk during the week of the conference.

The workshop was devoted to the topics Optimal Control, Filtering and Stochastic Analysis, with emphasis on Applications and on connections between Probability Theory and certain aspects of Quantum Physics. Both the survey lectures and those lectures where most recent research results were presented aimed to give an up-to-date account of these fields.

We wish to thank the scientific organisations in the home and/or guest countries for the financial support they provided conference participants. In particular, we are indebted to the Deutsche Forschungsgemeinschaft (German Science Foundation), which - via the Sonderforschungsbereich 72, a research program established at the Institute of Applied Mathematics of the University of Bonn - made this meeting possible. Our special thanks go to the members of the International Program Committee,

A.V. Balakrishnan
A. Bensoussan
M.H.A. Davis
A. Friedman
B. Grigelionis
F. Guerra
G. Kallianpur
H. Kunita
W. Runggaldier
J. Zabczyk

whose assistance helped us to bring such a fine group of researchers together. We appreciate also the splendid translating work done during the conference by Professors I. Gyöngy and J. Zabczyk. Last, but not least least, we would like to thank Ms. Barbara Chapman for such an excellent job in helping to organise the conference.

Bonn, June 1985

Norbert Christopeit
Kurt Helmes
Michael Kohlmann

CONTENTS:

Some points of interaction between stochastic analysis and quantum theory

by

Sergio Albeverio

Mathematisches Institut

Ruhr-Universität

4630 Bochum 1

and

Bielefeld-Bochum Stochastics

Research Centre

In this talk I will try to give a survey of some aspects of interrelations between stochastic analysis and quantum theory, emphasizing particularly some recent developments. It will be, by necessity, a partial account mainly influenced by developments in which I participated and many important topics will be only briefly mentioned, fortunately some are being discussed in other talks, to which I hope mine can be in some sense complementary. I shall also try to supply omitted topics by some appropriate references.

First of all I want to mention that there are two main ways probability, and more particularly stochastic analysis, enters quantum mechanics. The first is a basic, conceptual one, which goes back to the very origins of the probabilistic interpretation of the wave function (Schrödinger, Born,...), and has a modern extremely interesting version in Nelson's stochastic mechanics and the work around it, a very illuminating report on this has been given by Francesco Guerra at this Meeting, so I will only touch briefly these aspects (let me mention however an important paper by Nelson [1] which has appeared after this conference and will be published in the Proceedings of the 1. International Ascona - Como Meeting; incidentally these Proceedings [2] will contain also several other contributions of great relevance to this point).

Besides this "conceptual" relation between probability and quantum mechanics, there is a more "technical one", in which probability theory, in particular stochastic analysis, is used as a mathematical tool, to obtain results about quantum mechanical objects. Viceversa, concepts and methods forged in connection with quantum mechanics influence new developments in stochastic analysis. In this talk I will mainly concentrate on these aspects. I like however to emphasize that in many instances the distinction between "conceptual" and "technical" is in this context somewhat artificial. E.g. the approach to quantum mechanics by Dirichlet forms, to be discussed below, is in many ways (connection with stochastic mechanics, alternative formulation of quantum field theory) more than a technical tool.

Content

I. Dirichlet quantum mechanics

This is an approach to the formulation of a Hamiltonian, and hence a dynamics, for quantum mechanics which has its roots in the Schrödinger formulation of quantum theory and in the canonical approach to quantum mechanics and quantum field theory, and has received a strong interest in recent years in connection with questions of quantum field theory (e.g. [3]), nuclear physics [4] , [5] and solid state physics [5], on one hand, and specific questions of the theory of symmetric Markoff processes on

the other hand [6], [7], see also e.g. [8]. Let us schematically illustrate the relations between "Dirichlet quantum mechanics" and "Schrödinger quantum mechanics". Whereas the basic Hilbert space in quantum mechanics is $L^2(\mathbb{R}^s,dx)$ (with s the dimension of space, and dx Lebesgue measure), it is $L^2(\mathbb{R}^2,\nu)$ in "Dirichlet quantum mechanics", where $d\nu \equiv \varphi^2 dx$, $\varphi \in L^2_{loc}$, $\varphi \neq 0$ Lebesgue a.e.
Multiplication by $1/\varphi$ gives a unitary mapping from $L^2(\mathbb{R}^2,dx)$ onto $L^2(\mathbb{R}^2,\nu)$: for $f \in L^2(\mathbb{R}^s,dx)$ we have in fact $U_\varphi f \equiv \varphi^{-1} f \in L^2(\mathbb{R}^s,\nu)$, with $\| U_\varphi f \|_{L^2(\nu)} = \|f\|_{L^2(dx)}$.

The dynamics in Schrödinger quantum mechanics is given by a self-adjoint lower bounded operator H in $L^2(\mathbb{R}^2,dx)$.
To H there corresponds in $L^2(\mathbb{R}^s,\nu)$ the unitarily equivalent operator $H_\nu \equiv U_\varphi H U_\varphi^{-1}$. If $\varphi \in L^2(dx)$ we have that 1 is in the domain $D(H_\nu)$ of H_ν and in fact $H_\nu 1 = 0$ iff $H\varphi = 0$. Situations in which $\varphi \in L^2(dx)$ and $H\varphi = 0$ occur easily, e.g. H of the form $-c\Delta + V +$ constant, with $c > 0$, V multiplication by a real smooth function, with "not too negative behaviour" at infinity. In fact for any function of this type $-c\Delta + V$ has lower bounded spectrum [9], [10], hence by adding a constant we get the infimum of the spectrum to be 0; under general conditions then "by ellipticity" 0 is an isolated eigenvalue, with positive eigenfunction φ, the so called "ground state", [9] (Th. C. 81.), [11]. The case $c \neq 1/2$ can obviously be reduced to the case $c = 1/2$ by a suitable change of coordinates, so that it is enough to consider the case $-\frac{1}{2}\Delta + V$. In this case then H is non negative, hence H_ν is also non negative. Moreover on smooth functions of compact support, it has the form $-\frac{1}{2}\Delta - \beta_\nu \cdot \nabla$, with β_ν the gradient vector field $\beta_\nu \equiv \nabla \ln \varphi = -\frac{1}{2} \nabla^* 1$, where 1 is the function identically 1 in $L^2(\mathbb{R}^s,\nu) \otimes \mathbb{R}^s$ and $*$ is the adjoint from $L^2(\mathbb{R}^s,\nu) \otimes \mathbb{R}^s$ into $L^2(\mathbb{R}_s,\nu)$; the second equality is easily proven by using $\nabla^* = (\partial_i^*, i=1,\dots,s)$, $\partial_i^* = -\partial_i - 2(\beta_\nu)_i$, where i denotes the i-th component and $\partial_i \equiv \frac{\partial}{\partial x_i}$, $*$ denoting here the adjoint in $L^2(\mathbb{R}^s,\nu)$. We also remark that, H_ν being non negative, $H_\nu^{1/2}$ is a well defined non negative self-adjoint operator, hence the quadratic form $\mathcal{E}_\nu(f,g) \equiv (H_\nu^{1/2} f, H_\nu^{1/2} g)_\nu$, is well defined for all $f,g \in D(H_\nu^{1/2})$, with $(,)_\nu$ the scalar product in $L^2(\mathbb{R}^s,\nu)$.
An easy integration by parts yields, at least for $f,g \in C_o^2(\mathbb{R}^s)$,

$$\mathcal{E}_\nu(f,g) = \frac{1}{2} \int_{\mathbb{R}^s} \nabla f \cdot \nabla g d\nu. \tag{I.1}$$

The relation between V,β_ν,φ is expressed (with the shift by a constant alluded above, so that $H\varphi = (-\frac{1}{2}\Delta + V)\varphi = 0$) by:

$$V = \frac{1}{2}\,\Delta\varphi/\varphi = \frac{1}{2}\,[\beta_\nu^2 + \mathrm{div}\,\beta_\nu]. \tag{I.2}$$

In this way we see how, starting from the Schrödinger quantities $L^2(\mathbb{R}^s,dx)$, φ,H (s.t. $H = -\frac{1}{2}\Delta + V$, say on $C_0^\infty(\mathbb{R}^s)$, $V = \frac{1}{2}\,\Delta\varphi/\varphi$) we can get at the corresponding Dirichlet quantities $L^2(\mathbb{R}^s,\nu)$, $d\nu = \varphi^2 dx, H_\nu$ (s.t. $H_\nu = U_\varphi H U_\varphi^{-1}$ and $H_\nu = -\frac{1}{2}\Delta - \beta_\nu \cdot \nabla$, say on $C_0^\infty(\mathbb{R}^s)$, with $\beta_\nu = \nabla \ln\varphi$) and we can use H_ν to construct the energy form $\mathcal{E}_\nu(f,g) \equiv (H_\nu^{1/2}f,\ H_\nu^{1/2}\ g)_\nu = \frac{1}{2}\int \nabla f \cdot \nabla g d\nu$ associated with φ, hence with ν.

$\mathcal{E}_\nu$ is what is known as an "energy form" (e.g. [6], [7], [12] - [14]) this is a (special) local regular Dirichlet form in the sense of the general theory of Dirichlet forms and symmetric Markov processes, developed by Beurling, Deny, Fukushima and Silverstein, see [15]. $\mathcal{E}_\nu$ is in 1-1 correspondence with H_ν.

Thus the Schrödinger quantum mechanics gives us a basic quantity, the "energy form", viceversa, given an energy form $\frac{1}{2}\int \nabla f \cdot \nabla g d\nu$ defined for a positive Radon measure $d\nu = \varphi^2 dx$, $\varphi \neq 0$ a.e., $\varphi \in L^2_{loc}(\mathbb{R}^s)$ as a closed quadratic (positive) form, one can associate uniquely a self-adjoint positive operator in $L^2(\mathbb{R}^s,\nu)$ by $\mathcal{E}_\nu(f,g) = (H_\nu^{1/2}f,\ H_\nu^{1/2}g)$ for all $f,g \in D(H_\nu^{1/2})$. H_ν can be used then to define by

$$H = U_\varphi^{-1}\ H_\nu\ U_\varphi \tag{I.3}$$

a "generalized Schrödinger Hamiltonian". H is formally of the form $-\frac{1}{2}\,\Delta + V$, with $V = \frac{1}{2}\,\Delta\varphi/\varphi$, however both Δ and V might be meaningless as operators or even as forms on the domain of H.

As an example, let us take $s = 3$, $\varphi(x) \equiv e^{\alpha|x|}/(2\pi|x|)$, with $\alpha \in \mathbb{R}$. Then H_ν as well as H are well defined, V is however not well defined (in some sense, which can be made precise by non standard analysis, V is actually of the form $-\frac{\pi^2}{8\varepsilon^2}\,\chi_1(x|\varepsilon)(1-\frac{8\alpha}{\pi^2}\varepsilon) + \frac{\alpha^2}{2}$, with ε infinitesimal, χ_1 the characteristic function of the unit ball with center at the origin; thus it is correct to think of V as a "point interaction" at the origin [16]).

When does the method of defining the "generalized Schrödinger operator" H by Dirichlet forms like $\mathcal{E}_\nu$ work? As an example we mention conditions a), b), which are both separately sufficient:

a) $\varphi^2 > 0$ a.e., $\varphi \in L^2_{loc}$ with $|S| = 0$, where $|\ \ |$ means Lebesgue measure and

$S \equiv \{x \in \mathbb{R}^s | \int_{U_\varepsilon(x)} \varphi^{-2}(y)dy = \infty \}$, where U (x) is any neighbourhood of x, see [17].

b) $\varphi \neq 0$ a.e., $\varphi \in H^{1,2}_{loc}(\mathbb{R}^s)$.

For s=1 there is a necessary and sufficient condition related to a), see [12], [17], [18].

By Fukushima's method $\mathcal{E}_\nu$ is in a 1-1 way associated to a ν-symmetric diffusion process X_t, stationary with invariant measure ν and generator H_ν. This process is called [19]a "(ν-) distorted Brownian motion". It has the general form $X_t = X_o + N_t + b_t$, with b_t a Brownian motion in $\mathbb{R}^s$ and N_t a continuous additive functional, locally of zero energy. It solves, up to a killing time τ, the stochastic differential equation $dX_t = \beta_\nu(X_t)\, dt + db_t$, so that $N_t = \int_o^t \beta_\nu(X_s)ds$, roughly speaking whenever $\beta_\nu(X_t)$, is well defined, see [12], [20], [21], [24].

Sufficient conditions for the existence of the Girsanov functional

$\exp\,[\int_o^t \beta_\nu(x + b(s))\, db(s) - \frac{1}{2} \int_o^t \beta_\nu^2(x + b(s))ds]$

are also known, see [6], [7], [20], [21].

If V(x + b(s)) itself has a meaning (which is the case when φ has suitable second derivatives not only first ones, as required for the existence of β_ν), then also the "Schrödinger picture" Feynman-Kac functional $\exp\,(-\int_o^t V(x + b(s))ds)$, appropriate for discussing an exponentially killed (if $V \geq 0$) Wiener process, exists, see e.g. [9] (Th.A. 2.7), [22], [127], [129], [131]-[134];

Let us now consider an important example, of use also later, in the case of quantum field theory. Let us start from the classical Newton equation mass $\times$ acceleration = force for a 1-degree of freedom u(t) ("string") moving harmonically in $\mathbb{R}^s$:

$$\ddot{u}(t) = -A^2\, u(t),\ u(t) \in \mathbb{R}^s,\ t \in [0,\infty), \tag{I.4}$$

with initial condition $u(0) = x \in \mathbb{R}^s$, A a positive $s \times s$ symmetric matrix. The Schrödinger quantities φ, V, which we now denote by φ_o, V_o, are $\varphi_o(x) = \left(\frac{\det A}{\pi^s}\right)^{1/4}$ $\exp\left[-\frac{1}{2}(Ax,x)\right]$, $V_o(x) = \frac{1}{2}(A^2x,x)_{\mathbb{R}^s}$, the Dirichlet quantities are $d\nu_o(x) = \varphi_o^2(x)dx = N(0;(2A)^{-1})$, (the Gaussian on $\mathbb{R}^s$ with mean zero and covariance matrix $(2A)^{-1}$), $\beta_{\nu_o}(x) = -Ax$. X_t is in this case the Ornstein-Uhlenbeck process in $\mathbb{R}^s$ with covariance $E(X_i(t)X_j(s)) = \left(-\frac{d^2}{dt^2} + A^2\right)^{-1}_{i,j}(t,s)$, $t,s \in \mathbb{R}$, $i,j=1,\dots,s$. E is expectation with respect to the path-space measure μ_o, s.t. $\nu_o = \mu_o \upharpoonright \sigma(X(0))$, where $\sigma(Y)$ means the σ-algebra generated by Y. In physics all these quantities, related to Gaussian measures, are said to be "free" or "non interacting"; interaction arises when V_o is replaced by a non quadratic V and, equivalently, ν_o by a non Gaussian ν, β_{ν_o} by a non linear β_ν. Let us now mention in a very schematic way some recent work concerning the above topics:

a) For a stochastic calculus for processes X_t associated to (more general) Dirichlet forms see [6], [7], [12], [15], [20], [23]-[26], [142].

b) For criteria for explosions, transience, recurrence of the processes X_t see e.g. the work around "Ichihara's test": inf $\varphi^2(x) > 0$ on compacts, $\varphi^2(x) \leq c_1 e^{c_2|x|} \Rightarrow$ conservativeness of X_t (this criterium does not involve any smoothness of φ!): see [6], [7], [26].

c) Criteria for unattainability of zeros of φ and ergodic properties of X_t (whose correlate in quantum mechanics are tunneling/non-tunneling effects) have been discussed in the recent literature: typically if $Z \equiv \{x \in \mathbb{R}^s | \varphi(x) = 0\}$ then X_t does not reach Z for quasi-every $x \in \mathbb{R}^s$, and $\mathbb{R}^s - Z = \bigcup_i^\infty C_i$, with C_i connected, invariant, irreducible s.t. $X_t \upharpoonright C_i$ has stationary invariant measure $\nu \upharpoonright C_i$. The situation is related to capacity zero but is more involved, symmetry has to be exploited too (typically, for $s = 1$: let $\varphi(x)^2 \geq 2C|x|^{2\gamma}$, $0 < \gamma < 1/2$ to the right of zero, $\varphi(x)^2 \leq 2C|x|$ to the left of zero: then the $\mathcal{E}_\nu$-capacity of 0 is strictly positive, yet X_t does not go through 0 from right to left). For such results see [6], [7], [27], [28]. Applications have been given to biological systems [29] as well as several physical dynamical systems, see e.g. [30], [31]. Recently, a theory of time dependent Dirichlet forms is being developed and unattainability criteria, of relevance in the non stationary case of stochastic mechanics, have also been obtained there [32] - [35].

d) Uniqueness problems have been discussed: what is given primarily is actually not a closed form but rather a densely defined quadratic form, say $\frac{1}{2}\int \nabla f \cdot \nabla g d\nu$, for f,g in a suitable dense domain, say $C_0^\infty(\mathbb{R}^s)$. Does the restriction $\mathcal{E}_\nu \upharpoonright C_0^\infty(\mathbb{R}^s)$ of $\mathcal{E}_\nu$ determine uniquely $\mathcal{E}_\nu$? Do there exist more than one closed Dirichlet forms extending $\mathcal{E}_\nu \upharpoonright C_0^\infty(\mathbb{R}^s)$? On the operator side: is H_ν already uniquely determined by its restriction to $C_0^\infty(\mathbb{R}^s)$? Do there exist other self-adjoint extensions of $H_\nu \upharpoonright C_0^\infty(\mathbb{R}^s)$? If the answer is yes to the latter question then H_ν is essentially self-adjoint on $C_0^\infty(\mathbb{R}^s)$ and the closure of $H_\nu \upharpoonright C_0^\infty(\mathbb{R}^s)$ is unique. In this case there is only one self-adjoint bounded semigroup with generator extending $H_\nu \upharpoonright C_0^\infty(\mathbb{R}^s)$. We speak in this case of "strong uniqueness". We speak of "Markoff uniqueness" if there is only one self-adjoint Markoff semigroup with generator extending $H_\nu \upharpoonright C_0^\infty(\mathbb{R}^s)$. As shown by Fukushima [20] and Wielens [36] , see also [17], Markoff uniqueness is in general weaker than strong uniqueness (this is evident in the case where $\mathbb{R}^s$ is replaced by a bounded domain, in which case the semigroup corresponding to absorbing resp. reflecting boundary conditions with generators having domains the Sobolev spaces $H_0^{1,2}$ resp. $H^{1,2}$ are of course different and all different from the "maximal" Krein semigroup, see [15]). The following recent result is due to N. Wielens: $\varphi^2 > 0, \varphi \in Lip_{loc} \Rightarrow$ strong uniqueness [36]. An extension to manifolds is mentioned by Fukushima in [6], [7]. The uniqueness questions have great relevance also for physics, since H_ν determines the dynamics, and different realizations of it lead to different dynamics!

e) Let $\mathcal{E}_{\nu_n}$ be a sequence of energy forms associated to measures $d\nu_n = \varphi_n^2 dx$. Suppose in a suitable sense ν_n converge to a measure ν. $\mathcal{E}_{\nu_n}$ determines H_{ν_n} which in turn determines $H_n = U_{\varphi_n}^{-1} H_{\nu_n} U_{\varphi_n}$. All H_n are defined in the fixed Hilbert space $L^2(\mathbb{R}^s, dx)$. When do the corresponding semigroups e^{-tH_n} converge? This question, related e.g. to different uniqueness questions in Stroock-Varadhan's martingale problem approach, has been discussed recently in [13] , [37], [38].

f) A Donsker-Varadhan's type of asymptotics for "Dirichlet" processes X_t has been developed recently in [26] , [39]. It would be nice to find applications e.g. in problems involving "polymer measures".

g) Applications of the relation between Dirichlet and Schrödinger quantum mechanics can be given in non linear filter theory. This has been greatly stimulated by S. Mitter [40] , [41] and by work by Beneš [42] , Baras [44], Davis [43], Hazewinkel [47], Marcus [44], Ocone [44] and others.

It seems that Dirichlet theory should come to play an even greater role, allowing for more singular drift coefficients. In another direction, but still in accord with interplays between stochastic analysis and quantum mechanics, let us mention that the Duncan-Mortensen-Zakai equation for the non normalized conditional density can be solved for suitable drift coefficients using an explicit computation of Feynman-Kac functionals for potentials V which are Fourier transforms of bounded complex measures, a trick already exploited for Feynman-path integrals by K. Ito and S. Albeverio and R. Høegh-Krohn [45] . This application to filter theory is contained in recent work by T. Arede [46].

h) For some further work using Dirichlet forms in quantum mechanics see e.g. [119], [121], [135], [148].

II. Dirichlet processes in infinitely many dimensions - Generalized Markoff fields-Quantum fields

Let us return to our example (I.4). We shall replace formally $\mathbb{R}^s$ by $L^2(\mathbb{R}^s)$, so that u(t) is for any $t \in \mathbb{R}$ an element u(t,x) of $L^2(\mathbb{R}^s)$ (x runs over $\mathbb{R}^s$: u(t,x) can be thought as an "excitation field" at time t and place x). A should then be a symmetric operator in $L^2(\mathbb{R}^s)$ and we take A^2 to be the "Klein-Gordon operator" $-\Delta + m^2$ for some constant $m>0$, with Δ the s-dimensional Laplacian. In this case ν_o is the Gaussian measure with mean zero and covariance $(2A)^{-1}$ (which can easily, by Minlos theorem, be realized with support e.g. on $\mathcal{S}'(\mathbb{R}^s)$). The associated infinite-dimensional (e.g. $\mathcal{S}'(\mathbb{R}^s)$-valued) Ornstein-Uhlenbeck process $X_t(x)$ has mean zero and covariance

$$E(X_s(x)X_t(y)) = (-\frac{d^2}{dt^2} + A^2)^{-1}_{x,y}(s,t) = (-\Delta_d + m^2)^{-1}(t,y,s,x),$$

with Δ_d the $d \equiv s+1$-dimensional Laplacian. In the same way as the process X_t associated to (I.4) was stationary and time-reversal symmetric, i.e. homogeneous with respect to the 1-dimensional Euclidean group, the present process $X_t(x)$ is homogeneous with

respect to the d-dimensional Euclidean group (as seen by its covariance). Its path space measure μ_o, with respect to which the above expectation is taken, restricted to the σ-algebra Σ_o associated with X_o (in a suitable sense) can be identified with ν_o. (X_t,μ_o) is the so called "free Markov field" studied by Molchan, Pitt, Wong and especially Nelson [47]. Its "global Markov property" (Markov property with respect to arbitrary sets) has been exploited recently, in connection with interactions and a Dirichlet problem with distributional data, by Albeverio and Høegh-Krohn [48], Dynkin [49] , Röckner [50] , see also Dobrushin-Minlos [51] , Gielerak [58] , Kolsrud [120] , Rozanov [53] , Zegarlinski [54] . The Ornstein-Uhlenbeck generator of $X_t(x)$ is the operator H_{ν_o} s.t. $(H_{\nu_o}^{1/2}f, H_{\nu}^{1/2}g)_{\nu_o} = \frac{1}{2} \int_{\mathcal{S}'(\mathbb{R}^s)} \nabla f \cdot \nabla g d\nu_o$, the scalar product being now in $L^2(\mathcal{S}'(\mathbb{R}^s), \nu_o)$. One has here, as remarked in [24] , strong uniqueness. The sample path properties of $X_t(x) \equiv X(y)$, $y \equiv (t,x) \in \mathbb{R}^d$ (support properties of μ_o) have also seen studied, see e.g. the references in [3]: what is essential here is that they are, as easily seen from the form of the covariance, sufficiently bad not to allow formation without "renormalization" of multiplicative functionals of the type $d\mu^v(X) = \exp(-\int_{\mathbb{R}^d} v(X(y))dy)d\mu_o(X)$, which would yield in principle "interacting", "non Gaussian" path space measures $d\mu\,(X)=d\mu^v(X)/\int d\mu^v(X)$ (the case $v(\alpha) = \alpha^4$ is what is usually called the "φ_d^4-model", the case $v(\alpha)$ = polynomial in α is the so called $P(\varphi)_d$-model [55], [56]). Formally $d\mu$ is the path space measure of an "interacting" global Markoff field X (in an intuitive sense: Markoff property with respect to separating d-1-dimensional hyperplanes, e.g.), which is homogeneous with respect to the Euclidean group in $\mathbb{R}^d$. Such Markoff Euclidean fields are interesting in physics, inasmuch as they yield, by an analytic continuation procedure $X_t(x) \to X_{it}(x)$ "relativistic local fields" (i.e. the fields sought for in quantum field theory).Formally, the generator H_ν to a field X with formal path space measure $d\mu(X)$, has the form $(H_\nu f,g)_\nu = \frac{1}{2} \int_{\mathcal{S}'(\mathbb{R}^s)} \nabla f \cdot \nabla g d\nu$, with $\nu \equiv \mu \restriction \Sigma_o$. Note that this picture can be realized by non standard analysis [16](see also e.g. [57]), however it is an open problem whether the non standard analysis result can be exploited to yield results interpretable in the "standard world" for d = 4.
For d = 2 one has standard as well as non standard constructions of μ, ν , see e.g. [3], [16], [24], [58]. Let us note that a general theory of energy forms in

infinite dimensions exists [3], [58], [59], in much the same spirit as the above mentioned finite dimensional theory. One starts with a rather arbitrary Radon probability measure ν on $\mathcal{S}'(\mathbb{R}^s)$ (say) satisfying some weak positivity or regularity condition (e.g. $\beta_\nu \equiv -\frac{1}{2} \nabla^* 1 \in L^2(\nu)$, which corresponds in the finite dimensional case to $\nabla\varphi \in L^2(dx)$). Then one shows that the quadratic form $\frac{1}{2} \int \nabla f \cdot \nabla g d\nu$, well defined on smooth cylinder functions f,g, has a closure, which then is an energy form (local Dirichlet form, in an extended sense). Several potential theoretic, probabilistic and ergodic properties of such forms have been analyzed by Høegh-Krohn and myself, Paclet and Kusuoka. In particular they lead, under smoothness and positivity assumptions on ν, to nice associated diffusion processes. It is quite remarkable that all assumptions are satisfied for the known models of measures ν associated with d = 2-quantum fields, see [3], [58]. For a subset of models also the global Markoff property of the associated Markov fields (as well as the uniqueness property of the associated Gibbs fields, a strong version of "absence of phase transitions") has been shown [48], [52], [54]. However, even in this case, the question of "strong uniqueness" resp."Markoff uniqueness" of the semigroups associated with an $\mathcal{E}_\nu$ restricted to smooth cylinder functions (the analoque of the questions discussed under I d) in the finite dimensional case) are not completely settled. Høegh-Krohn and myself proved [24], [58] the equality of all generators on smooth cylinder functions. Takeda [130] has recently proven Markoff uniqueness for the case where ν is replaced by a measure absolutely continuous with respect to an abstract Wiener measure with regular positive density. If the density is tame by an extension of Wielens method he obtains strong uniqueness. Kusuoka has defined the analogue of the Sobolev space $H^{1,2}$ for the infinite dimensional Dirichlet forms under consideration. He has recently proven that Markov uniqueness is equivalent with the Dirichlet form corresponding to $H^{1,2}$ being equal to the Dirichlet one discussed previously in [24], [58]. This has been verified [60] in models with exponential interactions [61], see also [52], [54].

Remark: Construction of the process has also been achieved in some models of gauge fields with regularized interactions, see [62], [63].

Remark: Let us finally mention that another approach to the construction of the above path space measure have been pursued recently. It is the analogue of the construction of the Ising Gibbs fields as equilibrium measures of time dependent interacting models, a well known approach, mentioned also in the talk by Hans Föllmer, see e.g. the references in [16], [128]. For mathematical work along these lines, in the case of quantum fields see e.g. [65] - [67]. Mainly this work involves a "Girsanov-type" approach: it would be nice to develop a Dirichlet approach to it, along above lines.

III. Stochastic Schrödinger operators

This is a topic in which in the last few years there has been an intensive interaction between stochastic analysis and quantum theory. Fortunately there are already quite a few excellent surveys, see e.g. [68], [69], so that I can find some justification in being rather short here. My main purpose is to mention, for readers coming from other domains, at least some of the beautiful mathematical work being developed in this area. The typical object of study is a stochastic Schrödinger operator of the form $H_\omega = -\frac{1}{2}\Delta + V(\omega,x)$ in $L^2(\mathbb{R}^s,dx)$, where the potential V is a random field, i.e. depends on $x \in \mathbb{R}^s$ and on the sample point ω in some probability space $(\Omega,\mathcal{A},P)$. Such a "stochastic Hamiltonian" gives the dynamics for a quantum mechanical particle moving in a disordered medium. A class of important examples is $V(\omega,x) = \sum_i \lambda_i(\omega)\, f(x - \xi_i(\omega))$, with λ_i "random charges" and $\xi_i(\omega)$ "random sources", f a fixed (real-valued) function. One is particularly interested in random potentials V which are random fields ergodic on $\mathbb{R}^s$ (or sometimes only on some lattice like $\mathbb{Z}^s$), in this case namely important quantities attached to H_ω, also of physical relevance, are almost sure independent of ω i.e. one is able to make almost sure statements. In particular for such potentials one has the following type of results:

a) the spectrum $\sigma(H_\omega)$ of H_ω and its relevant parts (essential, absolutely continuous, discrete etc. spectrum) are all almost surely independent of ω i.e. non random, and the discrete spectrum of H_ω is void. For these results see e.g. [68] - [71], [125], [143].

b) The integrated density of states associated with H_ω exists. It is defined by introducing for any cube $\Lambda \subset \mathbb{R}^s$ the Hamiltonian $H_\Lambda(\omega)$ defined as $-\frac{1}{2}\Delta + V(\omega,x)$, with Δ having Dirichlet (absorbing) boundary conditions on $\partial\Lambda$ and looking at $\lim_{\Lambda\uparrow\mathbb{R}^s} |\Lambda|^{-1} \#\{k \in \mathbb{N} \mid \lambda_k(H_\Lambda(\omega)) \le E\}$, with $|\Lambda|$ the volume of Λ, $\#(A)$ denoting the number of elements in A, λ_k being the k-th eigenvalue of $H_\Lambda(\omega)$. Kirsch and Martinelli [72] have shown that for a large class of V the above limit exists for any $E \in \mathbb{R}$ and is independent of the fact that we have chosen Dirichlet boundary conditions for Δ (i.e. it is the same when one choses e.g. Neumann boundary conditions for Δ). Call this limit N(E): it is by definition the integrated density of states. The asymptotics of N(E) for E tending to the extrema of the spectrum have been determined and the so called "Lifshitz exponents", described heuristically by Lifshitz, have been recovered in important work by Fukushima, Kirsch, Martinelli, Nakao, Simon and others, see e.g. [72], [144]. It would be nice to extend such results on one hand to the case where $\mathbb{R}^s$ is replaced by some fractal set and, on the other hand, to the case of manifolds.

c) Important results on the localization problem ("Andersons's transition") have been obtained. The most complete results are for the case s = 1, let us quote e.g. Kotani' result that the spectrum of $H(\omega)$ in an interval (a,b) is absolutely continuous iff the Ljapunov index γ attached to $H(\omega)$ is 0 on (a,b). For "most cases" H has pure point spectrum for s = 1, for these results see e.g. [73],[74]. Recently results in higher dimensions (s > 1) showing the presence of pure point spectrum ("localization") for large disorder or low energy have been obtained [75], see also [76] - [79].

d) Let me close by mentioning a possible connection with part I in an approach to stochastic Schrödinger operators vis "stochastic Dirichlet forms". In work of Fukushima [7] and Takeda [126] stochastic Dirichlet forms of the type $\frac{1}{2}\int_{\mathbb{R}^s} \nabla f \cdot \nabla g \; \nu_\omega(dx)$ appear, with some special stochastic measure ν_ω. E.g. for s = 1 the case where ν_ω is of the form $\nu_\omega(dx) = \exp(-b_x(\omega))dx$,with b_x Brownian motion on $\mathbb{R}$, has been studied and yields Brox's diffusion in a Wiener medium. Similar cases for $s > 1$, with b_x replaced by Levy Brownian motion have been studied [7], [126], as far as recurrence and transience are concerned, but e.g. spectral questions remain entirely open. I think this is an area where further work is called for.

IV. Local times - Intersection functionals of Brownian motion - Polymers and quantum fields

This is a further, large area of research in which there have been a very fruitful interaction between stochastic analysis and quantum theory. My aim here is only to mention a few problems and give a couple of references, it would take us too far to go into details here. It is a fact that a probabilistic modelling of long polymers chains, see e.g. [16], [81], has yielded important connections between the study of equilibrium properties of polymer chains (a subject from chemistry, biology and physics) and both relativistic quantum field theory and non relativistic quantum mechanics ("scattering by polymers"). Whereas for the physics of the polymer themselves the interesting dimensions are $s \leqq 3$ (with $s = 3$ the most interesting case), for the applications to quantum field theory the interesting dimensions are $s \leqq 4$ (with $s = 4$ the most interesting one). Typical objects of study are the Edwards "polymer measures"

$$d\mu_1(b) \equiv Z^{-1} \exp\left[-\lambda \int_0^t \int_0^t \delta(b_\sigma - b_{\sigma'})\, d\sigma\, d\sigma'\right] dP(b),$$

with $P(b)$ the Wiener measure for a Brownian motion b in $\mathbb{R}^s$, and

$$d\mu_2(b) \equiv \tilde{Z}^{-1} \exp\left[-\lambda \int_0^t \int_0^t \delta(b_\sigma - \tilde{b}_{\sigma'})\, d\sigma\, d\sigma'\right] dP(b)\, d\tilde{P}(\tilde{b}),$$

with $b, \tilde{b}$ independent Brownian motions in $\mathbb{R}^s$. Here $Z, \tilde{Z}$ are formal normalizations making the measures μ_1, μ_2 into probability measures, λ is a real parameter. Two main questions arise:

a) Existence of above measures: this was solved for $\lambda \geqq 0$, $s=2$ by Varadhan [80], for $s = 3$ by Westwater [81], and for $s = 4$ and λ negative infinitesimal by Albeverio, Fenstad, Høegh-Krohn, Lindstrøm in [16], see also [82], [122], as far as μ_2 is concerned. Let us also mention work by Dynkin [83], J. Rosen [84], M. Yor [85] particularly relevant in the above cases for $s \leqq 3$, and by A. Stoll [86], who provides an alternative non standard construction in these cases, with strong "invariance principles" type results.

The study of the μ_2 measures goes through the study of stochastic Schrödinger operators of the form $-\frac{1}{2}\Delta + \lambda N_t(x,\tilde{b})$, with $N_t(x,\tilde{b}) = \int_0^t \delta(x-\tilde{b}(\sigma))d\sigma$ local time at x of Brownian motion, an object worthwhile studying in itself (e.g., up to now spectral questions are open), (mind also that, formally, $\int N_t(x,b)^2 dx = \int_0^t \int_0^t \delta(b(\sigma)-b(\sigma'))d\sigma d\sigma'$, of relevance in the μ_1-measure).

b) Asymptotics of above measures: the heuristic "Flory formulae" $E(b(t)^2)_{t\to\infty} \sim t^{6/(2+s)}$, for $s \leqq 4$, t for $s \geqq 5$, obtained essentially by scaling arguments, (with some uncertainties, "logarithmic terms", for s = 4) are far from being mathematically justified. Essentially only the case s=1 is under control, by work of Kusuoka[87]and Westwater [81], [88], involving an extension of Donsker-Varadhan's asymptotics. Local behaviour results on the "Dirichlet process" associated with μ_1 for s=3 are obtained by Kusuoka in [145].

c) There is an important connection of the path space measure μ of φ_d^4 (and more generally $F(\varphi^2)_d$-models), described in Sect. II, with polymer measures μ_i of above type. This has been discovered by Symanzik in [80] and exploited recently, see [16] and references therein. In particular a non standard analysis representation of μ in terms of measures μ_i, partially under control also for s = 4, has been achieved [16] , [82] , [122], but much remains to be done.

V. Stochastic connections, Markov cosurfaces, gauge fields

It is well known that the Brownian motion b(t), $t \in \mathbb{R}$ on a Lie group G solves, in the sense of multiplicative stochastic integrals [89] , [90] an equation of the type $b(t)^{-1}db(t) = \xi(t)$, with $\xi(t)$ white noise in the Lie algebra g of G. Can one extend this theory to the case where $t \in \mathbb{R}$ is replaced by $x \in \mathbb{R}^s$ or e.g. x ∈(s-dimensional Riemannian manifold M) and $\xi(x)$ is the curvature 2-form F(x) of some connection 1-form a (x) of a principal fibre bundle over M, with group G, i.e. solve the stochastic equation for forms D a = F? This question has been discussed and answered positively by Albeverio, Høegh-Krohn and Holden[8], [92]. Before describing some of this work let us mention that, at about the same time, and independently, Wong and Zakai have been discussing in general stochastic differential forms (without entering however in specific questions of stochastic equations), see [91], and Moshe Zakai's report at this conference). For processes/manifolds see also e.g. [138], [140].

Albeverio, Høegh-Krohn and Holden's approach involves the consideration of stochastic multiplicative G-valued measures η on a measurable space $(M,\mathcal{B})$. By definition such measures associate to each measurable set $A \in \mathcal{B}$ a G-valued random variable $\eta(A)(\omega)$, in such a way that $\eta(\emptyset)(\omega)$ is the unit in G and $\eta(A)$, $\eta(B)$ are independent if $A \cap B = \emptyset$ and moreover $\eta(A \cup B) = \eta(A) + \eta(B)$ in law, in this case. This together with suitable assumptions on continuity, non triviality and ergodicity yield the result that the law p_A of $\eta(A)(\cdot)$ has an invariant density and forms a generalized Markov semigroup p on G, indexed by $(M,\mathcal{B})$, in the sense that $p_{A\cup B}=p_A * p_B$ (III.1) whenever $A \cap B = \emptyset$. Viceversa any random family of measures $p(\cdot)$ with (III.1) yields, by a Kolmogorov type theorem, a stochastic multiplicative G-valued measure η, see [92] for details. An example is provided by $p_A \equiv \eta_{|A|}$, with η a 1-parameter convolution semigroup of measures on G, with $|A|$ the volume of A, if M is a Riemannian manifold. If G is a real vector space (i.e. in the Abelian case) all η and p are classified by a Levy-Khinchine type formula, see [92]. This can be used to induce a classification of η ad p also for arbitrary Lie groups G, "by integration from the Lie algebra g". Schematically, if a G-valued stochastic multiplicative measure η is given then

$$\int_{\varphi(A)} \eta(\varphi^{-1}[0,t))^{-1} d\eta(\varphi^{-1}[0,t)) \equiv \xi(A)$$

defines, for any $A \in \mathcal{B}$, a g-valued stochastic multiplicative measure, independent of φ (for φ a Borel isomorphism from $(M,\mathcal{B})$ into $(\mathbb{R}^+,\mathcal{B})$). Viceversa, if ξ is a g-valued stochastic multiplicative measure, then $\eta(A) \equiv \eta_t^{\varphi(A)}|t = 1$, with $\eta_t^{\varphi(A)}$ the non-anticipating Markov invariant solution, with left and right independent increments, of the stochastic equation $(\eta_t^{\varphi(A)})^{-1} d\eta_t^{\varphi(A)} = \chi_{\varphi(A)}(t)\xi(\varphi < t)$, is a stochastic multiplicative G-valued measure. If $\dim M = d$ and c_i are suitable d-1-dimensional oriented hypersurfaces, a product $c_1 \cdot c_2$ can be defined by preserving orientation. We call a map m associating to such hypersurfaces an element of G a cosurface if $m(c_1 \cdot c_2) = m(c_1) \cdot m(c_2)$, the product on the right hand side being in G, and $m(c)^{-1} = m(c^{-1})$. For d = 2 a cosurface is simply a G-valued curve integral. When m is given as the solution of $m((\tilde{c}(s))^{-1} dm(\tilde{c}(s)) = d \int_{\tilde{c}(s)} a$, with a a g-valued 1-form on M, and $\tilde{c}(s)$, $s \in [0,1]$ describes a closed oriented loop c, then m describes the holonomy.

Interesting cases where a and m are random have been studied in [8], [92].
E.g. if η is a stochastic multiplicative G-valued measure then, whenever well defined, $m(\partial A) \equiv \eta(A)$ gives a cosurface which is Markov in a natural sense. For d = 2 the m(c) can be looked upon as realizations of Wilson operators for continuum quantum gauge fields, which in turn can be realized as limits of lattice quantized gauge fields [8], [92]. For $d > 2$ the Markov cosurfaces yield Euclidean invariant fields associated with d-1-dimensional hypersurfaces (if $M = \mathbb{R}^d$, this then yields relativistic quantum fields associated with d-1-hyperplanes), see [8], [92] for details. We look upon these results as the beginnings of a stochastic analysis for group-valued random fields.

VI. Other topics

1) Many other topics of relevance in the theme stochastic analysis versus quantum theory have been discussed in other lectures at this conference (in particular the ones by H. Föllmer, G. Del Grosso, F. Guerra, S. Mitter, N.J. Portenko,..., K. Yasue).
2) Stochastic methods have been used in recent years in quantum mechanics for controlling eigenvalues and eigenfunctions and their asymptotics for $h \downarrow 0$ (h = Planck's constant) ("semiclassical limit") and $t \to 0$, $t \to \infty$ (short and large time behaviour). See e.g. [9], [22], [93]-[98], [136] For most recent work on the asymptotics for $h \downarrow 0$ of solutions of the heat resp. Schrödinger equation see e.g. [93]-[98], [104] resp. [96]. For tunneling transition probability in the case of potentials with "potential wells" see Jona-Lasinio's work with Martinelli and Scoppola and Faris, adapting Wentzell and Freidlin's methods, and extended recently in various directions [98], [131], [133].
The asymptotics for $t \to \infty$, $t \to 0$ as e.g. nice connections with hypercontractivity (supercontractivity, ultracontractivity) theory on one hand ([99]-[101]) and index theory [102] on the other hand. We also like to mention studies on stochasticity in

classical versus quantum mechanics [123], [146].

3) A stochastic spatial inverse problem for processes is being studied [103].

4) There has been a large amount of work involving discrete processes, instead of diffusion processes. It has been to some extent a basic restriction we put ourselves here to concentrate on diffusion type processes. E.g. in the representation of solutions of the heat and Schrödinger's equations the Feynman-Kac formula uses the $e^{\frac{1}{2}\Delta}$ - part of $e^{-t(-\frac{1}{2}\Delta + V)}$ (think of Lie-Trotter formula!) to provide the Brownian motion with respect to which expectations are taken, we could however have exploited the e^{-tV} part, considered as characteristic function of a process (e.g. a Poisson process) with independent increments. Such a viewpoint is basic to another approach to functional integration for quantum theory, developed e.g. in [104], [105], [139], [141] (and references therein). Also the Dirac equation has been handled by using Poisson processes, see e.g. [141].

5) In Ch. V we have mentioned a probabilistic approach to group-valued random fields, Another one, connected with the representation theory of the infinite dimensional Lie groups of mapping from a manifold M into a compact Lie group, has been pursued in recent years [106], [107], [109]-[111]. It has contacts with quantum field theory (representations of gauge groups), the identification problem in system theory (cfr. e.g. the Sobolev-Lie groups arising in [108]), the construction of a theory of non commutative distributions [109]. The proof of irreducibility for dim M ≧ 3 3 (or 2) of the representation uses orthogonality properties of Gaussian measures in infinite dimensional spaces [3], [110], the reduction theory for dim M = 1 (loop groups) uses properties of Brownian motion on Lie groups [110], [111]. Relations with the theory of Kac-Moody algebras and higher dimensional analogues have been found [107].

6) Work using stochastic analysis to prove new results of Atiyah-Singer indexes [102] and, more on the functional analytic side, [112]. Stochastic processes have also been discussed in connection with supersymmetric theories, see e.g. [113], [114], [137].

7) On the basic side mentioned in the introduction, we like to mention investigations which give classical mechanical models of diffusion processes, cfr. D. Dürr at the Ascona conference [115], and e.g. [147].

8) We should also like to mention work on quantum statistical mechanics as well as on "non commutative probability theory", where important developments towards an operator-valued stochastic analysis have been achieved [116]. This includes work concerning Fermi particles and fields, as well as dynamical semigroups and diffusion processes on C^*-algebras. For an introduction we advise the reader to look at P.A. Meyer's exposition [117]. For further recent references see e.g. [116], [118].

I am sure I have omitted many topics I should have mentioned and I apologize for it. My main purpose was to mention a few areas where important interaction between stochastic analysis and quantum theory has been going on in recent years. There is no doubt the interaction has been very fruitful and many more results can be expected in the near future.

Acknowledgements. I thank heartily the organizers for giving me the opportunity to speak at a most exciting meeting. It is a great pleasure to acknowledge here the great debt I have towards many friends with whom I collaborated on topics touched by this talk, in particular Raphael Høegh-Krohn and T. Arede, Ph. Blanchard, Ph. Combe, D. Dürr, J.E. Fenstad, M. Fukushima, F. Gesztesy, H. Holden, W. Karwowski, W. Kirsch, S. Kusuoka, T. Lindstrøm, J. Marion, F. Martinelli, M. Mebkhout, D. Merlini, R. Rodriguez, M. Sirugue-Collin, M. Sirugue, L. Streit, D. Testard, A. Vershik. To all of them my hearty thanks. I am grateful to J. Brasche, Dr. W. Kirsch, Dr. W. Loges, U. Spönemann for useful discussions. I also gratefully acknowledge the hospitality and financial support of the ZiF, University of Bielefeld (Proj. No. 2), of the Centre de Physique Théorique, CNRS, Marseille-Luminy and the Mathematics Institute of Oslo University as well as the Norwegian Science Foundation and the Volkswagenstiftung (Project BiBoS).

I thank Mrs. Mische and Richter for their skilful typing.

References

[1] E. Nelson, Field theory and the future of stochastic mechanics, Princeton Preprint (1985), to appear in [2].

[2] S. Albeverio, G. Casati, D. Merlini, Edt., "Stochastic Processes in Classical and Quantum Systems", Proc. 1. Intern. Ascona-Como Meeting, June 24-29, 1985, subm. Lect. Notes Phys. Springer Verlag, Berlin (1985)

[3] S. Albeverio, R. Høegh-Krohn, Diffusion fields, quantum fields and fields with values in Lie groups, pp. 1-98 in M.A. Pinsky, Edt., Stochastic Analysis and Applications, Adv. Probability and Related Topics, Vol.7, M. Dekker Inc., New York

and Basel (1984)

[4] R. Vilela-Mendes, Reconstruction of dynamics from an eigenstate, Preprint (1985)

[5] S. Albeverio, F. Gesztesy, R. Høegh-Krohn, H. Holden, Solvable models in quantum mechanics, book in preparation

[6] M. Fukushima, Energy forms and diffusion processes, pp. 65-97 in Mathematics + Physics, L. Streit Edt., World Publ. Co., Singapore (1985)

[7] M. Fukushima, Dirichlet space theory and its applications, to appear in Proc. 2. BiBoS-Symposium, Stochastic Processes - Mathematics and Physics, Edts. S. Albeverio, Ph. Blanchard, L. Streit, Lect. Notes Maths., Springer (1985)

[8] S. Albeverio, R. Høegh-Krohn, H. Holden, Markov processes on infinite dimensional spaces, Markov fields and Markov cosurfaces, BiBoS-Preprint (1985) to appear in Proc. Bremen Conf., Edt., L. Arnold, P. Kotelenez, D. Reidel (1985)

[9] B. Simon, Schrödinger semigroups, Bull. AMS 7, 447-526 (1982) (e.g. Theor. A.2.7)

[10] M. Reed, B. Simon, Methods of modern mathematical physics II, IV, Acad. Press, New York (1975);

[11] S. Albeverio, F. Gesztesy, W. Karwowski, L. Streit, On the connection between Schrödinger- and Dirichlet forms, BiBoS-Preprint, to appear in J. Math. Phys.

[12] S. Albeverio, R. Høegh-Krohn, L. Streit, Energy forms, Hamiltonians and distorted Brownian paths, J. Math. Phys. 18, 907-917 (1977)

[13] S. Albeverio, R. Høegh-Krohn, L. Streit, Regularization of Hamiltonians and processes, J. Math. Phys. 21, 1636-1642 (1980)

[14] L. Streit, Energy forms, Schrödinger theory, processes, Phys. Repts. 77, 363-375 (1981)

[15] M. Fukushima, Dirichlet Forms and Markov Processes, North Holland/Kodansha Amsterdam/Tokyo (1980)

[16] S. Albeverio, J.E. Fenstad, R. Høegh-Krohn, T. Lindstrøm, Non standard methods in stochastic analysis and mathematical physics, Acad. Press, to appear.

[17] M. Röckner, N. Wielens, Dirichlet forms - Closability and change of speed measure, pp. 119-144 in "Infinite dimensional analysis and stochastic processes", Ed. S. Albeverio, Res. Notes in Mathematics, Pitman, London (1985)

[18] K. Rullkötter, U. Spönemann, Dirichletformen und Diffusionsprozesse, Diplomarbeit, Bielefeld, 1982 and publication in preparation by U. Spönemann

[19] H. Ezawa, J.R. Klauder, L.A. Shepp, A path space picture for Feynman-Kac averages, Ann. Phys. 88, 588-620 (1974)

[20] M. Fukushima, On a stochastic calculus related to Dirichlet forms and distorted Brownian motion, Phys. Repts. 77, 255-262 (1981)

[21] M. Fukushima, On absolute continuity of multidimensional symmetrizable diffusions, pp. 146-176 in Functional analysis in Markov processes, M. Fukushima Edt., Lect. Notes Maths. 923, Springer Berlin (1982)

[22] B. Simon, Functional integration and quantum physics, Acad. Press, New York (1979)

[23] M. Fukushima, On a representation of local martingale additive functionals of symmetric diffusions, pp. 110-118 in Stochastic Integrals, Proc., Ed. D. Williams, Lect. Notes Maths. 851, Springer (1981)

[24] S. Albeverio, R. Høegh-Krohn, Quasi-invariant measures, symmetric diffusion processes and quantum fields, pp. 11-59 in "Les Méthodes Mathématiques de la Théorie Quantique des Champs", F. Guerra, D.W. Robinson, R. Stora, Edts., Ed. CNRS, Paris (1976)

[25] S. Nakao, Stochastic calculus for continuous additive functionals of zero energy, Z. Wahrscheinlichkeitsth. verw. Geb. 68, 557-578 (1985)

[26] Y. Oshima, M. Takeda, to appear in Ref. [7].

[27] S. Albeverio, M. Fukushima, W. Karwowski, L. Streit, Capacity and quantum mechanical tunneling, Comm. Math. Phys. 81, 501-513 (1981)

[28] M. Fukushima, A note on irreducibility and ergodicity of symmetric Markov processes, pp. 200-207 in "Stochastic processes in quantum theory and statistical physics", Edts. S. Albeverio, Ph. Combe, M. Sirugue-Collin, Lect. Notes Phys. 173, Springer, Berlin (1982)

[29] M. Nagasawa, Segregation of a population in an environment, J. Math. Biology 9, 213-235 (1980); and in Ref. [2].

[30] S. Albeverio, Ph. Blanchard, R. Høegh-Krohn, A stochastic model for the orbits of planets and satellites. An interpretation of Titius-Bode law, Exp. Math. 4, 365-373 (1983)

[31] Ph. Blanchard, Trapping for Newtonian diffusion processes, pp. 185-210 in "Stochastic Methods and Computer Techniques in Quantum Dynamics", H. Mitter and L. Pittner, Edts., Acta Phys. Austr. Suppl. XXVI, Springer Wien (1984)

[32] E.A. Carlen, Existence and sample path properties of the diffusions in Nelson's stochastic mechanics, BiBoS-Preprint (1985), to appear in Proc. BiBoS-Symposium I, Edts. S. Albeverio, Ph. Blanchard, L. Streit, Lect. Notes Maths., Springer, Berlin

[33] E. Nelson, Quantum Fluctuations, Princeton University Press (1985)

[34] P.A. Meyer, W.A. Zheng, Construction de processus de Nelson reversibles, pp. 12-26 in Sém. Prob. XIX, Edts. J. Azema, M. Yoy, Lect. Notes Maths. 1123, Springer, Berlin (1985)

[35] W.A. Zheng, to appear in Ref. [7]

[36] N. Wielens, On the essential self-adjointness of generalized Schrödinger operators, J. Funct. Anal. 61, 98-115 (1985)

[37] J. Brasche, Perturbation of Schrödinger Hamiltonians by measure-self-adjointness and lower semiboundedness, J. Math. Phys. 26, 621-626 (1985)

[38] S. Albeverio, S. Kusuoka, L. Streit, Convergence of Dirichlet forms and associated Schrödinger operators, BiBoS-Preprint (1984), to appear in J. Funct. Anal.

[39] M. Fukushima, M. Takeda, A transformation of a symmetric Markov process and the Donsker-Varadhan theory, Osaka J. Math. 21, 311-326 (1984)

[40] S.K. Mitter, On the analogy between mathematical problems of nonlinear filtering and quantum physics, Ricerche di Automatica 10, 163-216 (1980)

[41] S.K. Mitter, Nonlinear filtering and stochastic mechanics, in M. Hazewinkel, J.C. Willems, Edts., "Stochastic systems: the mathematics of filtering and identification and applications", D. Reidel, Dordrecht (1981)

[42] V.E. Benes, Exact finite dimensional filters for certain diffusions with nonlinear drifts, Stochastics 5, 65-92 (1981)

[43] M.H.A. Davis, Pathwise nonlinear filtering, in Ref [41]

[44] D.L. Ocone, J.S. Baras, S.I. Marcus, Explicit filters for diffusions with certain nonlinear drifts, Stochastics 8, 1-16 (1982)

[45] S.A. Albeverio, R.J. Høegh-Krohn, Mathematical theory of Feynman path integrals, Lect. Notes Maths. 523, Springer, Berlin (1976)

[46] T. Arede, A class of solvable non linear filters, in preparation

[47] E. Nelson, The free Markoff field, J. Funct. Anal. 12, 211-227 (1973)

[48] S. Albeverio, R. Høegh-Krohn, Uniqueness and the global Markov property for Euclidean fields. The case of trigonometric interactions, Comm. Math. Phys. 68, 95-128 (1979)

[49] E.B. Dynkin, Markov processes and random fields, Bull. AMS 3, 975-999 (1980)

[50] M. Röckner, A. Dirichlet problem for distributions and specifications for random fields, Mem. AMS 54, No 324 (1985)

[51] R.L. Dobrushin, R.A. Minlos, An investigation of the properties of generalized Gaussian random fields, Sel. Math. Sov. 1, 215-263 (1981)

[52] R. Gielerak, Verification of the global Markov property in some class of strongly coupled exponential interactions, J. Math. Phys. 24, 347-355 (1983)

[53] Y. Rozanov, Boundary problems for stochastic partial differential equations, BiBoS-Preprint (1985), to appear in Exp. Math.

[54] B.Zegarlinski,Verification of the global Markov property for Euclidean fields: the case of general exponential interactions, Comm. Math. Phys.; The Gibbs measures and partial differential equations I, II, BiBoS-Preprint (1985)

[55] B. Simon, The $P(\varphi)_2$ Euclidean (Quantum) Field Theory, Princeton Univ. Press (1974)

[56] A. Jaffe, J. Glimm, Quantum Physics, Springer, New York (1981)

[57] S. Albeverio, Nonstandard analysis; polymer models, quantum fields, pp. 233-254 in Ref. [31]

[58] S. Albeverio, R. Høegh-Krohn, Dirichlet forms and diffusion processes on rigged Hilbert spaces, Z. Wahrscheinlichkeitsth. verw. Geb. 40, 1-57 (1977); Hunt processes and analytic potential theory on rigged Hilbert spaces, Ann. I.H. Poincaré B XIII, 269-291 (1977)

[59] S. Kusuoka, Dirichlet forms and diffusion processes on Banach spaces, J. Fac. Sci. Univ. Tokyo 29, 79-95 (1982)

[60] S. Kusuoka, Lectures at Kyoto

[61] S. Albeverio, R. Høegh-Krohn, The Wightman axioms and the mass gap for strong interactions of exponential type in two-dimensional space-time, J. Funct. Anal. 16, 39-82 (1974)

[62] M. Asorey, P.K. Mitter, Regularized continuous Yang-Mills processes and Feynman-Kac functional, Comm. Math. Phys. 80, 43-58 (1981)

[63] B. Gaveau, Ph. Trauber, Une approche rigoureuse à la quantification locale du champ de Yang-Mills avec cut-off, C.R. Ac. Sci. Paris A291, 673-676 (1980)

[64] B. Haba, Instantons with noise. I Equations for two-dimensional models, BiBoS-Preprint (1985); and to appear in Ref. [2]

[65] S. Marcus, Stochastic diffusions on an unbounded domain, Pac. J. Math. 84, 143-153 (1979)

[66] G. Jona-Lasinio, P.K. Mitter, On the stochastic quantization of field theory, Roma Preprint (1985)

[67] Ph. Blanchard, J. Potthoff, R. Sénеor, A remark on perturbation expansion for unstable acting via stochastic quantization, BiBoS-Preprint (1985)

[68] a) W. Kirsch, F. Martinelli, Random Schrödinger operators: on recent results and open problems, to appear in Ref. [81]

b) W. Kirsch, Random Schrödinger operators and the density of states, pp. 68-102 in S. Albeverio, Ph. Combe, M. Sirugue-Collin, Stochastic aspects of classical and quantum systems, Lect. Notes Maths. 1109, Springer, Berlin (1985)

[69] R. Carmona, Random Schrödinger operators, Irvine Preprint (1984), to appear in Proc. Ecole d'Ete de Probabilites XIV, Saint Flour, 1984

[70] W. Kirsch, Über Spektren stochastischer Schrödingeroperatoren, Ph. Thesis, Bochum (1981)

[71] W. Kirsch, F. Martinelli, On the spectrum of Schrödinger operators with a random potential, Comm. Math. Phys. 85, 329-350 (1982)

[72] W. Kirsch, F. Martinelli, Large deviations and the Lifshitz singularity of the integrated density of states of random Hamiltonians, Comm. Math. Phys. 89, 27-40 (1983)

[73] S. Kotani, Ljapunov exponents and spectra for one-dimensional random Schrödinger operators, Kyoto Preprint (1984)

[74] N. Minami, to appear in Ref. [2]

[75] J. Fröhlich, F. Martinelli, E. Scoppola, T. Spencer, Anderson localization for large disorder or low energy, Preprint (1985)

[76] F. Martinelli, H. Holden, A remark on the absence of diffusion near the bottom of the spectrum for a random Schrödinger operator on $L^2(\mathbb{R}^\nu)$, Comm. Math. Phys. 93, 197-217 (1984)

[77] G. Jona-Lasinio, F. Martinelli, E. Scoppola, Multiple tunneling in D dimensions: a quantum particle in a hierarchical potential, Ann. Inst. H. Poincaré 42, No 1 (1985)

[78] F. Delyon, Y. Levy, B. Souillard, Anderson localization for multidimensional systems at large disorder or low energy, Preprint (1985)

[79] B. Simon, M. Taylor, T. Wolff, Some rigorous results for the Anderson model; Preprint

[80] K. Symanzik, Euclidean quantum field theory, in "Local Quantum Theory", Ed. R. Jost, Acad. Press, New York (1969)

[81] J. Westwater, On Edward's model for polymer chains, to appear in "Trends and developments in the eighties", Ed. S. Albeverio, Ph. Blanchard, World Publ., Singapore

[82] S. Albeverio, J.E. Fenstad, R. Høegh-Krohn, W. Karwowski, T. Lindstrøm, Perturbations of the Laplacian supported by null sets with applications to polymer measures and quantum fields, Phys. Letts. 104, 396-400 (1984)

[83] E.B. Dynkin, Polynomials of the occupation field and related random fields, J. Funct. Anal. 58, 20-52 (1984)

[84] J. Rosen, A local time approach to the self-intersections of Brownian paths in space, Comm. Math. Phys. 88, 327-338 (1983)

[85] M. Yor, to appear in Ref. [7]

[86] A. Stoll, Self-repellent random walks and polymer measures in two dimensions, Ph. D. Thesis, Bochum (1985); and to appear in Ref. [7]

[87] S. Kusuoka, Asymptotics of polymer measures in one dimension, pp. 66-82 in Ref. [17]

[88] J. Westwater, A variational principle for the Flory measure in 2 dimensions, ZiF-Preprint (1984)

[89] H.P. McKean, Stochastic Integrals, Acad. Press, New York (1969)

[90] M. Ibero, Intégrales stochastiques multiplicatives et construction de diffusions sur un groupe de Lie, Bull. Sci. Math. 100, 175-191 (1976)

[91] E. Wong, M. Zakai, Markov processes on the plane, Preprint; Multiparameter martingale differential forms, Preprint

[92] S. Albeverio, R. Høegh-Krohn, H. Holden, Markov cosurfaces and gauge fields, pp. 211-232, in Ref. [31]; Stochastic multiplicative measures, generalized Markov semigroups and groupvalued stochastic processes and fields, BiBoS-Preprint (1985); and paper to appear in [2]

[93] S. Albeverio, T. Arede, The relation between quantum mechanics and classical mechanics: a survey of mathematical aspects, pp. 37-76 in G. Casati, Edt., "Chaotic behavior in quantum systems", Plenum Press, New York (1985)

[94] R. Azencott, H. Doss, L'equation de Schrödinger quand h tends vers zero: Une approche probabiliste, pp. 1-17 in Ref. [68 b]

[95] D. Elworthy, A. Truman, K. Watling, The semi-classical expansion for a charge particle on a curved space background, Warwick Preprint

[96] J. Rezende, The method of stationary phase for oscillatory integrals on Hilbert spaces, Comm. Math. Phys.; Remark on the solution of the Schrödinger equation

for anharmonic oscillators via the Feynman path integral, Lett. Math. Phys. 7, 75-83 (1983); Stationary phase method on Hilbert space and semi-classical approximation in quantum mechanics, to appear in Ann. I.H. Poincaré ; [2]

[97] G. Jona-Lasinio, F. Martinelli, E. Scoppola, New approach in the semiclassical limit of quantum mechanics I. Multiple tunneling in one dimension, Comm. Math. Phys. 80, 223-254 (1981)

[98] B. Simon, Semiclassical analysis of low-lying eigenvalues, IV. The flea on the elephant, J. Funct. Anal. 63, 123-136 (1985)

[99] D. Bakry, M. Emery, Diffusions hypercontractives, pp. 177-206 in Sém. Prob. XIX, Edts. J.A. Azéma, M. Yor, Lect. Notes Maths. 1123, Springer, Berlin (1985)

[100] E.B. Davies, B. Simon, Ultracontractivity and the heat kernel for Schrödinger operators and Dirichlet Laplacians, J. Funct. Anal. 59, 335-395 (1984)

[101] S. Kusuoka, D. Stroock, Some boundedness properties of certain stationary diffusion semigroups, J. Funct. Anal. 60, 243-264 (1985)

[102] J.M. Bismut, The Atiyah-Singer theorems; a probabilistic approach I. The index theorem, J. Funct. Anal. 57, 56-99 (1984)

[103] S. Albeverio, S. Kusuoka, L. Streit, in preparation

[104] Ph. Blanchard, Ph. Combe, M. Sirugue, M. Sirugue-Collin, Estimates of quantum deviations from classical mechanics using large deviations, to appear in Proc. Heidelberg, Conf. Quantum Probability, Ed. L. Accardi et al

[105] Ph. Combe, to appear in Ref. [2], [7]

[106] J. Marion, to appear in Ref. [2]

[107] D. Testard, to appear in Ref. [7]

[108] P.S. Krishnaprasad, S.I. Marcus, M. Hazewinkel, Current algebras and the identification problem, Stochastics 11, 65-101 (1983)

[109] S. Albeverio, R. Høegh-Krohn, J. Marion, D. Testard, book in preparation

[110] S. Albeverio, R. Høegh-Krohn, D. Testard, Irreducibility and reducibility for the energy representation of the group of mappings of a Riemannian manifold into a compact semisimple Lie group, J. Funct. Appl. 41, 378-396 (1981); Factoriality of representations of the group of paths of SU(n), F. Junct. Anal. 57, 49-55 (1984)

[111] S. Albeverio, R. Høegh-Krohn, D. Testard, A. Vershik, Factorial representations of path groups, J. Funct. An. 52, 115-131 (1983)

[112] H. Cycon, R. Froese, W. Kirsch, B. simon, Recent results on Schrödinger operators, in preparation, Springer, Berlin

[113] Z. Haba, Instantons with noise. I. Equations for two-dimensional models, BiBoS-Preprint (1985)

[114] L. Girardello, to appear in Ref. [7]

[115] D. Dürr, to appear in Ref. [2]

[116] L. Accardi, V. Hemmen, V. Waldenfels, Edts., Ref. [115]

[117] P.A. Meyer, Eléments de probabilités "quantiques", Sém. Prob. XX, Lect. Notes Math., Springer, Berlin (1985)

[118] G. Ekhaguere, to appear in Ref. [2]

[119] R. Figari, S. Teta, The Laplacian in regions with many small obstacles: fluctuations around the limit operator, BiBoS-Preprint (1984)

[120] T. Kolsrud, On the Markov property for certain Gaussian random fields, BiBoS-Preprint (1984)

[121] S. Albeverio, W. Karwowski, M. Röckner, L. Streit, Capacity, Green's functions and Schrödinger equation, pp. 197-215 in Ref. [17]

[122] T. Lindstrøm, Nonstandard analysis and perturbations of the Laplacian along Brownian path, to appear in Ref. [32]

[123] J. Bellissard, Stability and instability in quantum mechanics, to appear in Ref. [81]

[124] T. Hida, L. Streit, Generalized Brownian functionals and the Feynman integral, Stoch. Proc. Appl. 16, 55-69 (1983)

[125] W. Kirsch, S. Kotani, B. Simon, Absence of absolutely continuous spectrum for some one dimensional random but deterministic potentials, Preprint (1985), Ann. I.H. Poincaré

[126] M. Takeda, On the recurrence of diffusions in random media, in preparation

[127] M. Demuth, Scattering by singular potentials, Akad. DDR, Berlin Preprint

[128] H. Föllmer, Von der Brownschen Bewegung zum Brownschen Blatt: einige neuere Richtungen in der Theorie der stochastischen Prozesse, pp. 159-190 in Perspectives in Mathematics, Anniversary of Oberwolfach 1984, Birkhäuser Verlag, Basel (1984)

[129] R. Carmona, Regularity properties of Schrödinger and Dirichlet semigroups, J. Funct. Anal. 33, 259-296 (1979)

[130] M. Takeda, On the uniqueness of Markovian extensions of diffusion operators on infinite dimensional spaces, ZiF-Preprint (1984), Osaka Math. J.

[131] G. Jona-Lasinio, Stochastic processes and quantum mechanics, Lect. given at the Colloque en l'Honneur de L. Schwartz, to appear in Astérique

[132] A.M. Berthier, B. Gaveau, Critère de convergence des fonctionelles de Kac et application en mécanique quantique et en géometrie, J. Funct. Anal. 29, 416- (1978)

[133] G. Jona-Lasinio, Qualitative theory of stochastic differential equations and quantum mechanics of disordered systems, Helv. Phys. Acta 56, 61- (1983)

[134] K.L. Chung, Probabilistic approach to boundary value problems for Schrödinger's equation, Exp. Math. 3, 175-179 (1985)

[135] Ju.G. Kondratsev, V.D. Kosmanenko, The scattering problem for operators associated with Dirichlet forms, Sov. Math. Dokl. 26, 585-589 (1982)

[136] G.F. Dell'Antonio, Large time, small coupling behaviour of a quantum particle in a random field, Ann. I.H. Poincaré 39, 339-384 (1983)

[137] L. Streit, Stochastic processes - quantum physics, pp. 3-52 in Ref. [31]

[138] K. Helmes, A. Schwane, Levy's stochastic area formula in higher dimensions, J. Funct. Anal. 54, 177-192 (1983)

[139] D. DeFalco, in preparation, to appear in Ref. [2]

[140] F. De Angelis, in preparation, to appear in Ref. [2]

[141] M. Sirugue-Collin, in preparation, to appear in Ref. [2]

[142] Y. Oshima, T. Yamado, On some representations of continuous additive functionals locally of zero energy, J. Math. Soc. Jap. 36, 315-339 (1984)

[143] H. Englisch, One dimensional Schrödinger operator with ergodic potential, Zeitschr. für Anal. u. Anw. 2, 411-426 (1983)

[144] W. Kirsch, B. Simon, Lifshitz tails for random plus periodic potentials, Bochum-Pasadena Preprint (1985)

[145] S. Kusuoka, The path property of Edward's model for long polymer chains in three dimensions, pp. 48-65 in Ref. [17]

[146] G. Casati, Ed., Chaotic behavior in quantum systems, Plenum Press (1983)

[147] Y. Okabe, A generalized fluctuation-dissipation theorem for the one-dimensional diffusion process, Commun. Math. Phys. 98, 449-468 (1985)

[148] S. Albeverio, Ph. Blanchard, F. Gesztesy, L. Streit, Quantum mechanical low energy scattering in terms of diffusion processes, pp. 207-227 in Ref. [68b].

[149] J.R. Klauder, Measure and support in functional integration, to appear in "progress in Quantum Field Theory", Eds. H. Ezawa, S. Kamefuchi, North Holland, Amsterdam (1986)

ON A CLASS OF STOCHASTIC DIFFERENTIAL EQUATIONS [1] WHICH DO NOT SATISFY LIPSCHITZ CONDITIONS

A. V. Balakrishnan

Electrical Engineering Department
University of California at Los Angeles
Los Angeles, California
U.S.A.

We study a class of nonlinear stochastic equations of the rotational motion of a rigid body subject to random torques. Nonlinear equations do not satisfy the Lipschitz conditions; we show the existence of "pathwise" unique solutions, which display some unexpected properties such as the first order distributions being asymptotically Gaussian.

1. INTRODUCTION

We study (existence of, and properties of, solutions) a class of nonlinear stochastic equations characterizing rotational motion of a rigid body subject to random torques, arising in Satellite Attitude Control Dynamics [1]. The novelty consists in that the nonlinearity does not satisfy the Lipschitz conditions usually imposed for existence of solutions [2], so that the construction of a solution is not obvious. We show the existence of "pathwise" unique solutions, so that in particular both the Ito and the finitely additive White Noise formulations yield the same solution. What is more interesting, the solution displays some unexpected properties such as the first order distributions being asymptotically Gaussian, even though the equations are nonlinear.

2. THE PROBLEM

Letting $x(t)$ denote (3×1 matrix) angular velocity,[2] and $\otimes$ denoting the usual vector cross-product, the kinematic (Euler) attitude motion equation [1] can be written:

$$M\dot{x}(t) + x(t) \otimes Mx(t) + Kx(t) = N(t) , \quad \text{a.e. } 0 < t , \tag{2.1}$$

where M is the moment of inertia matrix, real symmetric and positive definite, $\otimes$ denotes vector cross-product, and K is the damping matrix which is again a real symmetric and positive definite matrix, and finally $N(t)$ is the (random) input torque (3×1 vector):

$$N(t) = \begin{vmatrix} N_1(t) \\ N_2(t) \\ N_3(t) \end{vmatrix} .$$

We have thus a stochastic differential equation and we shall consider the case where $N(t)$ is white Gaussian (or more precisely, "finitely additive" white Gaussian noise, in order not to confuse with other uses of this term). The spectral density will be denoted by D, a symmetric nonnegative definite matrix. We recall that in the finitely additive White Noise theory [3], the sample paths of the noise are in $L_2[0,T]$, $T < \infty$ for every T and the process is defined by the characteristic function

$$E\left| e^{i\int_0^T [N(t),h(t)]dt} \right| = \exp -\frac{1}{2}\int_0^T [Dh(t),h(t)]\,dt \qquad (2.2)$$

for any $h(\cdot)$ in $L_2[0,T]$.

We can also consider an Ito model which takes the form:

$$M dx + f(x)dt = dW \qquad (2.3)$$

where

$$f(x) = x \otimes Mx + Kx \qquad (2.4)$$

and $W(\cdot)$ is a 3×1 Wiener process with

$$E[dW\, dW^*] = D\, dt\ .$$

We shall take an "input-output" (or "simulation") point of view. We need a notion of "sample pathwise" solution to either (2.1) and (2.2): by this we mean that in each finite interval $0 \le t \le T < \infty$, we have a unique (continuous) solution (2.1) for given sample function $N(\cdot)$ in $L_2[0,T]$. Similarly (2.2) has a unique continuous solution for every $W(\cdot)$ in $C[0,T]$. We emphasize that this requirement (of pathwise existence and uniqueness) for the Ito equation (2.2) is different from the usual (see, e.g., [4]) in that we are not satisfied with such a solution for <u>almost all</u> (with respect to Wiener measure) sample paths in $C[0,T]$. In our particular problem, the coefficients are time-invariant and the diffusion term is a constant. In this case it is well known (see [4]) that if the drift term satisfies a global Lipschitz condition, then the usual Picard construction does yield a unique solution $x(\cdot)$ in $C[0,T]$ for <u>each</u> specified $W(\cdot)$ in $C[0,T]$. However in our drift term the nonlinear part is:

$$x \otimes Mx = \tilde{x} M x$$

where

$$x = \begin{vmatrix} 0 & -x_3 & x_2 \\ x_3 & 0 & -x_1 \\ -x_2 & x_1 & 0 \end{vmatrix}$$

where

$$x = \begin{vmatrix} x_1 \\ x_2 \\ x_3 \end{vmatrix}$$

and is a second degree homogeneous polynomial in x. Hence it is not globally Lipschitz. However, as can be easily verified, it is locally Lipschitz: for all x in $S(x_0;R)$ where $S(x_0;R)$ denotes a sphere of radius R centered at x_0, we can find a constant K_R such that

$$\|f(x)-f(y)\| \le K_R \|x-y\|\ , \qquad x,y \in S(x_0;R)\ . \qquad (2.5)$$

This does guarantee the existence and uniqueness of a strong solution for (2.2), as shown for example in [4], but does <u>not</u> yield a solution for <u>every</u> $W(\cdot)$ in $C[0,T]$, only for $W(\cdot)$ in a set of Wiener measure one, up to an explosion time. In our particular problem however it is possible to do better by taking advantage of some special features of the problem, as we shall now proceed to show. We shall begin with an important special case of (2.1) -- of interest on its own -- the Axially Symmetric case, where the arguments are simpler.

First, it is convenient to change (2.1) by the substitution:

$$y = Mx$$

yielding

$$\dot{y} - y \otimes My + (KM^{-1})y = N(t) \qquad \text{a.e. } 0 < t \qquad (2.6)$$

and the corresponding Ito version

$$dy = (y \otimes My)dt - (KM^{-1})y\,dt + dW \, . \qquad (2.7)$$

It is convenient to replace KM^{-1} by K, as we shall in what follows. Also, we shall omit "a.e." in the equations such as (2.6), as obvious.

3. THE AXIALLY SYMMETRIC CASE

This is the special case where M is diagonal and has the form

$$M = \text{diag.}\ [a,a,2a] \, . \qquad (3.1)$$

If we specialize K also to be diagonal with no real loss in generality, and normalize a to $a=1$, we may rewrite (2.6) as

$$y = \begin{vmatrix} y_1 \\ y_2 \\ y_3 \end{vmatrix}$$

$$\left.\begin{aligned} \dot{y}_1 + y_3 y_2 + k_1 y_1 &= N_1(t) \\ \dot{y}_2 - y_3 y_1 + k_2 y_2 &= N_2(t) \\ \dot{y}_3 \qquad\quad + k_3 y_3 &= N_3(t) \end{aligned}\right\} \qquad (3.2)$$

where

$$k_1, k_2, k_3 \quad \text{are all} > 0 \, .$$

From (3.2) we obtain immediately that

$$y_3(t) = e^{-k_3 t} y_3(t) + \int_0^t e^{-k_3(t-s)} N_3(s)\, ds$$

which can be expressed

$$= e^{-k_3 t} y_3(0) + \tilde{W}_3(t) + k_3 \int_0^t e^{-k_3(t-s)} \tilde{W}_3(s)\, ds \qquad (3.3)$$

where

$$\tilde{W}_3(t) = \int_0^t N_3(s)\, ds$$

(and we shall denote:

$$\tilde{W}(t) = \int_0^t N(s)\, ds \quad . \,)$$

It is clear that we have a unique pathwise solution for each $N_3(\cdot)$ in $L_2[0,T]$. Next it is obvious that we may consider the first two equations in (3.2) as time-varying linear equations, by fixing $y_3(\cdot)$. The (2×2) matrix equation

$$\dot{\Phi}(t) = A(t)\, \Phi(t)$$

where

$$A(t) = \begin{vmatrix} -k_1 & -y_3(t) \\ y_3(t) & -k_2 \end{vmatrix} \qquad (3.4)$$

for fixed $y_3(\cdot)$ in $C[0,T]$ has a unique C_1-solution for $t \geq 0$, with

$$\Phi(0) = I$$

and writing

$$Y(t) = \begin{vmatrix} y_1(t) \\ y_2(t) \end{vmatrix}$$

we have the unique continuous solution of (3.2) as:

$$Y(t) = \Phi(t)\,Y(0) + \int_0^t \Phi(t)\Phi(s)^{-1} \begin{vmatrix} N_1(s) \\ N_2(s) \end{vmatrix} ds$$

which can be rewritten as:

$$Y(t) = \Phi(t)\,Y(0) + \begin{vmatrix} \tilde{W}_1(t) \\ \tilde{W}_2(t) \end{vmatrix} + \int_0^t \Phi(t)\Phi(s)^{-1}A(s) \begin{vmatrix} \tilde{W}_1(s) \\ \tilde{W}_2(s) \end{vmatrix} ds \,. \qquad (3.5)$$

This establishes a unique pathwise solution, as required, for each $N(\cdot)$ in $L_2[0,T]$. But as (3.5) and (3.3) show, the solution continues to be defined for each $\tilde{W}(\cdot)$ in $C[0,T]$. In particular, replacing $\tilde{W}(\cdot)$ by $W(\cdot)$ in (3.3), (3.5), it is readily seen that we get the solution to the Ito equation (2.7) specialized to the Axially Symmetric case:

$$y_3(t) = y_3(0) + \int_0^t k_3 y_3(s)ds + W_3(t)$$

$$y_2(t) = y_2(0) + \int_0^t (y_3(s)y_1(s) - k_2 y_2(s))ds + W_2(t)$$

$$y_1(t) = y_1(0) - \int_0^t (y_3(s)y_1(s) + k_1 y_1(s))ds + W_1(t)$$

$$0 < t \qquad (3.6)$$

Next let us examine the statistical properties. Of course $y_3(\cdot)$ is a Gaussian process, and if we fix $y_3(\cdot)$, the process $Y(\cdot)$ is "conditionally Gaussian." More precisely, let $B_3(t)$ denote the sigma-algebra generated by the process $y_3(s)$, $s \le t$. Then the conditional probability distribution of

$$Y(t_i)\,, \qquad i = 1,\dots,n, \quad t_i \le t$$

conditioned on $B_3(t)$ is Gaussian. Such processes have been treated in detail by Liptser-Shiryayev in [5] and more recently by [6]. An immediate consequence of the conditionally Gaussian property is that it proves that the Markov process $Y(\cdot)$ defined by (3.3), (3.5) possesses transition densities, as follows by applying Fubini's theorem. Further the forward Kolmogorov of Fokker-Planck equations [2] hold.

We can do more in this particular case. In fact to make explicit analysis easier, we shall set

$$k_1 = k_2 = k$$

$$D = \text{diag.}\ (d,d,d_3)\,.$$

Which from the physical point of view is not much of a limitation. In that case we exploit the fact that we are dealing with rotation dynamics and can write for the solution of (3.4):

$$\Phi(t) = e^{-kt}\psi(t)$$

$$\psi(t) = \begin{vmatrix} \cos \Omega_3(t) & \sin \Omega_3(t) \\ -\sin \Omega_3(t) & \cos \Omega_3(t) \end{vmatrix}$$

$$\Omega_3(t) = \int_0^t y_3(s)\, ds \,,$$

and

$$\psi(t)^* = \psi(t)^{-1}\,.$$

Moreover:

$$\Phi(t)\Phi(s)^{-1} = e^{-k(t-s)} \begin{vmatrix} \cos(\Omega_3(t) - \Omega_3(s)) & \sin(\Omega_3(t) - \Omega_3(s)) \\ -\sin(\Omega_3(t) - \Omega_3(s)) & \cos(\Omega_3(t) - \Omega_3(s)) \end{vmatrix} . \qquad (3.7)$$

Hence, letting, for $s \leq t$:

$$R(t,s) = E[Y(t)Y(s)^* \mid B_3(t)]$$

we have

$$R(t,s) = \Phi(t)\Phi(s)^{-1}(\Phi(s)\,R(0,0)\,\Phi(s)^* + \frac{d}{2k}(1-e^{-2ks})I) \quad . \tag{3.8}$$

Let

$$\bar{R}(t,s) = E[R(t,s)] \quad . \tag{3.9}$$

Then

$$\lim_{t\to\infty} \bar{R}(t,t) = \frac{d}{2k} I$$

and

$$\lim_{s\to\infty} \bar{R}(t+s,s) = I \frac{d}{2k} e^{-t(k+\frac{r(t)}{t})} \tag{3.10}$$

where

$$r(t) = \lim_{s\to\infty} E\left[\cos \int_s^{s+t} y_3(\sigma)d\sigma\right] = \frac{1}{2} \int_0^t \int_0^t R_3(\sigma - s)\, d\sigma\, ds \tag{3.11}$$

where, in our case:

$$R_3(\sigma) = \lim_{s\to\infty} E[y_3(\sigma+s)\, y_3(s)] = \frac{d_3}{2k_3} e^{-k_3\sigma} \quad .$$

We can conclude from (3.10) that the bandwidth of the process $Y(\cdot)$ is larger than that indicated by the damping matrix K. The spectral density is not rational and the spectral spreading is proportional to the noise spectral density. In the limiting case where, when $y_3(\cdot)$ is white noise, (3.11) yields

$$\frac{r(t)}{t} = \frac{d_3}{2} \tag{3.12}$$

and (3.10) is consistent with the white noise theory for the "bilinear" system case [7]. In fact the matrix $A(t)$ in that case satisfies the commutativity condition in [7].

From (3.8) it follows that $R(t,t)$ converges with probability one to

$$\frac{d}{2k} I$$

and it is easy to see that the first order distribution is asymptotically Gaussian with zero mean and covariance matrix

$$\text{diag.} \left(\frac{d}{2k}, \frac{d}{2k}, \frac{d_3}{2k_3}\right) . \tag{3.13}$$

and the process covariance:

$$E[Y(t)y(s)^*] = \begin{vmatrix} r(t-s) & 0 \\ 0 & R_3(t-s) \end{vmatrix} , \qquad t \geq s \; . \tag{3.14}$$

4. THE GENERAL CASE

Let us now go on to consider the general case: The White Noise version

$$\dot{y} - y \otimes My + Ky = N(t) \tag{4.1}$$

or the Ito version:

$$dy = (y \otimes My)\, dt - Ky\, dt + dW(t) \quad . \tag{4.2}$$

We are primarily interested in the White Noise version, and our object is to show that (4.1) has a unique continuous solution satisfying:

$$y(t) = y(0) + \int_0^t (y \otimes My - Ky)\,ds + \int_0^t N(s)\,ds\,, \qquad 0 \le t \le T < \infty \tag{4.3}$$

for every $N(\cdot)$ in $L_2[0,T]$. First, let us consider the case where $N(t)$ is continuous in $0 \le t \le T$. Since the function

$$y \otimes My$$

is locally Lipschitz, it follows that for $N(\cdot)$ fixed, we can find a "local" solution satisfying (4.1) in $0 \le t < b$, for some $b < \infty$, and in fact a "maximal" such solution valid up to the escape time $b \le \infty$, (by elementary theory of differential equations, as in [8] for example). We shall now show that b cannot be finite. From (4.1) it follows that

$$[y(t),\dot{y}(t)] = -[Ky(t),y(t)] + [N(t),y(t)] \tag{4.4}$$

Hence

$$\frac{1}{2}\frac{d}{dt}(\|y(t)\|^2) \le [N(t),y(t)] - \gamma\|y(t)\|^2 \tag{4.5}$$

where γ is the smallest eigenvalue of K.

From (4.5) we obtain that

$$\frac{d}{dt}\|y(t)\| \le [N(t),e(t)] - \gamma\|y(t)\| \tag{4.6}$$

where

$$e(t) = \frac{y(t)}{\|y(t)\|} = 0 \qquad \text{if } \|y(t)\| = 0\,.$$

From (4.6) we have that

$$\|y(t)\| \le e^{-\gamma t}\|y(0)\| + \int_0^t e^{-\gamma(t-\sigma)}[N(\sigma),e(\sigma)]\,d\sigma \tag{4.7}$$

in particular

$$\|y(t)\| \le e^{-\gamma t}\|y(0)\| + \int_0^t e^{-\gamma(t-\sigma)}\|N(\sigma)\|\,d\sigma\,, \tag{4.8}$$

showing that b cannot be finite. Moreover, if $\{N_k(\cdot)\}$ is any sequence in $C[0,T]$ such that

$$\sup_k \int_0^t \|N_k(t)\|^2\,dt < \infty\,,$$

the corresponding solutions $\{y_k(\cdot)\}$ would be such that

$$\sup_k \sup_{0\le t\le T} \|y_k(t)\| < \infty$$

and by the local Lipschitzian property (2.5) it would follow that $\{y_k(\cdot)\}$ are actually equicontinuous. This is enough to establish that for each $N(\cdot)$ in $L_2[0,T]$, we have a unique solution of (4.3) in $C[0,T]$, so that we can define

$$y(\cdot) = \Phi(H(\cdot)) \tag{4.9}$$

where the mapping $\Phi(\cdot)$ takes $L_2[0,T]$ into $C[0,T]$ and is of course compact.

Let $\mathcal{L}$ denote the Volterra operator defined by:

$$h = \mathcal{L}g\,; \qquad h(t) = \int_0^t g(s)\,ds$$

mapping $L_2[0,T]$ into itself. In fact $\mathcal{L}$ maps $L_2[0,T]$ into $C[0,T]$. We shall now show that we can express $\Phi(\cdot)$ as:

$$\Phi(N(\cdot)) = \mathcal{L}N(\cdot) + \psi(\mathcal{L}(N(\cdot)) \tag{4.10}$$

where $\psi(\cdot)$ is a continuous mapping of $L_2[0,T]$ into itself. For this purpose let

$$Z(\cdot) = y(\cdot) - \mathcal{L}N(\cdot)$$

or,

$$Z(t) = y(t) - \tilde{W}(t)$$

where

$$\tilde{W}(t) = \int_0^t N(s)\, ds$$

$$Z(t) = Z(0) + \int_0^t [(Z(s)+\tilde{W}(s)) \otimes M(Z(x)+W(s))]ds - \int_0^t K(Z(s)+\tilde{W}(s))ds \quad (4.11)$$

and since $Z(t)$ is absolutely continuous, we can rewrite (4.11) as:

$$\dot{Z}(t) = (Z(t)+\tilde{W}(t)) \otimes M(Z(t)+\tilde{W}(t)) - K(Z(t)+\tilde{W}(t)) \quad (4.12)$$

Hence

$$[Z(t),\dot{Z}(t)] = +[Z(t), (Z(t)+\tilde{W}(t)) \otimes M(Z(t)+\tilde{W}(t))] - [K(Z(t)+\tilde{W}(t), Z(t)] \quad (4.13)$$

Now the first term on the right in (4.13) is

$$= [Z(t), \tilde{W}(t) \otimes M(Z(t)+\tilde{W}(t))]$$

$$= [Z(t), \tilde{W}(t) \otimes MZ(t)] + [Z(t), \tilde{W}(t) \otimes M\tilde{W}(t)] \ .$$

We make use now of the elementary equality:

$$\|x \otimes My\| \le \sqrt{2}\, \|x\| \lambda \|y\|$$

where λ is the largest eigenvalue of M. Letting

$$W_T = \max \|\tilde{W}(s)\| \ , \qquad 0 \le s \le T$$

we obtain

$$\frac{d}{dt}\|Z(t)\| \le \tilde{W}_T \cdot \|Z(t)\| + \lambda(\tilde{W}_T)^2 - \gamma\|Z(t)\| \quad (4.14)$$

and hence

$$\sup_{0\le t\le T} \|Z(t)\| \le c(T, \tilde{W}_T, \|Z(0)\|) \quad (4.15)$$

where $c(\cdot,\cdot)$ is a continuous function of its arguments. Let,

$$N_1, N_2 \in L_2[0,T]$$

$$\tilde{W}_1 = \mathcal{L}N_1$$

$$\tilde{W}_2 = \mathcal{L}N_2 \ .$$

By (4.15) the corresponding solutions are bounded and from (4.12), exploiting the local Lipschitzian property (2.5), we obtain a local Lipschitzian property

$$\|Z_1(t) - Z_2(t)\| \le m \max_{0\le t\le T} \|\tilde{W}_1(t) - \tilde{W}_2(t)\| \quad (4.16)$$

and

$$\int_0^t \|Z_1(t) - Z_2(t)\|^2 dt \le m\int_0^T \|\tilde{W}_1(t) - \tilde{W}_2(t)\|^2 dt \ . \quad (4.17)$$

It follows further that a unique continuous (4.11) is defined for any $\tilde{W}(\cdot)$ in $C[0,T]$. Finally, the solution $Z(\cdot)$ of (4.11) defines a mapping $\psi(\cdot)$ such that

$$Z(\cdot) = \psi(\mathcal{L}(N(\cdot)))$$

where $\psi(\cdot)$ is continuous on $L_2[0,T]$ into $L_2[0,T]$, establishing (4.10). Moreover this is enough to show that $\Phi(N)$ is a physical random variable.

The fact that we have a unique continuous solution of (4.11) for each $\tilde{W}(\cdot)$ in $C[0,T]$ allows us to also conclude that the Wiener process version (4.2) has a unique continuous solution for every sample path in $C[0,T]$, the explosion time being infinite.

5. SOME PROPERTIES OF THE SOLUTION PROCESS

Let us discuss briefly some (statistical) properties of interest of the solution process for (4.1) and (4.2). First of all we note that we have a conditionally Gaussian process for $y_1(\cdot)$ if we fix $y_2(\cdot)$ and $y_3(\cdot)$. This is enough to yield that the process $y(\cdot)$ has conditional densities which may be characterized by Fokker-Planck (or Kolmogorov forward) equations. Let

$$p(y; s \mid x,t)$$

denote the conditional density in the usual notation (see [2]) of $y(s)$ given $y(t) = x$, $s \le t$. Then we have:

$$\frac{\partial p}{\partial s} = -\nabla\cdot((Ky - y \otimes My)p) + \frac{1}{2} \sum_1^3 \sum_1^3 d_{ij} \frac{\partial^2 p}{\partial y_i \partial y_j} \tag{5.1}$$

where $D = \{d_{ij}\}$, and since

$$\nabla\cdot(y \otimes My) = 0 \tag{5.2}$$

by our assumption that M is symmetric, (5.1) can be simplified to:

$$\frac{\partial p}{\partial s} = -[\nabla p, Ky] - p\nabla\cdot Ky + \frac{1}{2} \sum_1^3 \sum_1^3 d_{ij} \frac{\partial^2 p}{\partial y_i \partial y_j} + [\nabla p, y \otimes My] \tag{5.3}$$

and here the fourth term constitutes the non-Gaussianness. It is easy to deduce that the steady state (i.e., asymptotic) first order distribution of $y(\cdot)$ is Gaussian if and only if

$$[R^{-1}y, y \otimes My] = 0 \qquad \text{for all } y$$

or, equivalently,

$$R^{-1} = (\alpha I + \beta M) \tag{5.4}$$

where I is the 3×3 Identity and α and β are real numbers such that

$$\alpha + \beta\gamma > 0$$

and R (the covariance) is defined by:

$$D = KR + RK^* \,. \tag{5.5}$$

This generalizes (3.13) derived for the Axially Symmetric Case.

FOOTNOTES

1. Research supported in part by the Air Force Office of Scientific Research under Grant No. 83-0318.

2. Usually ω is used to denote angular velocity but obviously not usable to stochastic context for danger of confusion with ω as sample point! similarly we cannot use I for moment of inertia for fear of confusion with Identity Matrix.

REFERENCES

[1] Wertz, J.R., Spacecraft Attitude Determination and Control (D. Reidel Publishing Co., 1980).

[2] Gikhman, I. and Skorokhod, A.V., Stochastic Differential Equations (Springer-Verlag, 1972).

[3] Kallianpur, G. and Karandikav, R.L., A Finitely Additive White Noise Approach to Nonlinear Filtering, J. Appl. Math and Optimization, 10 (1983) 159-185.

[4] Ikeda, N. and Watanabe, S., Stochastic Differential Equations and Diffusion Processes (North-Holland, 1981).

[5] Liptser, R.S. and Shiryayev, A.N., Statistics of Random Processes, II (Springer-Verlag, 1978).

[6] Kleptsina, M.L. and Veretennikou, A. Yu., On Filtering and Properties of Conditional Laws of Ito-Volterra Processes, Steklov Seminar (1984).

[7] Balakrishnan, A.V., On Abstract Stochastic Bilinear Equations with White Noise Inputs, J. of Appl. Math and Optimization, 10 (1983) 351-366.

[8] Birkhoff, G. and Rota, G.C., Ordinary Differential Equations, 3rd edition (John Wiley, 1978).

CURRENT RESULTS AND ISSUES

IN STOCHASTIC CONTROL

A. BENSOUSSAN

University Paris Dauphine and INRIA

GENERAL INTRODUCTION.

Since the basic review of W. FLEMING [1], several surveys and books have appeared during the fifteen past years, relating substantial progress of the theory of stochastic control. Among the books, let us mention W. FLEMING - R. RISHEL [1], A. FRIEDMAN [1], A.N. SHYRYAEV [1] , D. BERTSEKAS - S.E. SHREVE [1], I.I. GIHMANN - A.V. SKOROKHOD [1], N.V. KRYLOV [1], M.J. KUSHNER [1], E.B. DYNKIN - A.A. YUSHKEVICH [1], A. BENSOUSSAN [1], A. BENSOUSSAN - J.L. LIONS [1], [2], G. KALLIANPUR [1].

In addition several monographs and lecture notes provide an additional important material. Among many, let us mention J.M. BISMUT [1], N. ELKAROUI [1], M. NISIO [1], M. HAZEWINKEL - J.C. WILLEMS [1], S.K. MITTER - A. MORO [1], H. KOREZLIOGLU - G. MAZZIOTTO - J. SZPIRGLAS [1], M.A.H. DEMPSTER [1].

The Bad Honnef conference proceedings since 1979 represent a good source for the evolution of the art. Let us also mention the three volumes published by the french CNRS, under the responsability of I. LANDAU [1]. Finally the survey of M.H.A DAVIS [1] has been an important guide for the present paper.

This abundant litterature showed the importance of the development of the field. The mathematical techniques which are used are quite diversified (partial differential equations, stochastic processes, control theory, algebraic methods, numerical analysis...) and the applications are numerous (Engineering, Economics, Operations Research,...).

We emphasize in this review the control of diffusions with full or partial observation, for one decision maker and give some hints to other problems. Needless to say, this part of the theory, although of basic importance, does not cover everything. Even for it, we do not pretend here to be complete, which would have been impossible anyway in one article.

I. STOCHASTIC CONTROL WITH COMPLETE INFORMATION.

1. Introduction

This kind of problem has been studied from the very beginning. The diversity arises from the type of dynamics of the state, the type of control and the approach (Dynamic Programming or Maximum Principle).

Dynamic Programming leads to the study of the value function. According to the nature of the state equation, the value function is solution of a functional equation (Bellman equation) which can be more or less easily studied by Functional Analysis methods.

The difficulty arises from the lack of regularity of the value function (leaving aside the situation of non degenerate diffusions). When there is no regularity it is necessary to introduce weak solutions (connected to semigroup approaches) or purely probabilistic methods.

The stochastic Maximum Principle can be formulated. However its applicability is much more questionable than in the deterministic case.

The diversity of the control possibilities is illustrated by the following cases (continuous control, optimal stopping, impulse control, bounded variations control, constraints on the state etc...).

2. Stochastic Control for diffusions

2.1. *Statement of the problem*

An excellent survey can be found in P.L. LIONS [1]. Let U be a separable metric space and U_{ad} a compact subset of U. Consider

(2.1) $g(x,v) : R^n \times U_{ad} \to R^n$

$\sigma(x,v) : R^n \times U_{ad} \to L(R^m ; R^n)$

continuous and bounded ; g_x, σ_x bounded.

Let $(\Omega, \mathcal{A}, P)$ be a probability space equipped with a Filtration F^t, and a F^t Wiener process $w(t)$ with values in R^m.

An admissible control is a process $v(t)$ adapted to F^t, with values in U_{ad}. For each $v(.)$ we can solve the stochastic differential equation

(2.2) $$dx = g(x(t), v(t))dt + \sigma(x(t), v(t))dw(t)$$

$$x(0) = x$$

Let $\mathcal{O}$ be a bounded smooth domain of R^n. Assume that $x \in \bar{\mathcal{O}}$, and let τ be the exit time of the process $x(.)$ from the domain $\mathcal{O}$.

Consider functions

(2.3) $$f(x,v),\ c(x,v),\ \phi(x)$$

continuous and bounded with values in R

Let us define the cost function

(2.4) $$J(x;v(.)) = E\ \{ \int_0^\tau f(x(t), v(t))(\exp - \int_0^t c(x(s),v(s))ds)dt + $$
$$+ \phi(x(\tau))\exp - \int_0^\tau c(x(s),v(s))ds\}$$

and the value function

(2.5) $$u(x) = \operatorname*{Inf}_{v(.)} J(x;v(.))$$

Consider the family of 2nd order differential operators

(2.6) $$A^v = - a_{ij}(x,v) \frac{\partial^2}{\partial x_i \partial x_j} - g_i(x,v) \frac{\partial}{\partial x_i} + c(x,v)$$

in which the matrix

$$a = \frac{1}{2} \sigma\sigma^*$$

Bellman equation reads

(2.7) $$\sup_{v \in U_{ad}} \{A^v u - f^v\} = 0 \quad \text{in } \mathcal{O}$$

where $f^v(x) = f(x,v)$

2.2. *Regularity results*

It is possible to interpret (2.7) in a classical sense, when there is regularity and non degeneracy. Assume

(2.8) $$\psi(.v) \in W^{2,\infty}(R^n), \quad \|\psi(.,v)\|_{W^{2,\infty}} \le C \quad , \forall v$$

where $\psi = \sigma_{ij}$, g_i, c, f

(2.9) $$\phi \in W^{3,\infty}(R^n)$$

(2.10) $$\lambda = \inf \{c(x,v), x \in R^n, v \in U_{ad}\} \ge 0$$

(2.11) $$a(x,v) \ge \nu I , \quad \nu > 0$$

Theorem 2.1. Under the assumptions (2.8) to (2.11), $u \in C^{2,\alpha}(\bar{\mathcal{O}})$, $(\alpha \in]0,1[)$, $u = \phi$ on $\Gamma(\partial\mathcal{O})$ and satisfies (2.7). It is the unique solution of (2.7) in that class of functions.

The final regularity result is due to L.C. EVANS [1]. The same result with $u \in W^{2,\infty}(\mathcal{O})$ has been previously given by P.L. LIONS [2], L.C. EVANS - P.L. LIONS [1]. The first results in this field have been obtained by N.V. KRYLOV [1], [2]. Of course, when σ does not depend on v, then (2.7) becomes a quasi linear elliptic equation, and the result of Theorem 2.1 is classical.

2.3. *Degeneracy cases*

It is possible to prove a regularity result in cases of degeneracy, provided the constant λ is large enough. The type of regularity result which can be expected is the following

(2.12) $$u \in W^{1,\infty}(\mathcal{O}),$$

$$\partial^2_\xi u \le C , \text{ in } \mathcal{D}'(\mathcal{O}), \forall |\xi| = 1$$

$$A^v u \in L^\infty(\mathcal{O}), \quad \| A^v u\|_{L^\infty} < C$$

The Bellman equation (2.7) then holds a.e. x. We refer to P.L. LIONS [1] for details.

Note that the assumption λ large enough is necessary. Indeed consider $\mathcal{O} =]-1, +1[$, $\sigma = 0$, $g(x,v) = x$, $f = 0$, $\phi = 1$, $c = \lambda$, then $u(x) = |x|^{\lambda}$.

The 2nd property (2.12) requires $\lambda \geq 2$ (in particular). Concerning uniqueness, a solution of (2.12), (2.7) satisfying the boundary condition $u = \phi$ on Γ, is necessarily unique. This follows from the following uniqueness result (P.L. LIONS [1]).

<u>Theorem 2.2</u>. <u>Let</u> $\tilde{u} \in C(\bar{\mathcal{O}}) \cap W^{1,n}_{loc}(\mathcal{O})$ <u>such that</u> $\tilde{u} = \phi$ on Γ <u>and</u>

(2.13) $\quad A^v\tilde{u} \leq f^v$ <u>in</u> $\mathcal{D}'(\mathcal{O})$ $\forall v$

(2.14) $\quad \sup_v (A^v\tilde{u} - f^v) = 0$ <u>in the sense of measures</u>

(2.15) $\quad \exists\, \theta \in L^N_{loc}(\mathcal{O})$, <u>such that</u> $\Delta\tilde{u} \leq \theta$ <u>in</u> $\mathcal{D}'(\mathcal{O})$ <u>then</u> $\tilde{u} = u$ <u>in</u> $\mathcal{O}$

□

A condition such as (2.15) is necessary to ensure uniqueness. Indeed, consider the Bellman equation

(2.16) $\quad |Du| + \lambda u = 1 \quad , \quad$ a.e. in R^n

then $\forall\, \beta \geq 0$, $x_o \in R^n$,

$$u(x) = \frac{1}{\lambda}\,(1 - \beta \exp - \lambda|x - x_o|)$$

is a solution in $W^{1,\infty}$, satisfying (2.13), (2.14) but not (2.15).

2.4. *Characterization of the value function when the Bellman equation is not satisfied.*

Assume only (2.8), (2.9) and

(2.17) $\quad \lambda$ (defined by (2.10)) > 0

No specific assumption on the size of λ is made. In that case no regularity result for the value function holds. We have the following characterization (see P.L. LIONS - J.L. MENALDI [1] , P.L. LIONS [3]).

Theorem 2.3. Assume (2.8), (2.9), (2.17). Then $u \in L^\infty(\mathcal{O})$, and $\forall$ B bounded open set such that $\bar{B} \subset \mathcal{O}$, one has $\| a^{1/2} Du\|_{L^2(B)} \le C_B$ (independant of v)

(2.18) $\quad A^v u \le f^v$ in $\mathcal{D}'(\mathcal{O}) \quad \forall v.$

Moreover if $v \in C(\bar{\mathcal{O}})$, satisfies (2.18) and

$$\overline{\lim_{y\in\mathcal{O},y\to x}}\ v(y) \le u(x) \qquad \forall x \in \Gamma$$

then $v \le u$ in $\mathcal{O}$

□

The value function appears as the maximum subsolution of the Bellman equation (2.7). Another useful concept, introduced by M.G. CRANDALL and P.L. LIONS [1] for non linear 1st order P.D.E and extended by P.L. LIONS [1] for non linear 2nd order P.D.E. is that of viscosity solution.

If ϕ is a continuous function on $\mathcal{O}$, one defines the super differential of order 2 of ϕ at point x as

(2.19) $$D_2^+ \phi(x) = \{(\xi \in R^n, \eta \in L(R^n;R^n)) \ /$$

$$\limsup_{y\to x, y\in\mathcal{O}} [\phi(y)-\phi(x) - \xi.(y-x) - \frac{1}{2}\eta(y-x)^2]|y-x|^2 = 0\}$$

and the subdifferential of order 2,

(2.20) $$D_2^- \phi(x) = - D_2^+(-\phi)(x).$$

Consider a non linear 2nd order P.D.E. of the form

(2.21) $$H(x,u,Du,D^2u) = 0 \quad \text{in } \mathcal{O}$$

Then $u \in C(\mathcal{O})$ is a viscosity solution of (2.21) if $\forall x \in \mathcal{O}$

(2.22) $$\forall (\xi,\eta) \in D_+^2 u(x), \quad H(x,u,\xi,\eta) \le 0$$

$$\forall (\xi,\eta) \in D_-^2 u(x), \quad H(x,u,\xi,\eta) \ge 0$$

Then one has the following result (P.L. LIONS [1])

Theorem 2.4. If the value function $u \in C(\mathcal{O})$, it is a viscosity solution of (2.7). Moreover if $v \in C(\bar{\mathcal{O}})$, satisfies $v = u$ on Γ and is a viscosity solution of (2.7), then $v = u$

3. Stochastic control of discontinuous processes.

3.1. *Piece wise deterministic Markov processes*

After some earlier work of R. RISHEL [1], important recent contributions to the theory are due to M. DAVIS [2], and D. VERMES [1]. We describe here the results of VERMES.

Let $\mathcal{O}$ be a bounded smooth domain of R^n, and $\Gamma = \partial\bar{\mathcal{O}}$. Let

(3.1) $g(x)$ Lipschitz function from R^n into R^n

and

$$\Gamma^+ = \{x \in \Gamma \mid g(x).\nu(x) > 0, \text{ where } \nu \text{ denotes the unit outward normal}\}.$$

If one solves the differential equation

$$\frac{dx}{ds} = g(x) \qquad x(t) = x \in \bar{\mathcal{O}} \tag{3.2}$$

then the trajectory denoted by $X_{xt}(x)$ exists from $\bar{\mathcal{O}}$ at points belonging to Γ^+. We write

$$E = \bar{\mathcal{O}}, \; E_\partial = \Gamma^+ \;, \; E_o = E - E_\partial .$$

Let

$$\lambda : E \to [0,\infty), \quad p : E_\partial \to M^1(E_o), \tag{3.3}$$

$$q : E \to M^1(E_o), \text{ continous functions}$$

where $M^1(E_o)$ denotes the set of probability measures on E_o.

A piece wise deterministic process (PDI) has the following characteristics.

i) The trajectories $x(t)$ are right continuous functions with left hand limits. Their discontinuities do not accumulate

ii) in between two jumps the process evolves according to the dynamics (3.2)

ii) Let $\beta(t) = \inf \{t < s \mid x(s-) \neq x(s)\}$

$= $ time of the first jump after t.

One has

$$P(\beta(t) \geq t + h \mid X_{x(t),t}(s) \in E_o,\ s \in [t,t+h])$$
$$= \exp - \int_t^{t+h} \lambda(X_{x(t),t}(s))ds$$

$$P(\beta(t) \geq t+h \mid X_{x(t),t}(.) \text{ contains points of } E_\partial \text{ on the interval } [t,t+h]) = 0$$

iii) Let B be a Borel subset of E_o, then

$$P(x(\beta) \in B \mid x(\beta-) \in E_o) = \int_B q(x(\beta-),d\xi)$$

$$P(x(\beta) \in B \mid x(\beta-) \in E_\partial) = \int_B p(x(\beta-),d\xi)$$

Let also

$$\alpha(t) = \sup \{0 \leq s \leq t \mid x(s-) \neq x(s)\}$$

$$\eta(t) = x(\alpha(t))$$

$z(t) = t-\alpha(t)$, length of time since the last jump

The triple $x(t)$, $\eta(t)$, $z(t)$ is also a PDP on the state space $\tilde{E} = E \times E \times [0,\infty)$

Given a point $x \in E$ (initial value $x(0-) = x$), then DAVIS has shown that it is possible to construct a unique Markov process $x(t)$, P^x which is a P.D.P. corresponding to the data (g,λ,p,q).

The next step is to introduce some control possibilities. Let $(Y_o,\mathcal{Y}_o)$ and $(Y_\partial,\mathcal{Y}_\partial)$ be compact Polish spaces equipped with their Borel σ algebras.

We assume now that

$$(3.3) \qquad g(x,y) : R^n \times Y_o \to R^n$$

$$\lambda(x,y) : E \times Y_o \to 0,\infty)$$

$$p(x,y,d\xi) : E_\partial \times Y_\partial \to M^1(E_o)$$

$$q(x,y,d\xi) : E \times Y_o \to M^1(E_o)$$

be continuous functions.

VERMES [1] considers <u>relaxed strategies</u>, following ideas of VINTER and LEWIS [1] for deterministic control.

A relaxed strategy is a pair

$$u^o : E_o \times [0,\infty) \to M^1(Y_o)$$

$$u^\partial : E_\partial \to M^1(Y_o).$$

The two arguments in u^o will represent the position of the state at the time of last jump, and the time elapsed since the last jump.

Define

$$g^u(x,\eta,z) = \int_{Y_o} g(x,y)\ u^o(\eta,z;dy)$$

the corresponding trajectory $X^u_\eta(s)$ is the solution of

$$(3.4) \qquad \frac{dX}{ds} = g^u(X,\eta,s)$$

$$X(0) = \eta$$

Given cost functions $\ell_o(x,y)$, $\ell_\partial(x,y)$, one defines similarly $\ell^u_o(x,\eta,z)$ and $\ell^u_\partial(x)$ with

$$\ell^u_\partial(x) = \int_{Y_\partial} \ell_\partial(x,y) u^\partial(x;dy).$$

The trajectory (3.4) being well defined, it is possible to define a controlled P.D.P. corresponding to a relaxed strategy and thus to introduce a payoff

(3.5) $$J_x(u) = E_x^u \{ \int_0^T \ell_o^u(x,\eta,z)ds + \sum_{t\leq T \,:\, x(t-)\in E_\partial} \ell_\partial^u(x(t-)) \}$$

Under suitable assumptions VERMES proves the existence of an optimal relaxed strategy and gives a characterization of the value function as the upper envelope of the smooth subsolutions of the Bellman equation.

Moreover a necessary and sufficient condition of optimality is given, which extends the theory of VINTER and LEWIS for the deterministic case.

3.2. *Stochastic control of diffusions with jumps*

A more classical approach that the one discussed in § 3.1 consists in studying the Bellman equation. This is possible in the case there is sufficient regularity. This is the case for diffusions with jumps. We describe here the work of BENSOUSSAN-LIONS [2]. The Bellman equation belongs to the following class. Consider

(3.6) $$A = - \sum_{i,j} \frac{\partial}{\partial x_i} a_{ij} \frac{\partial}{\partial x_j} + \sum_i a_i + a_o$$

where

(3.7) $$a_{ij} = a_{ji} \in W^{1,\infty}(R^n),\ a_i \in L^\infty(R^n)$$

$$a_o \in L^\infty(R^n),\ a_o \geq \beta > 0$$

$$\Sigma a_{ij}\, \xi_i\, \xi_j \geq \alpha|\xi|^2,\quad \forall\ \xi \in R^n,\ \alpha > 0.$$

Let $\mathcal{O}$ be a bounded domain of R^n, such that $\Gamma = \partial\mathcal{O}$ is of class C^2. Consider an operator B such that

(3.8) $$B \in L(W_o^{1,p}(\mathcal{O})\ ;\ W^{-1,p}(\mathcal{O})),\ \forall\ p \geq 2,\ p < \infty$$

$$B = B_r^1 + B_r^2 \quad ,\ \forall\ r\ ,\ \text{with}$$

$$B_r^1 \in L(W_o^{1,p}\ ;\ W^{-1,p}(\mathcal{O})),\ \ ||B_r^1|| \leq O(r)$$

with $o(r) \to 0$ and $r \to 0$

$$B_r^2 \in L(W_o^{1,p}\ ;\ L^p)$$

Consider next a non linear operator H verifying

(3.9) $$H : W_0^{1,p}(\mathcal{O}) \to L^p(\mathcal{O}), \quad \forall\, p \geq 2,\ p < \infty$$

$$H(0) \in L^\infty \ ; \text{ if } \quad \phi \in W_0^{1,p}\ ,\ \psi \in W_0^{1,q},\ q \geq p$$

$$H(\phi+\psi) - H(\phi) \in L^q \text{ and } |H(\phi+\psi) - H(\phi)|_{L^q} \leq C\|\psi\|_{W_0^{1,q}}$$

(3.10) $$\langle B\phi + H(\phi+\psi) - H(\phi),\ (\psi-k)^+\rangle \quad \leq C|(\psi-k)^+|\ \|(\psi-k)^+\| + k\beta_1 \int_{\mathcal{O}} (\psi-k)^+ dx$$

where $\beta_1 > 0$, $\forall\ \phi,\ \psi \in H_0^1(\mathcal{O})$, $k \geq 0$; C independant of ϕ,ψ,k.

Consider the equation

(3.11) $$Au - Bu - H(u) = h, \quad u \in W_0^{1,p}(\mathcal{O})$$

where $h \in L^p(\mathcal{O})$.

One has the result

Theorem 3.1. Under the assumptions (3.7), (3.8), (3.9), (3.10) there is one and only one solution of (3.11)

□

In the applications we have in mind the operators B ang H are given by

(3.12) $$B\phi(x) = \int_{R^n} [\phi(x+z) - \phi(x) - z.\nabla\phi\ \chi_{|z|\leq 1}]c_0(x,z)m(dz)$$

where

(3.13) $$0 \leq c_0 \leq 1, \text{ measurable, } |D_x c_0| \leq C$$

and m(dz) is a positive measure on R^n, not necessarily bounded, with a singularity at 0, satisfying

(3.14) $$\int_{\{|z|\leq 1\}} |z|^2 m(dz) + \int_{\{|z|>1\}} |z| m(dz) < \infty$$

whereas $H(\phi)$ is defined by

(3.15) $$H(\phi)(x) = \inf_{v \in U_{ad}} \{f(x,v) + D\phi . g(x,v) - \phi\, \tilde{a}_1(x,v) + \int_{R^n} [\phi(x+z) - \phi(x) - z\, \Delta\phi\, \chi_{|z| \leq 1}]\, c_o(x,z) c_1(x,v,z) m(dz)$$

where

(3.16) $$f,\ c_1,\ \tilde{a}_1 : R^n \times U_{ad} \to R$$

$g : R^n \times U_{ad} \to R^n$, bounded measurable

$\sup \tilde{a}_1^- < \beta$

$1 + c_1 \geq 0 \quad , \ |c_1| \leq C|z|$

The operators B and H defined by (3.12) and (3.15) satisfy the assumptions (3.8), (3.9), (3.10).

By a convenient redifinition of g, it is possible to assume that

(3.17) $$a_i = \sum_j \frac{\partial a_{ij}}{\partial x_j}$$

so that

$$A = - a_{ij} \frac{\partial^2}{\partial x_i \partial x_j}$$

Let $\Omega_o = D([0,\infty) ; R^n)$ and $x(t;\omega) = \omega(t)$ denote the canonical process. Let $M^t = \sigma(x(s),\ 0 \leq s \leq t)$, $A = M^\infty$. One can prove that for any x, there exits one and only one solution P^x of the martingale problem, i.e.

(3.18) $$P^x(x(0) = x) = 1$$

$$\forall\, \phi \in \mathcal{D}(R^n), \quad \phi(x(t)) - \phi(x) + \int_0^t (A-B)\phi\ (x(s))ds$$

is a P^x martingale with respect to M^t.

(this results is due to D. STROOCK [1], LEPELTIER - MARCHAL [2]).

An admissible control is a process, adapted to M^t, with values in U_{ad}. To any control, it is possible to associate the unique solution $Q^x_{v(.)}$ of the martingale problem

(3.19) $$Q^x_{v(.)}(x(0) = x) = 1$$

$$\forall \phi \in \mathcal{D}(R^n), \quad \phi(x(t)) - \phi(x) - \int_0^t \{a_{ij}(x(s)) \frac{\partial^2 \phi}{\partial x_i \partial x_j} +$$

$$+ g(x(s),v(s)).D\phi + \int_{R^n} [\phi(x(s)+z) - \phi(x(s)) - z\, D\phi \chi_{|z|\le 1}]$$

$$c_o(x(s)z)(1 + c_1(x(s),v(s),z))m(dz)\}\, ds$$

is a $Q^x_{v(.)}$, M^t martingale

Define next

(3.20) $$a_1(x,v) = a_o(x) + \tilde{a}_1(x,v)$$

and the cost function

(3.21) $$J^x(v(.)) = E^{Q^x_{v(.)}} [\int_0^\tau f(x(t), v(t))(\exp - \int_0^t a_1(x(\lambda),(\lambda))d\lambda)dt]$$

where τ represents the exit time of x(.) from the boundary of $\mathcal{O}$.

Note that the operators B and H require functions defined on the whole space. A function belonging to $W_o^{1,p}(\mathcal{O})$ is extented by 0 outside the domain, hence Bu and H(u) in (3.11) are well defined.

Consider (3.11) with h = 0, hence $u \in W_o^{1,p}(\mathcal{O})$, $\forall p < \infty$. One has the following result

Theorem 3.2. *Under (3.7), (3.13), (3.14), (3.16), (3.17), the solution of (3.11) (with h = 0) is given explicitly by the formula*

(3.22) $$u(x) = \text{Min } J^x(v(.))$$

An optimal control exists, defined by a feedback.

Remark 3.1.

The Bellman equation approach to the problem discussed in § 3.1. would lead the following equation

$$(3.23)\qquad \frac{\partial u}{\partial t} + \operatorname*{Inf}_y \{Du.g(x,y) + \int_{E_o} (u(\xi,t)-u(x,t))\lambda(x,y)p(x,y,d\xi) +$$

$$+ \ell_o(x,y)\} = 0 \quad \text{in } E_o$$

$$u(x,t) = \operatorname*{Inf}_y [\int_{E_\partial} u(\xi,t)q(x,y,d\xi) + \ell_\partial(x,y)] \quad , x \in E_\partial$$

$$u(x,T) = 0$$

Remark 3.2. For the control of random evolutions, see R. RISHEL [1], A. BENSOUSSAN - P.L. LIONS [1].

For the control of branching processes, see S. USTENEL [1].

4. Probabilistic methods - Semigroup methods

4.1. *Martingale methods*

Following the presentation of M. DAVIS [3], we explain these methods in the case of non generate diffusions, the control operating only on the drift. Let

$$(4.1)\qquad \sigma : R^n \to L(R^n;R^n)$$

$$\sigma_x \text{ bounded}$$

$$(4.2)\qquad a = \frac{1}{2}\sigma\sigma^* \geq \nu I$$

Let $(\Omega,\mathcal{A},P)$ be a probability space, equipped with a filtration F^t, $(\mathcal{A} = F^\infty)$ and a standardized Wiener process with values in R^n, denoted by $w(t)$.

We solve the stochastic differential equations

$$(4.3)\qquad dx = \sigma(x)dw \qquad x(0) = x_o$$

An admissible control is a process $v(.)$ adapted to F^t, with values in a compact subset U_{ad} of a metric space U.

Let

$$g(x,v) \; : \; R^n \times U_{ad} \to R^n, \tag{4.4}$$

define the Girsanov transform

$$\left.\frac{dP^{v(.)}}{dP}\right|_{F^t} = \exp \{ \int_0^t \sigma^{-1} \, g \, dw - \frac{1}{2} \int_0^t |\sigma^{-1} g|^2 ds \} \tag{4.5}$$

and the payoff

$$J(v(.)) = E^{v(.)} \{ \int_0^T f(x(t),v(t))dt + \phi(x(T)) \}. \tag{4.6}$$

The conditional remaining cost associated to the control v(.) is defined by

$$\psi^{v(.)}(t) = E^{v(.)} \{ \int_t^T f(x(s),v(s))ds + \phi(x(T)) \mid F^t \}. \tag{4.7}$$

The random variable $\psi^{v(.)}(t)$ belongs to $L^1(\Omega,F^t,P)$. The value function is defined by

$$J(t) = \underset{v(.)}{\text{ess inf}} \; \psi^{v(.)}(t), \tag{4.8}$$

and J(t) in an F^t adapted process. Of course, the problem really of interest is to compute J(o) and the corresponding optimal control.

For this problem, Bellman equation has a meaning and reads

$$\frac{\partial u}{\partial t} + Au = \inf_v \{f(x,v) + Du.g(x,v)\} \tag{4.9}$$

$$u(x,T) = \phi(x)$$

In fact, it can be shown that

$$J(t) = u(x(t),t) \tag{4.10}$$

However, the advantage of introducing J(t) directly by formula (4.8) is that a direct study can be conducted, without relying on Bellman equation (4.9). This approach can be generalized to cases where the Bellman equation is not easily studied. The main properties of J(t) are the following

$$(4.11) \qquad M^{v(.)}(t) = \int_0^t f(x(s),v(s))ds + J(t)$$

is for any v(.) a $P^{v(.)}$ submartingale.

It is a martingale if and only if the corresponding control is optimal.

By Doob Meyer decomposition theorem, one can write

$$(4.12) \qquad M^{v(.)}(t) = J(o) + A^{v(.)}(t) + N^{v(.)}(t)$$

where $A^{v(.)}(t)$ is an increasing predictable process (with $A^{v(.)}(0) = 0$), and $N^{v(.)}(t)$ is a $P^{v(.)}$, F^t martingale.

Considering next the process

$$(4.13) \qquad w^{v(.)}(t) = w(t) - \int_0^t \sigma^{-1} g(x(s),v(s))ds$$

then $w^{v(.)}(t)$ is a F^t, $P^{v(.)}$ Wiener process.

Assume for instance that $F^t = \sigma(w(s), s \leq t)$, then from KUNITA-WATANABE [1] representation theorem, one can prove that

$$(4.14) \qquad N^{v(.)}(t) = \int_0^t \gamma^{v(.)}(s).\sigma(x(s))dw^{v(.)}(s)$$

where $\gamma^{v(.)}$ is an F^t adapted process. From (4.11), (4.12), (4.14) it follows that

$$(4.15) \qquad J(t) = J(0) + A^{v(.)}(t) + \int_0^t \gamma^{v(.)}(s)\sigma(x(s))dw^{v(.)}(s) - \\ - \int_0^t f(x(s),v(s))ds.$$

If u(.) is another control, one deduces from (4.15) and (4.13)

$$M^{u(.)}(t) = J(0) + A^{v(.)}(t) + \int_0^t \gamma^{v(.)}(s) \ (x(s))dw^{v(.)}(s) + \\ + \int_0^t (f(x(s),u(s)) - f(x(s),v(s)))ds$$

and from (4.13)

$$= J(0) + A^{v(.)}(t) + \int_0^t \gamma^{v(.)}(s)\sigma(x(s))dw^{u(.)}(s) +$$

$$+ \int_0^t f(x(s),u(s)) + \gamma^{v(.)}(s)g(x(s),u(s)) -$$

$$- (f(x(s),v(s)) + \gamma^{v(.)}(s)g(x(s),v(s)))\ ds$$

$$= J(0) + A^{u(.)}(t) + \int_0^t \gamma^{u(.)}(s)\sigma(x(s))dw^{u(.)}(s)$$

Necessarily

$$\gamma^{v(.)}(s) = \gamma^{u(.)}(s) = \gamma(s)$$

and

$$A^{v(.)}(t) - \int_0^t [f(x(s),v(s)) + \gamma(s).g(x(s),v(s))]ds$$

is independant of the control

Since $A^{u(.)}(t) = 0$, when $u(.)$ is optimal, it follows that

$$A^{v(.)}(t) = \int_0^t [f(x(s),v(s)) + \gamma(s).g(x(s),v(s)) - \tag{4.16}$$

$$- (f(x(s),u(s)) + \gamma(s).g(x(s),u(s))]ds$$

Hence

$$f(x(s),v(s)) + \gamma(s).g(x(s),v(s)) \geq f(x(s),u(s)) + \gamma(s).g(x(s),u(s)) \tag{4.17}$$

The relation (4.17) is a necessary condition of optimality. In fact, it can be shown (cf DAVIS [1]), that the minimization of the Hamiltonian (4.17) yields an optimal control.

After the work of RISHEL [2], martingale methods have been used for stochastic control problems in some generality by STRIEBEL [1], BOEL-VARAYA [1], MEMIN [1], ELLIOT [1], BOEL-KOHLMANN [1], DAVIS-KOHLMANN [1], N.EL KAROUI [1], BREMAUD-PIETRI [1], GERTNER-RAPPAPORT [1].

Probabilistic methods are also used to obtain results on the existence of optimal control. After the initial work of V. BENES [1], a fairly complete theory has been developped by J.M. BISMUT [1] relying on potential theory methods.

4.2. Semigroup methods.

Semi group methods have been introduced by M. NISIO [2], [3]. She has proposed the concept of envelope of Markovian semi groups, and considered non linear semi groups associated to stochastic control. This approach has the advantage of weakening the formulation of the Dynamic Programming equation, in some analogy with probabilistic methods discussed in § 4.1. Several authors have contributed to the theory, J. ZABCZYK [1], L. STETTNER - J. ZABCZYK [1].

One nice property of this approach is that a purely analytic treatment is possible, independently of the probabilistic interpretation. Therefore one can proceed in the same spirit as in the P.D.E. case (cf section 2), develop the study of an analytic problem and interpret its solution as the value function of a stochastic control problem, see A. BENSOUSSAN [1] , A. BENSOUSSAN - M. ROBIN [1]. The analytic problems introduced in this way have motivated research on the theory of non linear semi groups in L^∞ spaces (see L. BARTHELEMY [1]).

We present here some ideas of this approach, following the presentation of A. BENSOUSSAN - M. ROBIN.

Let E be a topological space and $\mathcal{E}$ its Borel σ algebra.

Let

$$B \equiv B(E) = \text{space of bonded Borel functions on } E$$

$$C \equiv C(E) = \text{space of uniformily continuous and bounded functions on } E$$

We consider on B and C the sup norm, denoted by $\|\ \|$. We next consider a family $\Phi^m(t)$ of linear semi groups on B such that

$$\Phi^m(t+s) = \Phi^m(t)\Phi^m(s) \tag{4.18}$$

$$\Phi^m(0) = I$$

$$\|\Phi^m(t)\| \leq 1$$

$$\Phi^m(t)f \geq 0 \quad \forall\, f \geq B,\ f \geq 0$$

One also needs the following properties

(4.19) $\Phi^m(t) : C \to C$

(4.20) $t \to \Phi^m(t)f(x)$ is continuous from $]0,\infty[\to R$, $\forall$ x fixed, $\forall f \in C$.

Let us next consider a family L^m of elements of B, such that

(4.21) $L^m \geq 0, \quad L^m \in B$

$t \to \Phi^m(t)L^m$ is measurable from $]0,\infty[$ in C.

We introduce the following problem find

(4.22) $u \in B$, u is u.s.c.

$$u \leq \int_0^t e^{-\alpha s} \Phi^m(s)L^m ds + e^{-\alpha t} \Phi^m(t)u,$$

$\forall m = 1...N, \forall t.$

One can then prove the following

Theorem 4.1. Under the assumptions (4.18), (4.19), (4.20), (4.21), for $\alpha > 0$, the set of functions u satisfying (4.22) is not empty and has a maximum element

□

The formulation (4.22) is of course very weak. It is important for the applications to give conditions under which the maximum element u is continous. Let us assume that

(4.23) $|L^m(s) - L^m(y)| \leq K|x-y|^\delta, \ 0 < \delta \leq 1, \ \forall m$

(4.24) $\forall g \in C^{0,\delta}(E)$, then $\forall t \geq 0$,

$\Phi^m(t)g \in C^{0,\delta}(E)$ and

$$||\Phi^m(t)g||_{C^{0,\delta}} \leq e^{\lambda t} \ ||g||_{C^{0,\delta}} \lambda \geq 0$$

One can then prove the following

Theorem 4.2. Under the assumptions of Theorem 4.1, and (4.23), (4.24), then if $\alpha > \lambda$,

$u \in C^{o,\delta}$. <u>For</u> $\alpha > 0$, <u>one can guarantee</u> $u \in C$

□

One can then define an evolution problem, as follows. Let

$$(4.25) \qquad \begin{aligned} & u(.) \in C([0,T];C) \ u(0) = \bar{u} \\ & u(t) \le \int_0^{t-s} e^{-\alpha\sigma} \Phi^m(\sigma) L^m d\sigma + e^{-\alpha(t-s)} u(s) \\ & \forall\ s \le t \end{aligned}$$

where $\bar{u} \in C$.

In a way similar to Theorem 4.1, one can prove that $\forall\ T > 0$, the set of elements solutions of (4.25) is not empty and has a maximum element.

Write

$$(4.26) \qquad u(t) = S(t)\bar{u}$$

which defines a non linear contraction semi group, and

$$(4.27) \qquad u(t) \to u \quad \text{as} \quad t \to \infty,$$

where u is the maximum element of the set (4.22).

Let us then give the interpretation of the maximum element of (4.22). Let us set

$$\Omega_o = E^I \quad , \quad x(t;\omega) \text{ the canonical process}$$

$$M_t^s = \sigma(x(\lambda) \ ; \ t \le \lambda \le s) \quad , \ M^t = M_t^\infty$$

For any m, one can define a probability P_m^{xt} on Ω_o, M_t such that

$$(4.28) \qquad E_m^{xt} \ \phi(x(s)) = \Phi^m(s-t) \ \phi(s), \ \forall\ s \ge t.$$

Let us consider the class W of step processes adapted to M_o^t, with values in [1,...,N].

More precisely, if $V \in W$, there exists a sequence

$$\tau_o = 0 \le \tau_1 \ldots \le \tau_n < \ \ldots .$$

which is deterministic, increasing and convergent to ∞ and

$$v(t;\omega) = v_m(\omega), \quad t \in [\tau_n, \tau_{n+1}[$$

where v_n is $M_o^{\tau_n}$ measurable.

To any $V \in W$, one can associate a unique probability P_V^x on Ω_o, M_o, such that

(4.29) $$P_V^x(x(0) = x) = 1$$

$$E_V^x\, [\phi(x(t)) \mid M_o^{\tau_n}] = \Phi^{v_n}(t-\tau_n)\phi(x(\tau_n))$$

$$\forall\, t \in [\tau_n, \tau_{n+1}[$$

We next define

(4.30) $$J^x(V) = E_V^x \int_0^\infty e^{-\alpha t}\, L^{v(t)}(x(t))dt$$

and

(4.31) $$W_q = \{V \mid \tau_n = \frac{n}{q}\}.$$

One can prove the following

Theorem 4.3. Under the assumptions of Theorem 4.2, one has

(4.32) $$u(x) = \inf_{V \in \cup_q W_q} J^x(V)$$

where $\cup$ *denotes the maximum element*

5. Stochastic Maximum Principle

The stochastic maximum principle has been first considered by H.J. KUSHNER [2]. A general theory based on random convex Analysis has been given by J.M. BISMUT [2]. A different approach has been considere by U. HAUSSMANN [2]. We propose here a formulation due to A. BENSOUSSAN [2]. Let

(5.1) $$g(x,t)\ ;\ R^n \times [0,\infty) \to R^n$$

$$\sigma(x,t)\ ;\ R^n \times [0,\infty) \to L(R^m;R^n)$$

measurable.

(5.2) $(\Omega,\mathcal{A},P)$ is a probability space, equipped with a filtration F^t and, $w(t)$ is a standardised F^t Wiener process with values in R^m.

We assume that

(5.3) $$|g(x,t) - g(x',t)| + |\sigma(x,t) - \sigma(x',t)| \leq K|x-x'|$$

$$|g(x,t)|^2 + |\sigma(x,t)|^2 \leq K_o^2(1 + |x|^2)$$

Denote by $L_F^2(0,T;R^k)$ the space of processes adapted to F^t with values in R^k, and let

(5.4) U_{ad} non empty convex closed subset of R^k

To any admissible control, one can associate the state of the system, which is obtained by solving the stochastic differential equation

(5.5) $$dx = g(x(t),v(t))dt + \sigma(x(t),v(t))dw(t)$$

$$x(0) = x_o$$

where $x_o \in R^n$. It is well known that

(5.6) $$E \sup_{0\leq t\leq T} |x(t)|^2 \leq C(1 + |x_o|^2 + E\int_0^T |v(t)|^2 dt).$$

Let next

(5.7) $$\ell(x,v) \ ; \ R^n \times R^k \to R$$

$$h(x) \ ; \ R^n \to R$$

be continously diffentiable

$$|\ell_x|, |\ell_v| \leq \bar{\ell}(1 + |x| + |v|)$$

$$|h_x| \leq \bar{h}(1 + |x|)$$

$$|\ell(x,v)| \leq \bar{\ell}(1 + |x|^2 + |v|^2)$$

$$|h(x)| \leq \bar{h}(1 + |x|^2)$$

and consider the cost function

(5.8) $$J(v(.)) = E\left[\int_0^T \ell(x(t),v(t))dt + h(x(T))\right]$$

which is well defined by virtue of the assumptions (5.3), (5.6), (5.7).

We introduce the Hamiltonian defined by

(5.9) $$H(x,v,q,r) = \ell(x,v) + q.g(x,v) + \sum_{j=1}^{m} r^j.\sigma^j(x,v)$$

and r represents the matrix whose j^{th} column is r^j.

Let u(.) be an optimal control for (5.5), (5.8) and y(.) be the corresponding optimal state. The first issue is to make precise the corresponding adjoint process. This is more delicate than in the deterministic case, since in the present situation there is in some sense two adjoint processes, one associated to the drift and another associated to the diffusion.

The first result is

Theorem 5.1. Under the assumptions (5.1), (5.2), (5.3), (5.7), and

(5.10) $$F^t = \sigma(w(s),\ s \le t)$$

there is one and only one pair p(t), r(t) which are adapted stochastic processes with values in R^n and $L(R^n;R^n)$ verifying

(5.11) $$-\,dp = (g_x^*(y,u)p + \ell_x(y,u) + \sum_j \sigma_x^{j*}(y,u)r^j)dt - \sum_j r^j(t)dw_j$$

$$p(T) = h_x(y(T))$$

□

One can then express the stochastic maximum principle as follows

Theorem 5.2. Under the assumptions of Theorem 5.1, if u(.) is an optimal control one has the condition

(5.12) $$H_v(y(t),\ u(t),p(t),r(t)).(v - u(t)) \ge 0$$

a.e.t, a.s. , $\forall v \in U_{ad}$ □

The drawback of this approach is that the process r(.) is not easily computable. The existence and uniqueness proof relies on a representation theorem for martingales and thus is not constructive. One recovers a difficulty already mentionned in the section on martingale methods.

Nethertheless, it is possible to use the stochastic maximum principle, in particular cases in order to characterize optimal controls (cf. U.G. HAUSSMANN [1] for a systematic study of these cases). Let us also mention an approximate maximum principle due to R.T. ELLIOTT - M. KOHLMANN [1], based on EKELAND [1] variational principle.

6. Impulse control problems

6.1. *Standard impulse control problem*

Let $(\Omega,\mathcal{A},P)$ be a probability space, F^t be a filtration with $F^\infty = \mathcal{A}$, and w(t) a n dimensional standard Wiener process. Let us consider functions

(6.1) $$g(x) : R^n \to R^n \quad \text{such that} \quad |g_x| \le C$$

(6.2) $$\sigma(x) : R^n \to L(R^n;R^n), \qquad |\sigma_x| \le C$$

An impulse control is a sequence

$$W = (\theta^1, \theta^2, \ldots, \theta^n, \ldots$$
$$\xi^1, \xi^2, \ldots, \xi^n, \ldots$$

of F^t stopping times, such that

$$\theta^n \le \theta^{n+1}$$

$$\theta^n \uparrow +\infty \text{ a.s.} \quad , \theta^n = \not+ \infty \text{ is possible}$$

and random variables such that

$$\xi^n \text{ is } F^{\theta^n} \text{ measurable} \quad , \quad \xi^n \ge 0.$$

The state corresponding to the control W is solution of the equation

(6.3) $$dx = g(x(t))dt + \sigma(x(t))dw + \Sigma\, \delta(t-\theta^i)\xi^i$$

$$x(0) = x$$

Let next f be a bounded function and $\mathcal{O}$ a smooth bounded domain of R^n. Let also $a_o(x)$ be a positive bounded function, one defines the cost function

(6.4) $$J_x(W) = E\,[\int_0^\tau f(x(t))(\exp - \int_0^t a_o(x(s))ds)dt + $$
$$+ \sum_n (k+c_o(\xi^n))(\exp - \int_0^{\theta^n} a_o(x(s))ds)\chi_{\theta^n<\infty}]$$

In (6.4) τ denotes the exit time of $\mathcal{O}$ of the process x(t). Let then

(6.5) $$u(x) = \operatorname*{Inf}_W J_x(W).$$

Considering the 2nd order differential operator

$$A = - a_{ij} \frac{\partial^2}{\partial x_i \partial x_j} - g_i \frac{\partial}{\partial x_i}$$

where $a = \frac{1}{2}\sigma\sigma^*$, then dynamic programming leads to the following analytic problem

(6.6) $$Au + a_o u \le f$$

$$u \le Mu$$

$$(Au + a_o u - f)(u - Mu) = 0$$

$$u|_\Gamma = 0$$

where $\Gamma = \partial\mathcal{O}$ denotes the boundary of $\mathcal{O}$.

The problem (6.6) is called a quasi variational inequality, by analogy with the problem of variational inequality. In (6.6) the operator M is explicitly given by

(6.7) $$Mu(x) = k + \operatorname*{Inf}_{\xi\ge 0} \{c_o(\xi) + u(x+\xi)\}.$$

The optimal strategy for an impulse control problem is characterized by a set called the continuation set.

In the continuation region, no action is made. When the state reaches the boundary of the continuation set then an impulse is exerted, which carries the state of the system back into the continuation set. See A. BENSOUSSAN - J.L. LIONS [2], M. ROBIN [1], J.L. MENALDI [1], [2], LEPELTIER - MARCHAL [2], N. EL KAROUI [1].

6.2. *Singular control problems*

In (6.7), k is strictly positive. It represents a "so called" fixed cost. Because of this, the optimal impulse carry the state inside the continuation region, far from the boundary. Interesting situations arise when k is 0. The corresponding problems are called "singular control problems".

Several authors have contributed to the theory, following an initial work of BATHER - CHERNOFF [1]. Among the main works, one can mention BENES - SHEPP - WITSENHAUSEN [1], KARATZAS [1], [2], KARATZAS - SHREVE [1], [2], MENALDI - ROBIN [1]. These problems are more clearly understood in the one-dimensional case.

For instance following KARATZAS [1], consider the process

(6.8) $$x(t) = x + w(t) + v(t)$$

where v(t) is the control, and w(t) is a Wiener process. The control v(t) is a process with bounded variations,

(6.9) $$v(t) = v^+(t) - v^-(t)$$

where v^+, v^- are non decreasing. Write

(6.10) $$\hat{v}(t) = v^+(t) + v^-(t)$$

The cost function is written as

(6.11) $$J_x(v(.)) = E\int_0^\infty e^{-\alpha t}(d\hat{v}(t) + f(x(t))dt)$$

The value function

(6.12) $$u(x) = \text{Inf } J_x(v(.))$$

is solution of the following Q.V.I.

$$(6.13)\qquad \frac{1}{2}u''(x) - \alpha u + f \geq 0$$

$$|u'| \leq 1$$

$$(\frac{1}{2}u'' - \alpha u + f)(1 - |u'|) = 0$$

KARATZAS [1] has shown, by explicit calculations, the existence of a smooth solution of (6.13). He has also studied the equivalent parabolic problem.

In a subsequent work, KARATZAS-SHREVE [1] [2] have established an interesting connection between a problem of singular control, the Monotone Follower Stochastic Control Problem and a problem of optimal stopping. More precisely, consider

$$(6.14)\qquad x(t) = x + w(t) - v(t),$$

where $v(t)$ is non decreasing process. The Monotone Follower control problem is described as follows

$$(6.15)\qquad u(x,t) = \operatorname*{Inf}_{v(.)} E\,[\int_0^t f(x(s),s)ds + \int_0^t \alpha(s)dv(s) + \phi(x(t))].$$

On the other hand, consider the problem of optimal stopping

$$(6.16)\qquad \tilde{u}(x,t) = \operatorname*{Inf}_{0\leq\theta\leq t} [\int_0^\theta f_x(x + w(s),s)ds + \alpha(\theta)\chi_{\theta<t} + \phi'(x + w(t))\chi_{\theta=t}].$$

The following relation holds

$$(6.17)\qquad \tilde{u} = \frac{\partial u}{\partial x}$$

Several techniques have been used by KARATZAS (Analytic and probabilistic). Problems of singular control in the case of diffusions with jumps have been also studied by MENALDI-ROBIN [1].

7. Direct methods - Specific problems

7.1. *Explicit solutions of Bellman equation*

Except the linear quadratic case, very few problems can be solved by giving an explicit solution of Bellman equation. Significant examples have been given by BENES, SHEPP, WITSENHAUSEN [1]. Most of these examples are in fact related to singular control cases.

Another interesting example is that of a linear system, with an exponential of a quadratic form as cost functional. The Bellman equation can be explicitly solved in this case and the optimal feedback is also linear. This problem has been introduced by P. WHITTLE [2].

The dynamic allocation problem, also considered by P. WHITTLE [1], has again a nice solution.

Consider the Bellman equation

$$\alpha u(x) = \underset{1\leq j\leq n}{\text{Max}} \left[\frac{1}{2}\sigma_j^2(s_j)\frac{\partial^2 u}{\partial x_j^2} + g_j(s_j)\frac{\partial u}{\partial x_j} + h_j(x_j)\right] \tag{7.1}$$

where $x = (x_1,\dots,x_n)$.

It is possible (under adequate assumptions) to reduce the solution of (7.1) to solving one dimensional problems. Consider the family of problems, indexed by a parameter m, of stopping time type ;

$$u_j \geq \frac{1}{2}\sigma_j^2(x_j)\frac{\partial^2 u_j}{\partial x_j^2} + g_j(x_j)\frac{\partial u_j}{\partial x_j} + h_j(x_j) \tag{7.2}$$

$$u_j \geq m$$

$$(u_j - m)\left[u_j - \frac{1}{2}\sigma_j^2(x_j)\frac{\partial^2 u_j}{\partial x_j^2} - g_j(x_j)\frac{\partial u_j}{\partial x_j} - h_j(x_j)\right] = 0,$$

then one can express u(x) by the formula

$$u(x) = K - \int_k^K \prod_j \frac{\partial u_j}{\partial m}(x_j,m)\,dm \tag{7.3}$$

where k and K are adequate constants. It is possible to show that

$$0 \leq \frac{\partial u_j}{\partial m} \leq 1$$

which justifies the formula (7.3).

The optimal index is then given by the Gittins rule. Let

(7.4) $$M_j(x_j) = \text{Min}\,\{m > k \mid u_j(x_j,m) = m\}$$

then the optimal index is

(7.5) $i^*(t)$ = smallest index which realizes $\underset{1\le j\le n}{\text{Max}}\; M_j(x_j(t))$.

This problem has been solved rigorously by I. KARATZAS [4].

In a more recent paper BENES [2] has considered a Bellman equation of the type (Burgers' type)

(7.6) $$\frac{\partial u}{\partial t} + \frac{1}{2}\,\Delta u + g.Du - \frac{1}{2}\,|Du|^2 + f = 0$$

$$u(x,T) = \phi(x)$$

which corresponds to the problem

(7.7) $$dx(t) = g(x)dt + v(t)dt + dw(t)$$

$$x(0) = x$$

$$J_x(v(.)) = E[\frac{1}{2}\int_0^T (f(x(t)) + v(t)^2)dt + \phi(x(T))]$$

BENES has been able to give some explicit formulas namely in the case when $g = DF$, where F is some function.

7.2. *Specific problems*

There is a large litterature dealing with stochastic control problems involving small parameters. It has initiated in particular with the seminal paper of W. FLEMING [2].

In A. BENSOUSSAN [3], one can find a survey of some of these techniques for obtaining approximated optimal controls, which can be easily computed.

More recently, problems with small parameters have been considered, using the theory of viscosity solutions described in section 2, introduced by M.G. CRANDALL-P.L. LIONS [1].

Several results of Ventsel Freidlin type have been proved quite differently with these new analytic tools. We refer to FLEMING - SOUGANIDIS [1],FLEMING - TSAI [1], LIONS - SOUGANIDIS [1], among the main references. The theory of viscosity solutions is also quite useful in the context of problems involving state-space constraints. The control has to be picked in order to maintain the state of the system in a domain. J.M. LASRY and P.L. LIONS [1] have considered the situation of diffusions, (see also for previous work M. DAY [1]), H. SONER [1] has studied the problem of deterministic systems as well as piece wise deterministic systems.

Ergodic Control problems appear as limit problems of stochastic control problems with small parameters (small discount). A good survey can be found in M. ROBIN [2]. For more recent work see A. BENSOUSSAN [3], P. GIMBERT [1].

II. STOCHASTIC CONTROL WITH PARTIAL INFORMATION

8. Direct solutions

The problem of stochastic control under partial information has been first studied in the case of special structures. But of course, the results which can been obtained in such a way are limited.

Besides the well known separation principle in the linear quadratic gaussian case (more precisely a certainly equivalence principle in this case), a class of problems which can be attacked by direct methods has been identified by M.H. WONHAM [1]. It is the class where the so called "Wonham's separation principle" holds (see also N. CHRISTOPEIT, K. HELMES [1], U.G. HAUSSMANN [3]).

The main feature is that an optimal feedback can be derived as a function of the Kalman filter, which represents the conditional mean of the state. Thus there is a sufficient statistics, namely the conditional mean which permits to express the optimal feedback. This sufficient statistics has the advantage of having the same dimension as the state of the system.

An other example where the optimal feedback can be expressed as a function of a sufficient statistics whose dimension is the same as that of the state, is the LEG problem. More precisely the linear - Exponential - Gaussian stochastic control problem is the following

(8.1) $$ds = (Fx + Bv)dt + G\, dw$$

$$x(0) = x_0$$

$$dy = H\,x\,dt + R^{1/2}db \qquad y(0) = 0$$

where w, b are two independant standard Wiener processes, and x_o is a gaussian random variable, with mean $\bar{x}_o$ and covariance matrix P_o (positive definite). The cost function to minimize is

$$(8.2) \qquad J(v(.)) = E\,\{\mu \exp(\mu/2)\,[M\,x(T)^2 + \int_0^T (Qx^2 + Nv^2)dt]\}.$$

It has been shown by A. BENSOUSSAN - J.M. VAN SCHUPPEN [1], that the problem (8.1), (8.2) admits an optimal control of the form

$$(8.3) \qquad u(t) = -N^{-1}(t)B^*(t)S(t)r(t)$$

where r(t) is the solution of

$$(8.4) \qquad dr = [F - RH^*R^{-1}H + \mu PQ]\,r\,dt + Bu\,dt + PM^*\,R^{-1}dy$$

$$r(0) = \bar{x}_o$$

and P is the solution of

$$(8.5) \qquad \dot{P} - FP - PF^* + P(H^*R^{-1}H - \mu Q)P - GG^* = 0$$

$$P(0) = 0_o$$

Note that when Q = 0, (8.4), (8.5) reduces to the Kalman filter, which is consistant with the fact that Wonham's separation principle holds in this case. In general r(t) is different from the Kalman filter, but is still a sufficient statistics whose size is the same as that of the state. In (8.3), S(t) is a matrix which is the solution of a Riccati equation.

There are also cases where it is still possible to characterize the optimal feedback as a function of a sufficient statistics, with a dimension higher than that of the state. Examples of this sort have been given by BENES and KARATZAS [1]. Consider for instance the problem

$$(8.6) \qquad dx = v(t)dt + dw \qquad x(0) = x_o$$

$$dy = x(t)dt + db \qquad y(0) = 0$$

where x_o has a distribution which is not necessarily Gaussian, and in fact general F.

But x_o is still independant from w and b. Then the conditional mean $\hat{x}(t)$ can be expressed as

$$(8.7) \qquad d\hat{x} = v\,dt + g(t,z(t))d\nu -$$

$$\hat{x}(0) = \int_{-\infty}^{+\infty} x\,dF$$

$$dz = h(t,z)dt + \beta(t)d\nu \qquad z(0) = 0$$

where g, h and β are specific deterministic functions.

In (8.7) ν represents the innovation process

$$d\nu = dy - \hat{x}\,dt$$

Using this result, Benes and Karatzas have shown that for the cost function

$$(8.\ 8) \qquad J(v(.)) = E\,[\int_0^T x^2\,dt + (x(T))^2]$$

with the constraint $v(t) \in [-1, +1]$, the optimal feedback is of the form

$$(8.9) \qquad u(t) = -\,\text{sign}\,\hat{x}(t),$$

result previously proved by them when F is gaussian. See also for this problem CHRISTOPEIT - HELMES [2].

9. Non linear filtering

9.1. *Basic results*

To simplify matters, let us consider the following model

$$(9.1) \qquad dx = g(x(t),t)dt + \sigma(x(t),t)dw$$

$$x(0) = \xi$$

$$dy = h(x(t),t)dt + db$$

$$y(0) = 0$$

where w,b are independant Wiener processes, and

(9.2) $\quad g, h \quad$ bounded measurable

(9.3) $\quad \sigma \quad$ bounded measurable $\in L(R^n;R^n)$

$\sigma_x \quad$ bounded

Let $\Omega = C([0,\infty) ; R^{n+d})$, $\omega(t) = \begin{pmatrix} x(t) \\ y(t) \end{pmatrix}$, $M_s^t = \sigma(\omega(\lambda), s \le \lambda \le t)$, $Y_s^t = \sigma(y(\lambda), s \le \lambda \le t)$, $M = M_o^\infty$, $M^t = M_o^t$.

We consider a probability P on (Ω,M) such that, $y(t)$ is a standard Wiener process with values in R^d, and there exist a Wiener process $w(t)$ with values in R^n, independant of $y(.)$, adapted to M^t, a random variable ξ which is M^o measurable, independant of $y(.)$ and $w(.)$, such that $x(.)$, $y(.)$ are independant and $x(.)$ satisfies the first equation (9.1). Defining then the process

$$z(t) = \exp \{ \int_0^t h(x(s),s)dy - \frac{1}{2} \int_0^t |h(x(s),s)|^2 ds\} \tag{9.4}$$

and performing the transformation of measures

$$\frac{d\tilde{P}}{dP}\bigg|_{M^t} = z(t) \tag{9.5}$$

then for $\tilde{P}$, the relations (9.1) hold ; the process b which is defined by

$$b(t) = y(t) - \int_0^t h(x(s),s)ds$$

is a $\tilde{P}$, Mt Wiener process, independant of $x(.)$.

Let ψ be a Borel bounded function on R^n. Define the operator

$$\pi(t)(\psi) = \tilde{E}\, [\psi(x(t)) \mid Y^t] \quad , \quad \forall\, t \tag{9.6}$$

Clearly

$$\pi(0)(\psi) = \pi_o(\psi) \tag{9.7}$$

where π_o is the probability distribution of ξ.

By an easy calculation, one obtains the KALLIANPUR - STRIEBEL, formula

(9.8) $$\pi(t)(\psi) = \frac{E(\psi(x(t))z(t) \mid Y^t)}{E(z(t) \mid Y^t)} = \frac{p(t)(\psi)}{p(t)(1)}$$

where clearly

(9.9) $$p(t)(\psi) = E\,[\psi(x(t))z(t) \mid Y^t].$$

The operator $p(t)$ is called the unnormalized conditional probability

We shall make the following assumption

(9.10) Let $\bar{v} \in C_0^\infty(R^n)$, $\beta(t)$ bounded and smooth function with values in R^d, there exists a function (not necessarily unique) $v \in C_b^{2,1}(R^n \times [0,T])$ such that

$$-\frac{\partial v}{\partial t} + A(t)v - v(x,t)h(x,t)\,.\,\beta(t) = 0$$

$$v(x,T) = \bar{v}(x).$$

The operator $A(t)$ arising in (9.10) corresponds of course to the diffusion (9.1), i.e.

$$A(t)\phi(x) = -\sum_{i,j} a_{ij}(x,t)\frac{\partial^2\phi}{\partial x_i \partial x_j} - \sum_i g_i(x,t)\frac{\partial\phi}{\partial x_i}$$

One can prove the following existence and uniqueness result

Theorem 9.1. Assume (9.2), (9.3), (9.10) and the existence of a probability P on (Ω,M) with the properties stated above. There exists one and only one process $p(t)$ with values in the space of positive finite measures on R^n, such that

(9.11) $\forall\ \psi(x,t)$ Borel bounded $p(t)(\psi(t)) \in L_Y^2(0,T)$ (space of square integrable processes adapted to Y^t)

(9.12) $\forall\ \psi(x,t) \in C_b^{2,1}(R^n \times]0,\infty[)$,

$$p(t)(\psi(t)) - \pi_0(\psi(0)) + \int_0^t p(s)(-\frac{\partial\psi}{\partial s} + A(s)\psi(s))ds$$

$$= \int_0^t p(s)(h(s)\psi(s)).dy(s)$$ □

The assumption (9.10) is satisfied either in the case of strong ellipticity of the operator A(t), or in the case of smooth coefficients.

Equation (9.12) is called Zakai equation. The question of uniqueness of solutions has been considered by H. KUNITA [1], relying on a different technique. He gives an explicit representation of the solution of (9.12) using multiple Wiener Ito integrals.

For a general presentation of the theory of non linear filtering , cf. G. KALLIANPUR [1], M.H.A. DAVIS, S.I. MARCUS [1]. See also N. KRYLOV, B.L.ROVOVSKII [1] for a useful technical argument.

9.2. *Density*

Can the measure p(t) be defined by a density with respect to Lebesgue measure i.e.,

(9.13) $$p(t)(\phi) = \int_{R^n} \phi(x)p(x,t)dx.$$

This problem has first been solved by E. PARDOUX [1] in the coercive case. One considers the stochastic P.D.E.

(9.14) $$dp + A^*(t)\, p\, dt = p(t)h(t).dy$$

$$p(0) = \pi_0$$

The initial condition in (9.14) means that the probability of is given by a density, also denoted by π_0. We assume that

(9.15) $$\pi_0 \in L^2(R^n) \cap L^1(R^n), \quad \pi_0 \geq 0$$

(9.16) $$a_{ij}\, \xi_1\, \xi_j \geq \alpha|\xi|^2, \quad \forall\, \xi \in R^n,\ \alpha > 0.$$

The following result has been proved by PARDOUX.

Theorem 9.2. We assume (9.2), (9.3), (9.15), (9.16), then there exists one and only one solution of (9.14) such that

(9.17) $$p \in L^2(\Omega,M,P\ ;\ C(0,T\ ;\ L^2(R^n))) \cap L^2_Y(0,T\ ;\ H^1(R^n))$$ □

It can then he proved that the representation formula (9.13) holds.

When the function h has some regularity properties it is possible to express p as a function of the sample paths of y(.). This is accomplished by the "robust" theory (cf J.M.C. CLARK [1], M.H.A. DAVIS [4], E. PARDOUX [1]).

One can express p as

(9.18) $$p(x,t) = q(x,t)\exp y(t).h(x,t)$$

where q is the solution of an ordinary P.D.E. parametrized by the path of y(.).

9.3. *Finite dimensional sufficient statistics*

Let us consider for some given function $\psi(x)$, the conditional expectation $\pi(t)(\psi)$ defined by (9.6) or (9.8). The question at stake is the existence of a stochastic dynamical system

(9.19) $$d\xi = \alpha(\xi)dt + \beta(\xi)\circ dy$$

defined on a finite dimensional manifold, such that

(9.20) $$\pi(t)(\psi) = \gamma(\xi(t))$$

where α, β, γ are maps defined on the manfold.

It is convenient to interpret $\beta(\xi)\circ dy$ as a Stratonovitch integral, to preserve invariance under diffeomorphismes. Besides the Kalman filter, an example of such a possibility has been given by V. BENES [3].

Consider the model (9.1) in dimension 1, with

$$g(x,t) = g(x), \qquad \sigma(x,t) = 1$$

$$h(x,t) = x$$

and g satisfies the Riccati equation

$$g' + g^2 = ax^2 + bx + c,$$

a finite dimensional sufficient statistics exists.

See D. OCONE [1], D. OCONE, J.S. BARAS, S.I. MARCUS [1] for further considerations.

It has been shown in the litterature that questions of existence of finite dimensional filters are related to finite dimensionality of lie algebras. Since formula (9.20), can be interpreted as two representations of the same dynamical system, BROCKETT [1] has suggested that the two corresponding Lie algebras be homomorphic. This provides a key to obtain finite dimensional filters or to prove non existence results such as in the case of the cubic sensor problem

$$g = 0 \quad , \quad \sigma = 1 \quad , \quad h(x,t) = x^3$$

(see M. HAZEWINKEL - S.I. MARCUS [1], HAZEWINKEL - MARCUS - SUSSMANN [1], SUSSMAN [1]). A general presentation can be found in S.K. MITTER [1].

9.4. *Additional remarks*

The question of existence of a density in the degenerate case have been considered by J.M. BISMUT - D. MICHEL [1] using techniques like Malhavin's calculus and stochastic flows (ELWORTHY [1], KUNITA [1], BISMUT [3]).

W. FLEMING - S. MITTER [1] have considered the transformation

$$p(x,t) = \exp - S(x,t)$$

which leads to an interesting stochastic Bellman equation.

10. Control of the Zakaī equation

10.1. *Formulation of the problem*

Consider functions

(10.1) $$g(x,v) : R^n \times U_{ad} \to R^n$$

$$h(x) \; ; \; R^n \to R^n$$

bounded measurable, g_x bounded

$$\sigma : R^n \to L(R^n;R^n), \ \sigma_x \ \text{bounded}$$

where U_{ad} is a non empty subset of R^k.

Let Ω, M, as in § 9.1, and let P be a probability on Ω such that, y(t) is a standard Wiener process with values in R^d, and there exists a Wiener process w(t) with values in R^n, and a random variable ξ M^o measurable, such that y(.), w(.), ξ are mutually independant.

An admissible control is a process $v(.) \in L^2_Y(0,T;R^k)$ $\forall$ T finite (set of square integrable processes with values in R^k, adapted to the filtration Y^t).

Let us denote by π the probability distribution of ξ. For any admissible control we solve the Ito equation

$$\text{(10.2)} \qquad dx = g(x,v)dt + \sigma(x)dw$$

$$x(0) = \xi$$

and we perform the change of probability measure $P^{v(.)}$ defined by

$$\text{(10.3)} \qquad \frac{dP^{v(.)}}{dP}\Big|_{M^t} = \exp \{ \int_0^t h(x(s))dy - \frac{1}{2}\int_0^t |h|^2 \, ds\} = z(t)$$

Defining as usual

$$b(t) = y(t) - \int_0^t h(x(s))ds$$

on (Ω, M, $P^{v(.)}$, M^t) we have the model (10.1) and

$$\text{(10.4)} \qquad dy = h(x(t))dt + db, \quad y(0) = 0$$

the process b(.) becoming a standard Wiener process.

Let us consider next

$$\text{(10.5)} \qquad f(x,v) \ ; \ R^n \times R^k \to R,$$

bounded mesurable

and the cost function

$$(10.6) \qquad J(v(.)) = E^{v(.)} \int_0^\infty e^{-\beta t} f(x(t),v(t))dt$$

$$= \int_0^\infty e^{-\beta t} E\, z(t) f(x(t),v(t)).$$

Using the definition (9.9), one can reformulate the cost function as

$$(10.7) \qquad J(v(.)) = E \int_0^\infty e^{-\beta t} p^{v(.)}(t)(f_{v(t)})dt$$

where we have set

$$f_v(x) = f(x,v)$$

and

$$p^{v(.)}(t)(\psi) = E(z(t)\psi(x(t)) \mid Y^t),$$

emphasizing the dependance with respect to $v(.)$.

Let us define the family of operators

$$(10.8) \qquad A^v = - \sum_{i,j} a_{ij}(x) \frac{\partial^2}{\partial x_i \partial x_j} - \sum_i g_i(x,v) \frac{\partial}{\partial x_i}$$

According to Theorem 9.1, $p^{v(.)}(t)$ satisfies the equation

$$(10.9) \qquad d\, p\,(t)(\psi) + p(t)(A^{v(t)}\psi) = p(t)(h\psi).dy$$

$$p(0)(\psi) = \pi(\psi)$$

$$\forall\, \psi \in C_b^2(R^n)$$

The problem has been reduced to a problem of stochastic control with complete observation. However the state of the system does not belong to a finite dimensional space.

10.2. Existence of optimal controls

The first question of interest concerns the existence of an optimal control. This problem has been considered by FLEMING - PARDOUX [1]. They assume that

$$(10.10) \qquad U_{ad} \quad \text{convex, compact}$$

(10.11) $g(x,v) = g_o(x) + g_1(x)v$

where g_o, g_1 are Lipschitz functions.

A finite horizon version of (10.7) is considered. We shall thus consider

(10.12) $\ell : R^n \to R$, bounded continuous.

The cost function (10.7) is changed into

(10.13) $J(v(.)) = E\, p^{v(.)}(T)(\ell)$

In fact F.P. consider relaxed controls. The probabilistic set up is the following.

Define $G = C([0,T];R^d) \times L^2(0,T;U_{ad})$ and $G^t = \sigma(y(s), \int_0^s v(\lambda)d\lambda,\ s \leq t)$.

One defines

$\mathcal{A}$ = set of probability measures on (G, G^T) such that y is a G^t standard Wiener process. This set becomes the set of admissible controls. It is convex and compact under weak sequentiel convergence.

Let $\nu \in \mathcal{A}$, for each sample $y(.)$, $v(.)$ (a.s.ν), one can define $p(t)$ by solving the Zakaï equation (10.9). More precisely, assuming smoothness of h, one considers the pathwise solution $p^{y,v}(t)$. The cost function (10.13) is now changed into

(10.14) $J(\nu) = \int_G p^{y,v}(T)(\ell)\nu(dy,dv)$

Then F.P. prove the lower semi continuity of J, which implies the existence of an optimal ν^*.

However optimal controls in the sense of § 10.1 may fail to exist in general.

Other types of existence results have been given by CHRISTOPEIT [1], ELLIOT - KOHLMANN [2] relying on different models for the observation process (deterministic functional of the state on discrete time observations).

For related work, see V.S. BORKAR [1] , J.M. BISMUT [4], U. HAUSSMANN [3]. See also N.J. CUTLAND [1] for existence results relying on non standard analysis techniques.

10.3. Semi group considerations

Several authors have extended to the present problem of stochastic control under partial observation, the non linear semi group theory of NISIO [2] relative to the full observation case. Several possibilities exist, depending on the choice of the state space.

A first natural possibility has been presented by W. FLEMING [3] and by M.H.A. DAVIS and M. KOHLMANN [2]. The state space is the space of $\mathcal{M}$ of finite measures on R^n.

We shall consider (10.9) and write the solution as follows

$$p_\pi^{v(.)}(t)$$

to emphasize the dependance with respect to the initial condition. One can notice that in the definition (10.9), π can be a finite measure on R^n, not necessarily a probability. Define the set $\mathcal{K}$ fo functionals on $\mathcal{M}$ such that if $F \in \mathcal{K}$

(10.15) $$F(\lambda\pi) = \lambda F(\pi), \ \lambda \geq 0, \ \forall \ \pi \in \mathcal{M}$$

(10.16) $$F(\pi_1 + \pi_2) \geq F(\pi_1) + F(\pi_2)$$

(10.17) $$\sup_{\|\pi\| \leq 1} F(\pi) < \infty$$

If ℓ is a bounded borel function, one defines $\hat{\ell}$ as a functional on $\mathcal{M}$ by setting

$$\hat{\ell}(\pi) = \pi(\ell),$$

and $\hat{\ell} \in \mathcal{K}$.

If $F \in \mathcal{K}$, one can define

(10.18) $$S(t)(F)(\pi) = \operatorname*{Inf}_{v(.)} E\, F(p_\pi^{v(.)}(t)).$$

In particular the control problem (10.13) is interpreted as

(10.19) $$\operatorname{Inf} J(v(.)) = S(T)(\hat{\ell})(\pi)$$

DAVIS - KOHLMANN have proceeded as follows. Let us write $p_\pi^v(t)$ for the solution of (10.9) when v(.) is a constant v. Define next the family of semi groups on $\mathscr{K}$

$$\phi^v(t)(F)(\pi) = F(p_\pi^v(t))$$

and

$$\psi(t)(F)(\pi) = \operatorname*{Inf}_v \phi^v(t)(F)(\pi)$$

It is checked that $\psi(t)$ maps $\mathscr{K}$ into itself. Now for any t, let

$$k_{N,t} = [t\ 2^N]$$

and

$$S_N(t)(F)(\pi) = [\psi(2^{-N})]^{k_{N,t}}(F)(\pi)$$

as $N \to +\infty$, $S_N(t) \to S(t)$.

V. BORKAR [2] has considered as $\mathscr{M}$ the space of probability measures on R^n, provided with a certain metric, and considers as $\mathscr{K}$ the set of uniformuly continuous functionals on $\mathscr{M}$. Consider the conditional probability

$$\pi_\pi^{v(.)}(t)(\psi) = \frac{p_\pi^{v(.)}(t)(\psi)}{p_\pi^{v(.)}(t)(1)}$$

and for $F \in \mathscr{K}$

$$Q(t)F(\pi) = \operatorname*{Inf}_{v(.)} E^{v(.)} F(\pi_\pi^{v(.)}(t)) \tag{10.20}$$

It is proved that Q(t) is a semi group on $\mathscr{K}$. In the case of density, i.e. the situation (9.14), it is also natural to consider as state space functional spaces to which the density belongs. This is the approach of A. BENSOUSSAN [4]. Let

$$H = L^2(R^n)$$

B = space of Borel bounded functionals on H

C = space of uniformuly continuous bounded functional on H

Let us define by $p_\pi^{v(.)}(t)$, the solution of

$$(10.21) \qquad dp + (A^{v(.)})^* p\, dt = p\, h.dy$$

$$p(0) = \pi$$

If F belongs to C, one defines

$$S(t)(F)(\pi) = \operatorname*{Inf}_{v(.)} E[F(p_\pi^{v(.)}(t))]$$

which again defines a non linear semi group on C. For the stopping time problem with partial observation, see Y. FUJITA - M. NISIO [1].

10.4. Dynamic Programming

The Bellman equation corresponding to (10.21) and

$$(10.22) \qquad J_\pi(v(.)) = E \int_0^\infty e^{-\beta t}\, p_\pi^{v(.)}(t)(f_{v(t)})dt$$

(cf (10.7)) would be the following. If $\Phi(\pi)$ is a functional on H, twice continuously differentiable (Frechet) satisfying

$$(10.23) \qquad \frac{1}{2} D^2\Phi(\pi)(\pi h, \pi h) + \operatorname*{Inf}_{v \in U_{ad}} \{(f_v, \pi) - D\Phi(\pi)((A^v)^* \pi)\}$$

$$= \beta\, \Phi \quad , \quad \forall\, \pi$$

If such a functional exists then a natural conjecture is

$$(10.24) \qquad \Phi(\pi) = \operatorname*{Inf}_{v(.)} J_\pi(v(.)).$$

BENES - KARATZAS [1] have considered some analogous equation, with a different functional space, and proved a verification theorem of type (10.24) (see also the presentation of M. DAVIS [1]).

It is clear however that (10.23) leads to serious difficulties. A way to avoid these difficulties is to consider a problem of semi group envelope, in the spirit of NISIO, as it has been done in § 4.2. This has been done by A. BENSOUSSAN [1].

Consider H, B, C as in (10.21) and let $p_\pi^v(t)$ denote the solution of (10.21) corresponding to a constant control. Define the family of linear semi groups on

C_1 = space of uniformuly continuous functionals on H with linear growth.

The introduction of C_1 is necessary, since we need to consider functionals (π, f_v) which are linear and thus belong to C_1, but not to C.

Let us consider the family of semi groups on C_1 (also on C) defined by

$$\Phi^v(t)(F)(\pi) = E\ F(p_\pi^v(t)).$$

Consider next the following set of functionals

$$\begin{aligned} &S \in C_1 \\ &S \leq \int_0^t e^{-\beta s}\ \Phi^v(s) f_v\ ds + e^{-\beta t}\ \Phi^v(t) S, \quad \forall\, t \geq 0 \end{aligned} \tag{10.25}$$

It is proved that if β is sufficiently large, the set (10.25) is not empty and has a maximum element. Moreover, denoting also by $S(\pi)$ this maximum element, then $S(\pi)$ is Lipschitz and

$$S(\pi) = \text{Inf } J_\pi(v(.)) \tag{10.26}$$

where $J_\pi(v(.))$ is defined by (10.22).

This carries the weak theory of dynamic Programming to the partially observable stochastic control problem.

10.5. *Maximum principle.*

The possibility of writing a stochastic maximum principle has been initialy proposed by M. KWAKERNAAK [1] (see also the presentation of M. DAVIS [1]). This problem has been solved by A. BENSOUSSAN [2], [4].

Let us go back to (9.14) and consider the cost function

$$J(v(.)) = E\,[\int_0^T p^{v(.)}(t)(f_{v(t)})dt + p^{v(.)}(T)(\ell)]. \tag{10.27}$$

One can prove the following

Theorem 10.1. Assume (9.2), (9.3), (9.15), (9.16), (10.4), (10.11). Let $u(.)$ be an optimal control for (9.14), (10.26). There exists (uniquely defined) $\lambda \in L_F^2(0,T;H^1(R^n))$ $K_i(t) \in L_F^2(0,T;L^2(R^n))$, $i = 1..n$ such that

$$dp + (A^{u(.)})^*(t)p\ dt = p\ h.dy \tag{10.28}$$

$$p(0) = \pi_o$$

(10.29) $$- d\lambda + A^{u(.)}(t)\lambda\, dt = (f_{u(t)} + \sum_i h_i K_i(t))dt - \sum_i K_i\, dy_i$$

$$\lambda(T) = \ell$$

(10.30) $$[\int_{R^n} (\frac{\partial f}{\partial v}(x,u(t)) + \sum_i \frac{\partial \lambda}{\partial x_i} \frac{\partial g_i}{\partial v}(x,u(t)))p(x,t)dx](v-u(t)) \geq 0$$

a.e.t, a.s

□

11. Miscellaneous

11.1. *Stochastic control of infinite dimensional systems*

As we have seen, the problem of stochastic control with partial observation is equivalent to a problem of stochastic control for an infinite dimensional system. This and other applications have motivated the study of such systems (the case of linear systems has been treated much earlier, see in particular the survey of A. BENSOUSSAN [5], and R.F. CURTAIN - A.J. PRITCHARD [1] ; see also G. DAPRATO - A. ICHIKAWA [1], J. ZABCZYK [2] .

11.2. *Several decision makers*

The field does not seen to have much progressed since the seventies. It is now clear that the "signalling" effect, which consists in exchanging information between the decision makers induces very complicated effects even in the case of simple structures. Recently J. LEVINE [1] has attempted to give a dynamic programming formulation for dynamic team problems, which contains a "signalling" term. But the treatment remains quite formal.

For practical purposes, the only possibility to handle such problems lies in the use a priori feedbacks.

11.3 *Discrete time problems and numerical methods*

Most questions can be reformulated in discrete time, and this is a necessary step to design numerical algorithms. Let us just mention the recent work of R.T. ROCKAFELLAR - R.J.B. WETS [1] in stochastic programming.

In a series of papers G.B. DIMASI - W.J. RUNGGALDIER [1], [2] ... have proposed numerical algorithms for solving stochastic control problems (with full or partial observation). See also N.M. VAN DIJK [1], D. BERTSEKAS [1], H.J. KUSHNER [1], and the collective article THEOSYS.

11.4. *The Expert system of J.P. QUADRAT*

The theory of Bellman equations and related topics of stochastic control have now reached a state of the art which represents a substantial "knowledge base". It is possible to incorporate it into an expert system. This is the objective of J.P. QUADRAT and his team (see GOMEZ, J.P. QUADRAT, A. SULEM [1]). The applications are numerous, since it will save programming efforts.

REFERENCES

A.V. BALAKRISHNAN (1) Applied functional Analysis, Springer Verlag, 1976.

L. BARTHELEMY (1) Application de la théorie des semi groupes non linéaires dans L^∞ à l'étude d'une classe d'inéquations quasi-variationnelles, Besançon Université de Franche-Comté (1980)

J.A. BATHER, H. CHERNOFF (1), Sequential decisions in the control of a spaceship, Proc. 5th Berkeley Symposium on Mathematical Statistics and Probability, 3 (1966), pp. 181-207.

V.E. BENES (1) Existence of Optimal Stochastic Control Laws, SICON 9 (1971) 446-475.

(2) Using Hopf-Cole, Gauge, and Girsanov Transformation to reduce Stochastic Control Problems to Gaussian Integrations (preprint)

(3) Exact finite dimensional Filters for certain Diffusions with non linear drift, Stochastics, 5, pp. 65-92

V.E. BENES, I. KARATSAS (1) Filtering of Diffusions controlled through their Conditional Measures, Stochastics, 1984, Vol. 13, pp. 1-23

(2) Estimation and Control for linear Partially Observable Systems with Non-gaussian initial distribution, Stochastic Proc. and Appli. 14 (1983) 233-248.

V.E. BENES, L.A. SHEPP, H.S. WITSENHAUSEN (1), Some solvable stochastic control problems, Stochastics, 4 (1980), pp. 134-160.

A. BENSOUSSAN (1) Stochastic Control by Functional Analysis Methods North Holland, Amsterdam 1982.

(2) Stochastic Maximum Principle for Distributed Parameter Systems, Journ. Franklin Institute, Vol 315 N° 5/6, pp. 387-406, May/June 1983.

(3) Méthodes de perturbations en Contrôle Optimal, to be published

(4) Maximum Principle and Dynamic Programming Approaches of the Optimal Control of Partially Observed Diffusions Stochastics 1983, Vol 9, pp. 169-222.

(5) Control of Stochastic Partial Differential Equations, in Distributed Parameter Systems (ed. W.H. RAY, D.G. Lainiotis) M. Dekker, N.Y. 1978.

A. BENSOUSSAN, J.L. LIONS (1) Applications of Variational Inequalities to Stochastic Control North Holland 1982.

(2) Impulse Control and Quasi-Variational Inequalities, Gauthier-Villars Paris 1984

A. BENSOUSSAN, P.L. LIONS (1) Optimal Control of Random Evolutions, Stochastics

A. BENSOUSSAN, M. ROBIN (1) On the Convergence of the Discrete Time Dynamic Programming Equation for General Semi-groups SICON, 20, (1982), pp. 722-746.

A. BENSOUSSAN, J.H. VAN SCHUPPEN (1) Optimal Control of partially observable Stochastic Systems with an exponential of integral Performance Index, SICON

D. BERTSEKAS (1) Dynamic Programming and Stochastic Control, Academic Press New York 1976

D. BERTSEKAS - S. SHREVE (1) Stochastic Optimal Control : The Discrete Time Case. Academic Press New York, 1978

J. M. BISMUT (1) Théorie Probabiliste du Controle des Diffusions Memoir of AMS vol 4 n° 167 (Janvier 1976)

(2) An Introductory Approach to Duality in Optimal Stochastic Control, SIAM Rev. Vol 20, n° 1, Janvier 1978

(3) Martingales, The Malliavin Calculus and Hypoellipticity under general Hormander Conditions ; Z. Wahrscheilichkeit theorie 56 (1981) 469-505.

(4) Partially Observable Diffusions and their Control, SICON 20, March 1982

J.M. BISMUT, D. MICHEL (1) Diffusions conditionnelles 1, 2 ; J. Functional Analysis 44 (1981), 174-211, 45 (1982) 274-292

R. BOEL, M. KOHLMANN (1) Stochastic Control over double Martingales in Analysis and Optimization of Stochastic Systems, ed. O.L.R. Jacobs, Academic Press, New york, 1979

R. BOEL, P. VARAIYA (1) Optimal Control of Jump Processes, SICON 15 (1977) 92-119

V.S. BORKAR (1) Existence of optimal controls for partially observed Diffusions, Preprint, Tata Institute Bangalore, 1982

(2) The Nisio Semigroup for Controlled Diffusions with Partial Observations, Preprint, Tata Institute.

P. BREMAUD, J.M. PIETRI (1), The Role of Martingale Theory in Continuous time Dynamic Programming (1978)

R.W. BROCKETT (1) Remarks on Finite Dimensional Nonlinear Estimation, Analyse des Systèmes, Astérisque, vol. 75-76, pp. 199-205, 1980

CHALAYAT , MOREL, D. MICHEL (1),

M.G. GRANDALL, P.L. LIONS (1) Viscosity Solutions of Hamiltnn-Jacobi Equations, Trans. Amer. Math. Soc, 1983

N. CHRISTOPEIT (1) Existence of optimal stochastic controls under partial observation, Z. Wahrscheinlichkeittheorie 51 (1980) 201-213

N. CHRISTOPEIT, K. HELMES (1) The Separation Principle for partially observed linear control systems : a general framework, Lecture notes in Control and Information Sciences 61, Springer-Verlag 1984

(2) Optimal control for a class of partially observable Systems, Stochastics, 8 (1982), 17-38

J.M.C. CLARK (1), The design of robust Approximations to the stochastic differential equations of non linear filtering, Communic. Systems and Random Process Theory (ed. J.K. SKWIRZYNSKI, Nato Advanc. Study Institute, Sijthoff and Noorshoff, 1978

R. CURTAIN, A.J. PRITCHARD (1) Infinite Dimensional Linear Systems Theory Lecture Notes in Control and Information Sciences 8, 1978, Springer

N.J. CUTLAND (1) Infinitesimal Methods in Control Theory : Deterministic and Stochastic

G.DAPRATO, A. ICHIKAWA (1) Stability and Quadratic Control for Linear Stochastic equations with Unbounded Coefficients, Preprint Scuola Normale Pisa

M.H.A. DAVIS (1) Some Current issues in Stochastic Control Theory,Stochastics, Special Issue on Stochastic Optimization, August 1981, ed. M.A.H. Dempster

(2) Piece Wise Deterministic Markov Processes : A General Class of non diffusion Stochastic Models, J.R. Stast. Soc. B, (1984), 147, n° 2

(3) Martingale Methods in Stochastic Control, in Stochastic Control and Stochastic Differential Systems, Lecture Notes in Control and Information Sciences 16, Springer Verlag, Berlin, 1979

(4) Pathwise nonlinear filtering, in Hazewinkel-Willems, pp. 505-528

M.H.A. DAVIS, M. KOHLMANN (1) Stochastic Control by Measure Transformation :a general Existence Result, Institute fur Angewandte Mathematik der Universitat Bonn, (1978)

(2) On the nonlinear Semigroup of Stochastic Control under Partial Observations, Preprint

M.H.A. DAVIS, S.I. MARCUS (1) An Introduction to non linear filtering, in Hazewinkel-Willems

M. DAY (1) On a Stochastic Control Problem with Exit Constraints, Appl. Math. Optim. 6, 181-188 (1980)

M.A.H. DEMPSTER (1) (ed) Stochastic Programming, Academic Press, London, 1980

G.DIMASI, W. RUNGGALDIER (1) On measure transformations for combined filtering and parameter estimation in discrete time, Systems and Control Letters Vol 2, n° 1, July 1982

G.B. DIMASI, M. PRATELLI, W.J. RUNGGALDIER (1) An Approximation for the non linear filtering problem with error bound, Preprint

E.B. DYNKIN, A.A. YISHKEVICH (1) Controlled Markov Processes, Springer Verlag, New York, 1979

N.EL KAROUI (1) Aspects Probabilistes du controle Stochastique, Lecture Notes 876, Springer Verlag 1981

K.D. ELWORTHY (1) Stochastic Dynamical Systems and their Flows, in Stochastic Analysis Academic Press, 1978

I. EKELAND (1) On the Variational Principle, J. Math. Anal. Appl. 47, 324-353 (1974

R.J. ELLIOT (1) the Martingale Calculus and its Applications, in Stochastic Control Theory and Stochastic Differential Systems, Lecture Notes in Control and Information Sciences 16, Springer Verlag Berlin, 1979

R.J. ELLIOT, M. KOHLMANN (1) The Variational Principle and Stochastic Optimal Control Universitat Bonn, SFB 72, Preprint 295

(2) On the existence of Optimal Partially Observed Controls, Preprint

L.C. EVANS (1) Classical Solutions of the Hamilton-Jacobi-Bellman Equation for uniformly elliptic Operators, Trans. Amer. Math. Soc. to be published

L.C. EVANS, P.L. LIONS (1) Resolution des Equations de Hamilton-Jacobi-Bellman pour des Operateurs uniformément elliptiques, CRAS Paris 290 (1980) p. 1049-1052

W.H. FLEMING (1) Optimal continous parameter stochastic control, SIAM Review 11, 470-509 (1969)

(2) Stochastic Control for small noise intensities, SICON, Vol. 9 N° 3, August 1971

(3) Nonlinear Semigroups for Controlled Partially Observed Diffusions, SICON, 20, 2, March 1982

W.H. FLEMING, S.K. MITTER (1) Optimal Stochastic Control and Pathwise Filtering on Non-degenerate Diffusions, Stochastics

W. FLEMING, E. PARDOUX (1) Optimal Control for Partially Observed Diffusions SICON 20, 2, March 1982

W. FLEMING, R. RISHEL (1) Deterministic and Stochastic Optimal Control, Springer-Verlag, New York , 1975

W.H. FLEMING, P.E. SOUGANDIS (1) PDE-Viscosity solution approach to some problems of large Deviations, Brown University, Report 84-29

W.H. FLEMING, C.P. TSAI (1) Optimal Exit Probabilities and Differential Games, Appl. Math. Optim. 7 (1981), 253-282

A. FRIEDMAN (1), Stochastic Differential Equations, Vol 1-2, Academic Press New York 1976.

Y. FUJITA, M. NISIO (1) Non linear semigroups associated with optimal stopping of controlled diffusions under partial observation, Preprint

I. GERTNER, D. RAPPAPORT (1) Optimality Criteria for Controlled Discontinuous Processes, Information Sciences 17, 75-90, (1979)

I.I. GIHMAN - A.V. SKOROHOD (1) Controlled Stochastic Processes, Springer-Verlag, New York 1979

F. GIMBERT (1) Problèmes de Neumann quasi linéaires ergodiques, Journ. Funct. Analysis, to be published

C. GOMEZ, J.P. QUADRAT, A. SULEM (1) Premiers pas vers un système expert en controle stochastique, Preprint

M. HAZEWINKEL, S.I. MARCUS (1) On Lie Algebras and Finite Dimensional Filtering, Stochastics, 7, pp. 29-62, 1982

M. HAZEWINKEL, S.I. MARCUS, H.J. SUSSMANN (1) Non Existence of Finite Dimensional Filters for Conditional Statistics of the Cubic Sensor Problem, Preprint, Amsterdam 1983

M. HAZEWINKEL - J.C. WILLEMS (1) Stochastic Systems ; The Mathematics of Filtering and Identification and Applications, Nato Advanced Study Institutes Series 78, Reidel Dordrecht 1981

U.G. HAUSSMANN (1) Some Examples of Optimal Stochastic Control or : The Stochastic Maximum Principle at Work, SIAM Review Vol 23, n° 3, July 1983

(2) On the stochastic Maximum Principle, SICON 16 (1978) pp. 236-251

(3) Optimal Control of partially observed Diffusions via the Separation Principle, in Lect. Notes in Control and Inf. Sciences, vol 43 (1982), 302-311

(4) On the Existence of Optimal Controls for Partially Observed Diffusions, SICON 20, 3, May 1982

G. KALLIANPUR, Stochastic Filtering Theory, Springer-Verlag, New York, 1980

I. KARATZAS (1) A class of singular stochastic control problems, Adv. Appl. Prob. 15 (1983), pp. 225-254

(2) The monotone follower problem in stochastic decision theory, Appli. Math. Opti. 7 (1981), pp. 175-189

(3) Probabilistic Aspects of finite fuel stochastic Control Proceedings National Academy of Sciences

(4) Gittins indices in the dynamic allocation problem for diffusion processes, Annals of Probability 1984, n° 1, 173-192

I. KARATZAS, S.E. SHREVE (1) Connections between Optimal Stopping ans Singular Control 1. Monotone Follower Problems SICON, Vol 22, n° 6, 1984

(2) Connections between Optimal Stopping and Singular Control 2. Reflected follower problems SICON, Vol 23.

H. KOREZLIOGLU, G. MAZZIOTTO, J. SZPIRGLAS (1) (ed.) Filtering and Control of Random Processes, Lecture Notes in Control and Information Sciences Springer Verlag, Berlin 1984

N.V. KRYLOV (1), Controlled Diffusion Processes, Springer-Verlag, New York 1980

(2) Control of a Solution of a Stochastic Integral Equation, Th. Proba. Appl. 17 (1972), p. 114-131.

N.V. KRYLOV, B. ROSOVSKI (1) On the first integral and Liouville equations for diffusion processes

H. KUNITA (1) Densities of a measure-valued process governed by a Stochastic partial differential equation, Systems and Control Letters 1 (1981) 100-104

H. KUNITA, S. WATANABE (1) On Square Integrable Martingales, Nagoya Math. Journal, Vol 30, August 1967

H.J. KUSHNER (1) Probability Methods for Approximations in Stochastic Control, Academic Press, New York, 1977

(2) Necessary Conditions for Continuous parameter stochastic Optimization problems, SICON, Vol 10, pp. 550-565, 1972

(3) Existence results for optimal stochastic controls, JOTA (1975) pp. 347-359

H.KWAKERNAAK (1) A minimum principle for stochastic control problems with output feedback, Systems and Control Letters 1 (1981) 74-77.

I.LANDAU (1) (ed.) Outils et Modèles Mathématiques pour l'Automatique, l'analyse des Systèmes et le Traitement du Signal. Vol 1 (1981) Vol. 2 (1983) Editions du CNRS Paris

J.M. LASRY, P.L. LIONS (1) Equations elliptiques non linéaires avec conditions aux limites infinies et contrôle stochastique avec contraintes d'état CRAS, t. 299, I, 7, 1984.

J.P. LEPELTIER, B. MARCHAL (1) Problème de Martingale et Equations différentielles stochastiques associées à un opérateur intégro-différentiel, Ann. Inst. Henri Poincaré, 12, n° 1 (1976), 43-103.

(2) Théorie générale du Contrôle impulsionnel

J. LEVINE (1) Thèse Paris 1984

P.L. LIONS (1) Hamilton-Jacobi-Bellman Equations and the Optimal Control of Stochastic Systems, International Congress of Mathematicians, Varsaw 1982.

(2) Resolution Analytique des problèmes de Bellman-Drichet, Acta-Math, 146 (1981), p. 151-166.

(3) Optimal Control of Diffusion Processes and Hamiton-Jacobi-Bellman equations, Comm. P.D.E. 1983 (Part 1 & 2) ; Part 3 in Non Linear Differential Equations and their Applications, College de France Seminar, Vol. 5, Pitman, London (1983)

P.L. LIONS, J.L. MENALDI (1) Optimal Control of Stochastic Integrals and Hamilton-Jacobi-Bellman Equations 1, SIAM J. Cont. Opti. 20 (1982) P. 58-81, 2, SIAM J. Cont. Opti. 20 (1982) p. 82-95.

P.L. LIONS, P.E. SOUGANIDIS (1) Differential Games, Optimal Control and Directional derivatives of Viscosity Solutions of Bellman's and Isaacs' Equations, SICON.

J. MEMIN (1) Conditions d'optimalité pour un problème de controle portant sur une famille de probabilités dominées par une probabilité P, Université de Rennes, 1977.

J.L. MENALDI (1) On the Optimal Stopping Time Problem for Degenerate Diffusions SICON 18 (1980) pp. 697-721.

(2) On the Optimal Impulse Control Problem for Degenerate Diffusions SICON, 18 (1980), pp. 722-739.

J.L. MENALDI, M. ROBIN (1), On Singular Stochastic Control Problems for diffusions with jumps, IEEE Trans. Aut. Control AC - 29, pp. 991-1004, (1984)

S.K. MITTER (1) Geometric Theory of Nonlinear Filtering, In LANDAU (ed.)

S.K. MITTER - A. MORO (1), (ed.) Non Linear Filtering and Stochastic Control, Springer-Verlag, Lecture Notes in Math 972, Berlin 1982.

M. NISIO (1) Lectures on Stochastic Control Theory, ISI, Lecture Notes 9 Mac Milan India Delhi 1981.

(2) On a non Linear Semi-Group attached to Stochastic Control, Pub. RIMS Kyoto, 12 (1976) 513-537.

(3) Remarks on stochastic optimal controls, Japan. Jour. Math. 1 (1975), 159-183.

D. OCONE (1) Topics in Nonlinear Filtering Theory, MIT, 1980.

D. OCONE, J.S. BARAS, S.I. MARCUS (1) Explicit Filters for Diffusions with certain Nonlinear Drifts, Stochastics, 8, pp. 1-16, 1982.

E. PARDOUX (1) Equations du filtrage non linéaire de la prédiction et du lissage, Stochastics 6 (1982) 193-231.

R.W. RISHEL (1) Dynamic Programming and Minimum Principle for Systems with jump Markov Disturbances, SIAM, J. Control 13 (1975), 338-371.

(2) Necessary and Sufficient Conditions for Continuous - time Stochastic Optimal Control, SIAM J. Control, 8 (1970) 559-571.

M. ROBIN (1) Controle impulsionnel des Processus de Markov, Thèse Paris, 1978.

(2) Long Term Average cost Control Problems for Continuous Time Markov Processes, Acta Applicandae Mathematicae, 1983.

R.T. ROCKAFELLAR, J.J.B. WETS (1) A Lagrangian Finite Generation Technique for Solving Linear-quadratic Problems in Stochastic Programming, Preprint.

A. N. SHIRYAYEV (1) Optimal stopping Rules, Springer-Verlag New York, 1978.

H.M. SONER (1) Optimal control with State - space Constraints, Part 1, 2, Brown University reports 84-26 ; 84-33.

L. STETTNER, J. ZABCZYK (1) Strong Envelopes of Stochastic Processes and a Penalty Method, Institute of Mathematics Polish, Academy of Sciences, Warsaw.

C. STRIEBEL (1) Martingale Conditions for the Optimal Control of Continuous time Stochastic Systems, 5th Symposium on non linear Estimations and Applications, San Diego, CA, 1974.

D. STROOCK (1) Diffusion Processes Associated with Levy Generators, Z. Wahrscheinlichkeitstheorie, 32, (1975), 209-244.

H.J. SUSSMANN (1) Approximate Finite-Dimensional Filters for some Nonlinear Problems, Stochastics, 7, pp. 183-204, 1982.

THEOSYS (1) Commande optimale de systèmes stochastiques, RAIRO, 18, 2, 1984.

S. USTUNEL (1) Construction of Branching Processes and their Optimal Stochastic Control, Applied Mathematics & Optimization Vol. 7, n° 1, 1981.

N.M. VANDIJK (1) Controlled Markov Processes : Time - discretization , CWI, Tract, 11, 1980.

D. VERMES (1) Optimal control of Piecewise Deterministic Markov Process Stochastics, 1985, Vol. 14, pp. 165-207.

R.B. WINTER, R.M. LEWIS (1) A Necessary and Sufficient Condition for Optimality of Dynamic Programming Type, making non a priori Assumption on the Controls, SIAM J. Control 16 (1978), 571-583.

P. WHITTLE (1) Multi-armed Bandits and the Gittins Index. J. Roy. Stast. SOC ; Ser. B, 42, 143-149.

(2) Risk-sensitive linear quadratic gaussian control, Adv. Appl. Prob. 13, 764-777 (1981)

(3) Optimization over Time : Dynamic Programming and Stochastic Control. Wiley, New York (1982).

W.M. WONHAM (1) On the Separation Theorem of Stochastic Control, SICON 6, (1968), 312-326.

J. ZABCZYK (1) Semi-group Methods in Stochastic Control, CRM, University of Montreal.

(2) Linear Stochastic Systems in Hilbert Spaces ; Structural Properties and Limit Behaviour, Preprint 236, Inst. Math. Acad. Scien. Warsaw, 1981.

A METHOD FOR CONSTRUCTING ε- OPTIMAL CONTROLS IN PROBLEMS WITH PARTIAL OBSERVATION OF THE STATE.

A. BENSOUSSAN *)   W.J. RUNGGALDIER **)

I. INTRODUCTION AND PROBLEM DESCRIPTION

I.1 Scope

We consider a stochastic control problem with partial information of the following type

$$
\begin{aligned}
dx_t &= g(x_t,v_t)dt + \sigma(x_t)dw_t \; ; \; x_o = \bar{x}_o \\
dy_t &= h(x_t)dt + d\eta_t \; ; \; y_o = 0 \qquad (1.1) \\
y(v) &= E\left\{ \int_o^T f(x_t,v_t)dt + \ell(x_T) \right\}
\end{aligned}
$$

where x_t is the state of the system, controlled by the process v_t, and y_t is the observation process.

For problem (1.1) there exist in general only ε- optimal controls and the purpose of this study is to derive a constructive method for determining them.

It is well known that problem (1.1) can be reduced to a problem of stochastic control with full information, where the state at time t is the (unnormalized) conditional distribution of x_t given $(y^t,v^t) := \{y_s,v_s \; ; \; s \leq t\}$ which is infinite-dimensional. Our method to determine an ε- optimal control for (1.1) consists in successive steps that approximate problem (1.1) by a discrete-time, complexe-information stochastic control problem, where state and control take only a finite number of values. For this latter problem an optimal control $\bar{v} = \{\bar{v}_t\}$ can therefore be computed, e.g. by the method of dynamic programming, and it is shown that, by choosing a sufficiently good approximation, this $\bar{v}$ is ε-optimal for the original problem (1.1).

---

*) University Paris Dauphine and INRIA.

**) Seminario Matematico,
Université di Padova, Padova, ITALY.

1.2 Assumptions

A.1. $v_t \in U$ compact metric

A.2. The functions

$$g(x,v) : \mathbb{R}^{d_1} \times U \to \mathbb{R}^{d_1} \; ; \; \sigma(x) : \mathbb{R}^{d_1} \to \mathscr{L}(\mathbb{R}^{d_1}, \mathbb{R}^{d_1});$$
$$h(x) \quad : \mathbb{R}^{d_1} \to \mathbb{R}^{d_2}$$
$$f(x,v) : \mathbb{R}^{d_1} \times U \to \mathbb{R} \; ; \; \ell(x) : \mathbb{R}^{d_1} \to \mathbb{R}$$

are Borel, bounded and Lipschitz in X, uniformily in v, i.e. $\forall$ x,x'

$$|g(x,v) - g(x',v)| \le \bar{g}|x-x'| \; ; \; |\sigma(x)-\sigma(x')| \le \bar{\sigma}|x-x'| \; ;$$
$$|h(x) - h(x')| \le \bar{h}\,|x-x'|$$
$$|f(x,v) - f(x',v)| \le \bar{f}|x-x'| \; ; \; |\ell(x)-\ell(x')| \le \bar{\ell}|x-x'|$$

I.3 Formulation via measure transformation

Given a probability space $(\Omega,\mathscr{F},P,\mathscr{F}^t)$ with a filtration $\{\mathscr{F}^t\}$, consider the following :

(i) $\{y_t\},\{v_t\}$ are independent $(P,\mathscr{F}^t)$-standard Wiener processes with values in $\mathbb{R}^{d_2}$ and $\mathbb{R}^{d_1}$ respectively. The r.v. $\bar{x}_o$ is $\mathscr{F}^o$- measurable, independent of $\{y_t, w_t\}$.

ii) The control process v_t is admissible if it takes values in V and is adapted to $Y^t := \sigma\{y_s \; ; \; s \le t\}$.

iii) For all admissible v_t, the equation

(1.2.a) $$dx_t = g(x_t,v_t)dt + \sigma(x_t)dw_t \; ; \; x_o = \bar{x}_o$$

has by A.2 a unique strong solution.

iv) The process

$$z_t = z_t^v := \exp\left\{\int_o^t h(x_s)dy_s - \frac{1}{2}\int_o^t |h(x_s)|^2 ds\right\}$$

is a $(P,\mathscr{F}^t)$- martingale with mean 1 ; it is therefore possible to perferm the change of probability measures

$$\left.\frac{dP^v}{dP}\right|_{\mathscr{F}^t} = z_t$$

v) On $(\Omega,\mathscr{F},P^v)$ we have

(1.2.b) $$dy_t = h(x_t)dt + d\eta_t^v \; , \; \eta_o = 0$$

where η_v^t is an $(P^v,\mathscr{F}^t)$-Wiener process independent of $\{w_t\}$ and $\bar{x}_o$.

vi) The objective function is

$$(1.2.c) \qquad J(v) = E^{v}\{\int_{o}^{T} f(x_t,v_t)dt + \ell(x_T) = Ez_t\{\int_{o}^{T} f(x_t,v_t)dt + \ell(x_T)\}$$

On $(\Omega,\mathcal{F},P^{v})$ the problems (1.2.a-c) corresponds to the original problem (1.1).

II. TIME DISCRETIZATION

II.1 Class of approximate controls

Given an integer $N > 0$, onsider the partition of $[0,T]$ consisting of $[n^{T}/_{N},(n+1)^{T}/_{N}]$, $n = o,\dots,N-1$ and let W_N denote the class of admissible step controls corresponding to this partition, i.e.

$$(2.1) \qquad W_N := \{v \mid v_t(\omega) = v_n(\omega) \text{ for } t \in [n^{T}N,(n+1)^{T}/_{N}) \ ; \ v_n \in U \text{ and } y^{n^{T}/_{N}} - \text{adapted}\}$$

Furthermore, let

$$(2.2) \qquad W := \bigcup_N W_N$$

The main interest in the class W derives from the fact that for much larger classes W of controls we have (see [1], see also [4, Thm 9.3.1])

$$(2.3) \qquad \rho := \inf_{v\in W} J(v) = \inf_{v\in V} J(v)$$

Furthermore,

$$(2.4) \qquad \rho_N := \inf_{v\in W_N} J(v) \xrightarrow{N\to\infty} \rho$$

This allows us to limit our search for an ε-optimal to the class W. The fact that optimal value for (1.1) (i.e.(1.2)) can be arbitarily closely approximated by use of step controls, leads us to consider as first approximation stage for (1.2) a time discretization.

II.2 Discrete-time approximation

Given an integer $N > 0$, let the process x_t^N be defined by

$$(2.5) \qquad x_t^N = x_n^N \quad \text{for } t \in [n^{T}/_{N},(n+1)^{T}/_{N}], \text{ where}$$

$$(2.6) \qquad x_{n+1}^N = x_n^N + g(x_n^N,v_n)^{T}/_{N} + \sigma(x_n^N)w_{n+1}^N \quad \text{with}$$
$$x_o^N = \bar{x}_o, \ v \in W_N \text{ and } w_{n+1}^N := w((n+1)^{T}/_{N}) - w(n^{T}/_{N})$$

Furthemore, defining

(2.7) $$z_T^N := \exp\{\int_0^T h(x_t^N)\,dy_t - \frac{1}{2}\int_0^T |h(x_t^N)|^2 dt\} = \exp\{\sum_{n=0}^{n-1} h(x_n^N)y_{n+1}^N - \frac{T}{2N}|h(x_n^N)|^2\}$$

where

(2.8) $$y_{n+1}^N := y((n+1)\tfrac{T}{N}) - y(n\tfrac{T}{N})$$

let $(v \quad W_N)$

(2.9) $$J^N(v) := Ez_T^N\{\int_0^T f(x_t^N,v_t)\,dt + \ell(x_T^N)\} = Ez_T^N\{\sum_{n=0}^{N-1} f(x_n^N,v_n)T/N + \ell(x_N^N)\}$$

and

(2.10) $$\tilde{\rho}_N := \inf_{v\in W_N} J_N(v)$$

The following theorem is a consequence of [2 ; Prop. 3.1] (see [1])

Theorem 2.1 : Given assumptions A.2, we have for some $C > 0$

i) $$|y^N(v) - J(v)| \le C(T/N)^{1/2}, \quad v \in W_N$$

ii) $$|\tilde{\rho}_N - \rho_N| \le C(T/N)^{1/2}$$ □

Notice that z_T^N in (2.7) allows us to define on $(\Omega,\mathcal{F})$ a probability measure P_N^v

(2.11) $$dP_N^v / dP = z_T^N$$

on $(\Omega,\mathcal{F},P_N^v)$ we can now consider, for $n = o,\dots,N-1$, the following discrete-time stochastic control problem with partial information

(2.12)
$$x_{n+1}^N = x_n^N + g(x_n^N,v_n)T/N + \sigma(x_n^N)w_{n+1}^N \;;\; x_o^N = \bar{x}_o$$
$$y_{n+1}^N = h(x_n^N)T/N + \eta_{n+1}^N$$
$$J^N(v) = Ez_T^N\{\sum_{n=o}^{N-1} f(x_n^N,v_n)T/N + \ell(x_N^N)\} = E_N^v\{\sum_{N=o}^{N-1} f(x_n^N,v_n)T/N + \ell(x_N^N)\}$$

where, see (2.8), y_n^N are the increments of the optimal observation process y_t and η_n^N is a sequence of i.i.d. zéro-mean gaussian vectors with covariance $(T/N)I$.

Remark 2.1 : On (2.12), instead of requiring v_n to be measurable with respect to $\tilde{y}^n := \sigma\{y_1^N,\dots,y_n^N\}$, we require it to be measurable with respect to a bigger σ-algebra, namely $y^{nT/N}$ which in what follows we shall write as Y^n. On the next subsection II.3 we shall see (Corollary 2.1) that with either σ-algebra we obtain the some optimal value. □

II.3 Reformulation of the discrete-time problem (complete-observation equivalent)

II.3.1 Unnormalized conditional densities

Assume $\bar{x}_o$ has a density $p_o(x)$ and that $\sigma(x)$ is invertible with $\sigma^{-1}(x)$ bounded. Let

(2.13)
$$p^N(x,\xi,v,y) := \exp\{-\frac{N}{2N}(\sigma\sigma^*)^{-1}(x-\xi-g(\xi,v)\frac{T}{N})^2 + h(\xi)y - \frac{T}{2N}|h(\xi)|^2\}\frac{1}{(2\pi\frac{T}{N})^{d_1/2}\det\sigma(\xi)}$$

and define, recursively for $n = o,\dots,N$, the following sequence

$$q_o^N(x) = p_o(x)$$

$$q_{n+1}^N(x) = \int p^N(x,\xi;v_n,y_{n+1}^N)q_n^N(\xi)d\xi$$

Remark 2.2 : From (2.12) and (2.13) it is easily seen that $q_n^N(x)$ can be interpreted as an unnormalized conditional density of x_n^N, given $(v_o,\dots,v_{n-1} ; y_1^N,\dots,y_n^N)$; it also follows that $q_n^N(x)$ depends on the past of the original observation process y_t only through the increments y_j^N, $j \le n$. □

We now have the following theorem, whose proof is in [1]

Theorem 2.2 : For all $\phi(.)$ bounded we have

$$E\{\phi(x_{n+1}^N)z_T^N \quad Y^{n+1}\} = \int_{\mathbb{R}^{d_1}} \phi(x)q_{n+1}^N(x)\,dx$$

Proof (sketch only ; for the complete proof see [1]).

With (see (2.7)) $\quad z_n^N = \exp\{\sum_{j=o}^{n-1}[h(x_j^N)y_{n+1}^N - \frac{T}{2N}|h(x_j^N)|^2]\}$

we have

i) $$E\{\phi(x_{n+1}^N)z_T^N \quad Y^{n+1}\} = E\{\phi(x_{n+1}^N)z_{n+1}^N \mid Y^{n+1}\}$$

ii) $$E\{\phi(x_{n+1}^N)z_{n+1}^N \mid Y^{n+1}\} =$$

$$= E\{z_{n+1}^N \quad E\{\phi(x_{n+1}^N) \mid Y^{n+1}, x_n^N, z_n^N\} \mid Y^{n+1}\} =$$

$$= \int \phi(x)\, E\{p^N(x,x_n^N ; v_n,y_{n+1}^N)z_n^N \mid Y^{n+1}\}dx =$$

$$= \int \phi(x)\, E\{p^N(x,x_n^N ; v,y)z_n^N \mid Y^{n+1}\}\Big|_{y = y_{n+1}^N, v = v_n} dx =$$

$$= \int \phi(x)\, E\{p^N(x,x_n^N,v,y)z_n^N \quad Y^n\}\Big|_{y = y_{n+1}^N, v = v_n} dx =$$

$$= \int \phi(x) \int p^N(x,\xi;v,y)q_n^N(\xi)d\xi\Big|_{y = y_{n+1}^N, v = v_n} dx$$

where in the last equality we assume that the statement of the theorem holds at stage n. To conclude, it therefore suffices to start the induction, i.e.

$$E\{\phi(x_1^N) z_1^N \mid Y^1\} = \int \phi(x) q_1^N(x)\,dx$$

□

Theorem 2.2 allows us to rewrite $J^N(v)$ in terms of the functions $q_n^N(x)$; i.e.

(2.15)
$$J^N(v) = \frac{T}{N} \sum_{n=o}^{n=1} E \int f(x,v_n) q_n^N(x)\,dx + E \int \ell(x) q_N^N(x)\,dx$$

The stochastic control problem with (infinite-dimensional) state-equation (2.14) and objective function (2.15) is now the complete-information equivalent of problem (2.12).

II. 3.2. Dynamic Programming

Assume in addition to A.1. and A.2.

A.3. : $g(x,v)$ and $f(x,v)$ are continuous in v.

Let B denote the Banach space of continuous functionals on L^1 $(\mathbb{R}^{d1})$ with linear growth, equipped with the norm

(2.16)
$$\| u \| := \sup_{q \in L^1} \frac{|u(q)|}{1 + |q|}$$

and define recursively for $n = N, \ldots, O$

(2.17)
$$\begin{cases} u_N^N(q) = \int_{\mathbb{R}^{d_1}} \ell(x) q(x)\,dx \\ u_n^N(q) = \inf_{v \in V} \{ \frac{T}{N} \int_{\mathbb{R}^{d_1}} f(x,v) q(x)\,dx + \\ \qquad + E\, u_{n+1}^N \left(\int p^N(.,\xi; v, \sqrt{\tfrac{T}{N}}\eta) q(\xi)\,d\xi \right) \} \end{cases}$$

where η is zero-mean Gaussian in $\mathbb{R}^{d_2}$ with covariance-matrix I. We now have the following theorem whose proof is in [1] .

Theorem 2.3. : Under the given assumptions

i) (2.17) defines a sequence in B

ii) There exists an optimal feed back $\hat{v}_n^N(q)$ which is a Borel function on L^1 with values in V

iii)
$$u_o^N(p_o) = \inf_{v \in W_N} J^N(v)$$

□

Letting

(2.18)
$$\tilde{W}_N := \{ v \in W_N \mid v_n \text{ is } \tilde{Y}^n - \text{measurable} \}$$

and recalling (see Remark 2.2) that $q_n^N(x)$ depends on y_s, $s \le n$ only through the increments y_j^N, $j \le n$, we finally have.

Collary 2.1

$$\tilde{\rho}_N = \inf_{v \in W_N} J^N(v) = u_o^N(p_o) = \inf_{v \in \tilde{W}_N} J^N(v)$$

□

Remark 2.3 We have approximated the original continuous-time partial-information control problem (1.2) with controls in W (see(2.2)) by the discrete-time, complete-information problem (2.14), (2.15) with controls in W_N (see(2.18)). The convergence of the approximation is guaranteed by (2.4) an Theorem 2.1. The approximating problem (2.14), (2.15) admits (Theorem 2.3) an optimal feedback control obtainable from (2.17). However, since the state in (2.14) is infinite - dimensional, the dynamic programming equations (2.17) do not provide an actual computational algorithm for determining this optimal control.

Further approximation steps are needed to reduce the problem to one with finite states and controls. In the rest of the paper we briefly mention such further steps ; for the full description we refer to [1]. □

III. FURTHER APPROXIMATIONS

III. 1 Finite-dimensional state

After the time-discretization in Section II we now perform an additional approximation step which is equivalent , but different from spatial discretization.

III. 1.1. Approximation and convergence

Let $g^m(x,v)$, $\sigma^m(x)$, $h^m(x)$, $f^m(x,v)$, $\ell^m(x)$ be step functions in the variable x that approximate in L^∞, and uniformly with respect to v, the corresponding functions in (2.12) ; also let $a^m(x)$ be a (truncated) step function approximating the identity function (see [1], see also [2], [3]). Corresponding to (2.6), (2.7), (2.9), (2.10) define ,

$$x_{n+1}^{N,m} = a^m(x_n^{N,m}) + \frac{T}{N} g^m(x_n^{N,m}, v_n) + \sigma^m(x_n^{N,m}) w_{n+1}^N, \qquad x_o^{N,m} = \bar{x}_o \tag{3.1}$$

$$z_T^{N,m} := \exp\{\sum_{j=o}^{N-1} [h^m(x_j^{N,m})\, y_{j+1}^N - \frac{T}{2N} | h^m(x_j^{N,m})|^2]\} \tag{3.2}$$

$$J^{N,m}(v) = E\, z_T^{N,m}\{\sum_{n=o}^{N-1} f^m(x_n^{N,n}, v_n) \frac{T}{N} + \ell^m(x_N^{N,n})\} \tag{3.3}$$

$$\rho^{N,n} := \inf_{v \in \tilde{W}_N} J^{N,n}(v) \tag{3.4}$$

Again, the following theorem is a consequence of [2, Prop. 3.2.] (see [1]) and define

$$(3.14)\qquad \rho^{N,m,H} := \inf_{v \in \bar{W}_N} J^{N,m,H}(v)$$

For problem (3.11), (3.12) the number of states $\pi_n^H = [\pi_{i,n}^H]_{i=1,\ldots,m}$ at each stage n is now finite. The dynamic programming algorithm then allows us to actually compute an optimal control

$$(3.15)\qquad \bar{v}_n(\pi_n^H) = \bar{v}_n(\bar{v}_o,\ldots,\bar{v}_{n-L};\ z_1,\ldots,z_n) \in \bar{W}_N$$

By (3.10) this control can also be expressed as

$$(3.16)\qquad \bar{v}_n(\bar{v}_o,\ldots,\bar{v}_{n-L};\ \bar{y}(y_1^N),\ldots,\bar{y}(y_n^N))$$

thereby becoming a control in the class $\tilde{W}_N$ and thus also in W so that it can be applied to our original problem (1.2) (i.e. (1.1)). In the next conclusions-section we show that, by choosing N,m, H sufficienthy large, this control $\bar{v}_n$ is ε-optimal for problem (1.2)

From [1] we now have the following

Theorem 3.3 :

i) $$\lim_{H\to\infty} J^{N,m,H}(v) = J^{N,m}(v)\ ;\quad v \in W_N$$

ii) $$\lim_{H\to\infty} \rho^{N,m,H} = \rho^{N,m}$$ □

IV . CONCLUSIONS

From (2.4) and theorems 2.1, 3.1, 3.3, given $\varepsilon > 0$, there exist N,m,H such that

$$(4.1)\qquad |\rho^{N,m,H} - \rho| < \varepsilon/2$$

The same theorems 2.1, 3.1, 3.3 imply that for all $v \in W_N$

$$(4.2)\qquad |J^{N,m,H}(v) - J(v)| < \varepsilon/2$$

Considering that the control $\bar{v}$ in (3.16) belongs to $\tilde{W}_N \subset W_N \subset W$ and is such that

$$(4.3)\qquad J^{N,m,H}(\bar{v}) = \rho^{N,m,H}$$

we then have that

$$(4.4)\qquad J(\bar{v}) \le \rho + \varepsilon = \inf_{v \in W} J(v) + \varepsilon$$

i.e. the control $\bar{v}$ is ε-optimal for the original problem (2.1) (i.e. (1.1)) provided N,m,H are sufficiently large.

Theorem 3.1 Given A.2, if m is large enough so that

i) $$\sup_v \| g^m(.,v) - g(.,v) \|_\infty \le \frac{1}{3} (\frac{T}{N})^{1/2}$$

ii) $$\| \sigma^m(.) - \sigma(.) \|_\infty \le (1/3)^{5/4} (\frac{T}{N})$$

iii) $$m \ge 3 \exp [2 + T(\| g \|_\infty + 2 \|\sigma\|^2_\infty) 2^{1/4 (2+N)} (\frac{T}{N})^{3/2}$$

then $(v \in W_N)$

$$|J^{N,m}(v) - J^N(v)| \le C [(\frac{T}{N})^{1/2} + \| h^m(.) - h(.) \|_\infty +$$

$$+ \sup_v \| f^m(.,v) - f(.,v) \|_\infty + \|\ell^m(.) - \ell(.) \|_\infty$$

with the same upper bound also for

$$|\rho^{N,m} - \tilde{\rho}^N|$$ □

Remark 3.1 Using a measure transformation analogous to (2.11), (3.1) and (3.3) can be seen to correspond to state-equation and objective function respectively, of a discrete-time problem with partial information analogous to (2.12). □

III. 1.2. Complete-observation equivalent

Denoting by $\phi^m(x,v)$ the generic step function in (3.1) - (3.3), let $\{D_i\}$ i = 1,..m be a partition if $\mathbb{R}^{d_1}$ such that

(3.5) $$\phi^m(x,v) = \sum_{\ell=1}^{m} \phi_i(v) \quad I_{D_i}(x)$$

analogously to (2.13), (2.14) define the matrix function

(3.6) $$P_{ij}(v,y) := \int_{D_i} \frac{\exp\{-\frac{N}{2T}(\sigma_j\sigma_j^*)^{-1}(x - g_j(v))^2 dx.}{(\frac{2\pi T}{N})^{1/2} \det \sigma_j} . \exp\{ h_j y - \frac{T}{2N} |h_j|^2 \}$$

and the sequence of m - vectors

(3.7) $$\begin{cases} \pi_{i,o} = \int_{D_i} p_o(x)\,dx \\ \pi_{i,n+1} = \sum_j \pi_{j,n}\, P_{ij}(v_n, y^N_{n+1}) \end{cases}$$ (i, = 1, ...,m)

corresponding to theorem 2.2. we have (see [1])

Theorem 3.2. :

$$E \{I_{D_i}(x_n^{N,m})\ z_T^{N,m} \mid \tilde{y}^n\} = \pi_{i,n} \qquad \square$$

which, recall (3.5), allows us to rewrite (3.3) as

$$J^{N,m}(v) = \frac{T}{N} \sum_{n=o}^{n-1} E \sum_{i=1}^{m} f_i(v_n)\ \pi_{i,n} + E \sum_{i=1}^{m} \ell_i\ \pi_{i,N} \tag{3.8}$$

Remark 3.2 [3.1] The approximating complete-information problem (3.7), (3.8) now has a finite-dimensional state $\pi_n = [\pi_{i,n}]_{i=1,\dots,m}$ but its possible values are still infinite, due to the fact that (see(3.7)) v_n and y_{n+1}^N take an infinite number of values. By letting $g^m(x,v)$, $f^m(x,v)$ be step-functions also in v, theorem 3.1 still holds (with the obvious modifications) and v_n in (3.7) can be condidered as taking a finite number of values. We can now complete our program by performing an additional approximation step consisting in a discretization of the observation increments y_n^N. $\square$

III. 2 Finite-valued state (Discretization of the observations).

Consider (see [1] , [3]) a partition $\{Y_h\}_{h=1,\dots,H}$ of the space $\mathbb{R}^{d_2}$ where the observation y_n^N take their values and let $\bar{y}_h$ be a representative element of the set Y_h. Also let

$$\bar{y}(y) := \sum_{h=1}^{H} \bar{y}_h\ I_{Y_h}(y) \tag{3.9}$$

Assume now that instead of the process y_n^N we observe the "discretized" process Z_n where

$$z_n := \bar{y}(y_n^N) \tag{3.10}$$

Using the observations $\{z_n\}$, the state-dynamics (3.7) and objective function (3.8) become

$$\begin{cases} \pi_{i,o}^H = \int_{D_i} p_o(x)\,dx \\ \pi_{i,n+1}^H = \sum_{j=1}^{m} \pi_{j,n}^H\ p_{ij}(v_n, Z_{n+1}) \end{cases} \qquad (i = 1,\dots,m) \tag{3.11}$$

and

$$J^{N,m,H}(v) = \frac{T}{N} \sum_{n=o}^{N-1} E \sum_{i=1}^{m} f_i(v_m)\pi_{i,n}^H + E \sum_{i=1}^{m} \ell_i \pi_{i,N}^H \tag{3.12}$$

respectively. Also let

$$\bar{W}_N := \{v \in W_N \mid v_n \text{ is } \sigma\{z_1,\dots,z_n\} \text{ - measurable}) \tag{3.13}$$

References

[1] A. BENSOUSSAN, W.J. RUNGGALDIER, "An approximation method for stochastic control problems with partial observation of the state". To be published.

[2] G.B. DI MASI, M. PRATELLI, W.J. RUNGGALDIER, "An approximation for the nonlinear filtering problem, with error bound" Stochastics, 1985, Vol. 14, pp. 247 - 271.

[3] G.B. DI MASI, W.J. RUNGGALDIER, "An approach to discrete - time stochastic control problems under partial observation". To be published.

[4] H.J. KUSHNER, Probability methods for approximations in stochastic control and for elliptic equations. Academic Press 1977.

Overload control for SPC telephone exchanges

- refined models and stochastic control

R.K. Boel

Laboratorium voor Theoretische Elektriciteit, Rijksuniversiteit Gent
Grote Steenweg Noord 2, B9710 Gent-Zwijnaarde, Belgium

J.H. van Schuppen

Centre for Mathematics and Computer Science
P.O. Box 4079, 1009 AB Amsterdam, The Netherlands

In telephone networks the switching and connecting operations are performed by the exchanges. The Stored Program Control (SPC) exchanges which are nowadays installed are computer controlled. One of the problems with these exchanges is the severe performance degradation during periods in which the demand for service exceeds the design capacity. The problem of overload control is then to maximize the number of successfully completed calls. In this paper two models for overload control of an SPC exchange are proposed that are refinements of an earlier model. A stochastic control problem for one of these models is shown to have a bang-bang type of optimal solution.

1. Introduction

The purpose of this paper is to present refined models for the operation of SPC telephone exchanges and to consider a stochastic control problem for overload control.

Telephone exchanges are the operational units at the nodes of telephone networks. In the last few years computer-controlled *Stored Program Control (SPC)* exchanges have been installed. In such an exchange the operations are executed by a processor according to a stored program. The operations of such an exchange may be summarized as follows. If a customer picks up the receiver this action generates a signal that will be detected by the exchange. After some delay, the exchange answers by sending a dial tone. After the customer has dialed the desired number, the exchange establishes, with some delay and depending on availability, a connection with the requested phone. All these different tasks have to be executed sequentially by the processor.

The performance of an SPC exchange can degrade considerably during periods in which the demand for service exceeds the design capacity [5]. The response time of the exchange during such periods is relatively long. This may cause impatient customers to dial prematurely, before a dial tone is given, after which an incompletely received telephone number takes up processor capacity and ends up as an unsuccessful call. Other requests for connections, that have been transmitted properly to the exchange, may encounter long processing delays. This then causes customers to abandon the call request and, possibly, to redial soon after. In this case capacity of the exchange is also wasted. That this performance degradation is a serious problem may be concluded from the data of [5].

The *problem of overload control* is then to maximize the number of successfully processed call requests. A call request may either be given access or be refused access. This decision represents the control action. References on this problem are [1,5,6,7,8,9,10,13]. The overload control problem also arises in mobile automatic telephony, in PBX business exchanges and other communication equipment.

A model for overload control has been proposed by one of the authors [13], based on an approach developed by F.C. Schoute [8,9,10]. This model consists of a hierarchical queueing system representing calls-in-build-up and tasks for the processor. Weak points in this model are: 1. there is no model for the successfully processed call requests; 2. retrials are not modeled. The present paper gives two refined models of an SPC exchange and then considers a stochastic control problem for one of these models.

The terminology used for counting and jump processes may be found in [4]. For a survey of modeling, stochastic filtering and stochastic control of such processes see [3]. References on the control of queueing systems are [11,12].

The authors acknowledge useful discussions with F.C. Schoute of Philips Telecommunicatie Industrie on the problem of overload control. They also thank the governments of Belgium and The Netherlands which through their cultural exchange agreement have provided financial support for the cooperation of the two authors.

2. A HIERARCHICAL QUEUEING SYSTEM

In this section the model for an SPC telephone exchange of [13] is summarized and discussed.

The mathematical model

A brief description of the technical operation of an SPC exchange follows. A customer who picks up the receiver sends thus a signal to the telephone exchange, to be called a *call request.* Call requests, when detected by the exchange, are placed in a buffer by the central processor. These buffered requests will be termed *calls-in-build-up.* During its presence at the buffer a call-in-build-up generates *tasks* which are executed sequentially by the central processor. Examples of tasks are a request for a dial tone, detection and recognition of dialed digits, the establishment of a connection, and related actions.

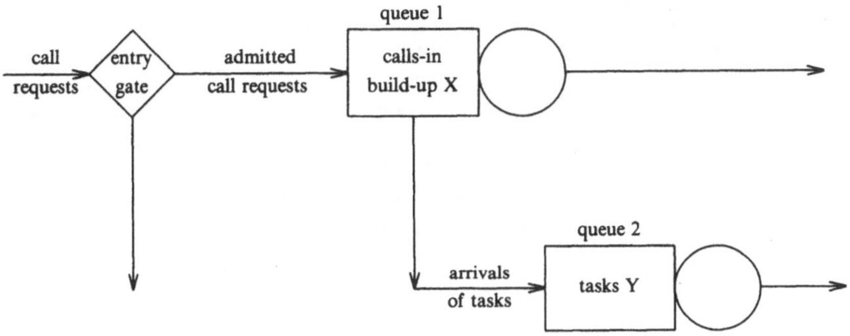

FIGURE 1. A hierarchical model for overload control.

The dynamics of the processor load may be modeled by a hierarchical queueing system as in figure 1. Call requests represented by an arrival process may or may not be admitted to the exchange, possibly based on the outcome of a toss of a coin. The probability of admission represents the control action. Call requests that have been refused access are assumed not to return. A call request that has been admitted to the exchange is placed in a buffer with an infinite number of servers. Such calls-in-build-up have independent identically exponentially distributed service times.

During its presence at queue 1 a call-in-build-up generates tasks that have to be executed by the central processor. The intensity of the arrival process of tasks is assumed to be proportional to the number of calls-in-build-up in queue 1. The task execution process is modeled by a single server queue M/M/1 operating on a first-in-first-out rule and with an infinite buffer.

Assume given a complete probability space (Ω, F, P) and a time index set $T = R_+$. Let

$$Z_+ = \{1,2,...\}, \quad N = \{0,1,2,...\}.$$

The construction of the hierarchical queueing system proceeds via a measure transformation indexed by a class of control policies U. For each admissable control policy $U(.)$ one obtains the following dynamic representation for the hierarchical queueing system,

$$dX(t) = [\lambda_0 U(t) - \mu_1 X(t)]dt + dM_1(t), \quad X(0), \tag{2.1}$$

$$dY(t) = [\lambda_2 X(t) - I_{(Y(t)>0)}\mu_2]dt + dM_2(t), \quad Y(0), \tag{2.2}$$

where $X:\Omega \times T \rightarrow R_+$ represents the number of calls-in-build-up, $Y:\Omega \times T \rightarrow R_+$ the number of tasks waiting or being served and M_1, M_2 local martingales. For details on this model see [13, sections 2 and 3].

Criticism and comments on the hierarchical queueing system

1. How to represent a successfully processed call request? A call request will be termed *successful* if it reaches a ringing or busy signal at the requested phone. The goal of overload control is to maximize the number of successfully processed call requests. It is preferable to exhibit successful call requests explicitly in the model rather than only in the cost function.

   Clearly a call request will be processed successfully if the delay in giving a dial tone and in establishing a connection is smaller than the time a customer is willing to wait. Thus one needs to model the time delays. How to do this is discussed at point 3 below.

2. The criticism may be voiced that in the hierarchical queueing system there is no connection between the server in queue 1 and the server in queue 2. Thus a call-in-build-up may leave from queue 1 before the tasks it has generated have been processed by the central processor in queue 2.

   To counter this criticism recall that a call request in queue 1 represents the active task generation phase during which tasks, such as a request for a dial tone, for a connection and for routing are generated. Should there then be a connection between the departure processes of queue 1 and queue 2, in particular should the time a call request is in the active task generation phase depend on the processing of its tasks? A little thought leads one to conclude that one has to distinguish call requests that are actively generating tasks and those that are merely waiting for the processing of these tasks. Queue 1 should include the former, another queue could represent the latter. Furthermore, there should be a connection between the processing of tasks and the waiting call requests. This the leads to the question what is the delay in processing a call request compared with the patience of a customer?

   Remark that in general the active task generation phase is longer than the period during which the customer dials the telephone number.

3. What is the time necessary to process a call request and how can one model the patience of a customer? The customer notices two types of delay, one in waiting for a dial tone and one in waiting for the connection. In the hierarchical model these delays are not explicitly represented. On the other hand, the time necessary to process the call request is not explicitly represented either. This period could be inferred from: 1. the time a call request is present in queue 1 actively generating tasks; 2. after a call attempt has left queue 1, the time it takes the processor to process the tasks generated by that call attempt.

   Notice that because of the memory in queue 2 the second period is sensitive to overload conditions. Thus in situations close to or in overload, the intensity of the arrival process of tasks is momentarily larger than the intensity of the server process of queue 2. Then queue 2 will increase rapidly and cause the waiting time necessary to process the tasks of a customer to grow too.

   The question is then how to refine the model such that the above mentioned time periods are exhibited explicitly?

4. In the hierarchical queueing system it is assumed that customers that have been refused access will not return. This is unrealistic. A fraction of customers will attempt to redial after some time. Such repeated call requests will be termed *retrials*. Although it is hinted at in [13] that retrials may be modeled, this has not been done yet. Forys [5] argues that retrials can be a very important cause of performance degradation.
5. In the model of [13] it is assumed that the number of calls-in-build-up and the number of tasks can be measured and used for control. In most exchanges this is not possible. In general one can observe only the number of calls-in-build-up and the idle time of the processor. The last measurement is not relevant for overload conditions. The full information case, in which one assumes knowledge of the past of all processes, is useful for theoretical analysis only. The ultimate goal is the partial information case, in which only practically available measurements are used. Solution of that problem will involve the solution of a filtering problem.

Based on the preceding comments, two new models are introduced in the next section. The aim is to represent all phenomena which cause the performance degradation under overload, while keeping the model analytically tractible.

3. Refined models

In this section two new models are proposed for the processor load in an SPC exchange. They differ from the hierarchical queueing system of section 2 in that a call request may be in an active or in a passive phase. In addition, there is an equation for the process of successfully processed call requests. In the first model retrials will be modeled.

Model 1

See figure 2. for the interconnections of the network of model 1.

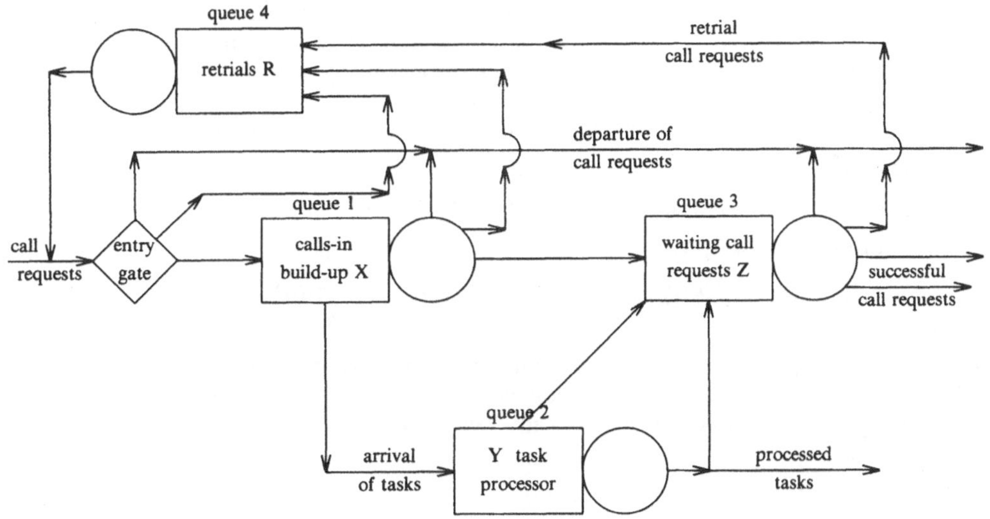

FIGURE 2. A refined model for overload control (model 1).

Assume given a complete probability space and a time index set $T = R_+$. An arrival process is denoted by A and a departure process by D. Super indices will be used to relate such processes to a queue. From now on all stochastic processes denoted by M with a certain index will denote local martingales.

The entry gate. The arrival process of original call requests is assumed to be a Poisson process with intensity λ_0 and with representation,

$$dA(t) = \lambda_0 \, dt + dM_1(t), \quad A(0). \tag{3.1}$$

As in the hierarchical queueing system of section 2 access to the exchange is controlled. The arrival process of queue 1 with the calls-in-build-up will be assumed to have the representation,

$$A^X(t) = \sum_{k=1}^{\infty} \bar{U}(k) I_{(\tau_k(A + D^R) \leq t)}, \tag{3.2}$$

where $\tau_k(A + D^R)$ is the stopping time of the k-th arrival of the process $(A + D^R)$ and $\bar{U}: \Omega \times N \to \{0,1\}$ is a random sequence that represents whether a call request is admitted or not. As in [13] one can then show existence of a stochastic process $U: \Omega \times T \to [0,1]$ such that A^X has the representation

$$dA^X(t) = [\lambda_0 + \mu_4 R(t)] \, U(t) \, dt + dM_2(t), \quad A^X(0). \tag{3.3}$$

Furthermore, as in [13], one can reformulate the model such that one only works with a control process U from some class $\underline{U}$ to be specified later. From such a control process U one can deduce the sequence $\bar{U}$ of (3.2).

Thus a fraction $U(.)$ of customers is admitted, and a fraction $(1 - U(.))$ is not admitted. Of the latter a fraction r_0 is assumed to redial after some time. This process will be modeled in the retrial queue, see below. There also the variable R will be defined.

The calls-in-build-up. A call request present in queue 1 will be termed a call-in-build-up. Its presence there signifies that it is actively generating tasks that have to be executed by the processor. It will be assumed that a call request is present at queue 1 for an exponentially distributed time with mean μ_1^{-1}. As in the hierarchical queueing system queue 1 will be taken to be a $. / M / \infty$ queue, thus with an infinite number of servers. The departure process from queue 1 is then represented by,

$$dD^X(t) = \mu_1 X(t) dt + dM_3(t), \quad D^X(0). \tag{3.4}$$

The equation for the number of calls-in-build-up $X: \Omega \times T \to R_+$ is,

$$X(t) = X(0) + A^X(t) - D^X(t), \tag{3.5}$$

$$dX(t) = [(\lambda_0 + \mu_4 R(t)) \, U(t) - \mu_1 X(t)] dt + dM_4(t), \quad X(0). \tag{3.6}$$

The tasks. During the presence of a call request in queue 1 it is assumed to generate tasks. The process of task generation will be assumed to be a Poisson process for each customer. The arrival rate at queue 2 is thus proportional to the number of calls-in-build-up. The representation of the arrival process of queue 2 is,

$$dA^Y(t) = \lambda_1 X(t) dt + dM_5(t), \quad A^Y(0). \tag{3.7}$$

The service times at queue 2 are independent and exponentially distributed with mean μ_2^{-1}. Tasks are executed in the order of their arrival.

$$dD^Y(t) = \mu_2 I_{(Y(t)>0)} dt + dM_6(t), \quad D^Y(0), \tag{3.8}$$

$$Y(t) = Y(0) + A^Y(t) - D^Y(t), \tag{3.9}$$

$$dY(t) = [\lambda_1 X(t) - \mu_2 I_{(Y(t)>0)}]dt + dM_7(t),\ Y(0). \tag{3.10}$$

Call requests waiting for the processing of their tasks. As mentioned in section 2, the presence of a call request in queue 1 represents the active task generation phase of the call request. However, after the active task generation phase there will be a period in which the customer has to wait for the processing of his tasks. This waiting time will be modeled by queue 3.

Only a fraction w_1 of the customers are assumed to be still waiting after their last task has been generated. The remaining $(1-w_1)$ fraction of customers is assumed to have departed. Of this a fraction r_1 goes to the retrial queue.

Let for $k \in Z_+$, $Z(k,.):\Omega \times T \rightarrow R_+$ be the number of tasks that have to be processed before the last task of the k-th customer leaving queue 1 is completed. The arrival process of $Z(k,.)$ is then taken to be,

$$A^Z(k,t) = W_1(k) I_{(\tau_k(D^X) \leq t)} Y_{\tau_k(D^X)-}, \tag{3.11}$$

where $\tau_k(D^X)$ is the stopping time at which the k-th customer departs from queue 1 and $W_1:\Omega \times N \rightarrow \{0,1\}$ is a sequence of independent random variables that determines whether a customer is still waiting or not. Assume that $P(W_1(k)=1)=w_1$ and that W_1 is independent of all other processes. The expression (3.11) is an approximation of the true waiting time for several reasons. For example, because it starts when the k-th customer leaves queue 1 rather than at the time this customer generates his last task. The departure process for $Z(k,.)$ must then be,

$$D^Z(k,t) = \sum_{s \leq t} I_{(Z(k,s-)>0)} \Delta D^Y(s), \tag{3.12}$$

$$Z(k,t) = Z(k,0) + A^Z(k,t) - D^Z(k,t), \tag{3.13}$$

$$dZ(k,t) = [\mu_1 w_1 X(t) Y(t) I_{(D^X(t)=k-1)} - \mu_2 I_{(Y(t)>0)} I_{(Z(k,t)>0)}]dt + dM_8(t),\ Z(k,0). \tag{3.14}$$

Summarizing, $Z(k,.)$ jumps to the value Y_{τ_k-} at $\tau_k(D^X)$, and subsequently jumps by -1 each time D^Y jumps by +1 until it becomes zero.

The patience of customers. Queue 3 will also model the patience of customers in waiting for the processing of their tasks. The total processing time of a customer consists of the time his call request generates tasks, which includes his dial time, and his waiting time after the generation of the last task. The task generation time is exponentially distributed by the assumptions for queue 1. This time does not depend on the state of the network, in particular not on overload conditions. It will be assumed that the waiting time of the customer after the last task in generated, is also exponentially distributed with mean μ_3^{-1}.

In accordance with the assumptions stated above concerning the waiting time of a customer after having left queue 1, one has the following representation. Here $P(k,t)=1$ represents that a customer is waiting and $P(k,t)=0$ that he is not waiting. For the k-th customer leaving queue 1,

$$P(k,t) = P(k,0) + A^P(k,t) - D^P(k,t), \tag{3.15}$$

$$A^P(k,t) = W_1(k) I_{(\tau_k(D^X) \leq t)}, \tag{3.16}$$

where $\tau_k(D^X)$ and W_1 are as defined below (3.11),

$$dA^P(k,t) = w_1 \mu_1 X(t) I_{(D^X(t)=k-1)} dt + dM_9(t),\ A^P(k,0), \tag{3.17}$$

$$dD^P(k,t) = \mu_3 I_{(P(k,t)>0)} dt + dM_{10}(t),\ D^P(k,0), \tag{3.18}$$

$$dP(k,t) = [w_1 \mu_1 X(t) I_{(D^X(t)=k-1)} - \mu_3 I_{(P(k,t)>0)}]dt + dM_{11}(t),\ P(k,0). \tag{3.19}$$

The successfully processed call requests. The call request of the k-th customer leaving queue 1 is successful if the processing of his last task is finished before his patience has run out. The successfully processed call requests may then be modeled by,

$$D^S(k,t) = \sum_{s \leq t} I_{(P(k,s-)>0)} I_{(Z(k,s-)=1)} \Delta D^Z(k,s), \tag{3.20}$$

$$= \sum_{s \leq t} I_{(P(k,s-)>0)} I_{((Z(k,s-)=1)} \Delta D^Y(s),$$

$$dD^S(t) = \sum_{k=1}^{\infty} dD^S(k,t) \tag{3.21}$$

$$= [\sum_{k=1}^{\infty} I_{(P(k,t)>0)} I_{(Y(t)>0)} I_{(Z(k,t)=1)}]\mu_2 dt + dM_{12}(t), \ D^S(0).$$

Retrials. Customers with a call request may be turned away by the exchange or loose their patience and terminate the call request. In the model these cases are represented by: 1. the call requests that have been refused access to the exchange by the entry gate; 2. the call requests that have been terminated by customers that are in the active task generation phase of queue 1; 3. the call requests that are unsuccessful because the customer's patience has run out before his last task has been processed. It is assumed that of the customers that have been turned away or that lost their patience, a fraction attempts to redial after an exponentially distributed time with mean μ_4^{-1}. In the model this will be represented by queue 4 that is in principle $./M/\infty$, with an infinite number of servers. A call request present in queue 4 will be termed to be in the *retrial mode.* The variable R represents the number of call requests that are in the retrial mode. The independent random sequences $Q_1, Q_3 : \Omega \times N \rightarrow \{0,1\}$ represent whether a call request goes to the retrial mode, if $Q_1(k)=1$, or not, if $Q_1(k)=0$. Assume that $P(\{Q_1(k)=1\})=r_1$, and $P(\{Q_3(k)=1\})=r_3$ and that the sequences Q_1, Q_2 are independent and independent of all other processes.

The process of retrials can then be modeled as,

$$dA^{R0}(t) = [\lambda_0 + \mu_4 R(t)] r_0 (1-U(t))dt + dM_{13}(t), \ A^{R0}(0), \tag{3.22}$$

$$A^{R1}(t) = \sum_{k=1}^{\infty} Q_1(k)(1-W_1(k)) I_{(\tau_k(D^x) \leq t)}, \tag{3.23}$$

$$dA^{R1}(t) = r_1(1-w_1)\mu_1 X(t)dt + dM_{14}(t), \ A^{R1}(0), \tag{3.24}$$

$$A^{R3}(t) = \sum_{k=1}^{\infty} Q_3(k) I_{(\tau_k(D^{SN}) \leq t)}, \tag{3.25}$$

where $\tau_k(D^{SN})$ is the stopping time of the k-th jump of the process D^{SN}, which process counts the number of call requests that leave queue 3 unsuccessfully,

$$D^{SN}(t) = \sum_{s \leq t} I_{(P(k,s-)=0)} I_{(Z(k,s-)=1)} \Delta D^Z(k,s),$$

$$dA^{R3}(t) = \sum_{k=1}^{\infty} r_3 \mu_2 I_{(P(k,t)=0)} I_{(Z(k,t)=1)} I_{(Y(t)>0)} dt + dM_{15}(t), \ A^{R3}(0), \tag{3.26}$$

$$A^R(t) = A^{R0}(t) + A^{R1}(t) + A^{R3}(t), \tag{3.27}$$

$$dA^R(t) = [r_0(1-U(t))(\lambda_0 + \mu_4 R(t)) + r_1(1-w_1)\mu_1 X(t), \tag{3.28}$$

$$+ \sum_{k=1}^{\infty} r_3 \mu_2 I_{(P(k,t)=0)} I_{(Z(k,t)=1)} I_{(Y(t)>0)}]dt + dM_{16}(t), \ A^R(0),$$

$$dD^R(t) = \mu_4 R(t)dt + dM_{17}(t), \ D^R(0), \tag{3.29}$$

$$R(t) = R(0) + A^R(t) - D^R(t), \tag{3.30}$$

$$dR(t) = [r_0(1-U(t))(\lambda_0+\mu_4 R(t)) + r_1(1-w_1)\mu_1 X(t) \tag{3.31}$$
$$+ \sum_{k=1}^{\infty} (\mu_2 r_3 I_{(P(k,t)=0)} I_{(Z(k,t)=1)} I_{(Y(t)>0)}) - \mu_4 R(t)]dt + dM_{18}(t), \; R(0).$$

The final stochastic dynamic system consists then of the formula's (3.6,3.10,3.11,3.14,3.16,3.19,3.31) with as controlled variable the successful departure process specified by (3.21). The specification of the stochastic control system is then completed by the definition of a class of admissable controls.

Model 2
Although model 1 answers the criticism of and comments on the hierarchical queueing model of section 2, it is rather complicated. Therefore a simplified model will be proposed below. Model 2 differs from model 1 in that the queues for the waiting call requests are aggregated to just one queue in which the distinction between customers disappears. Moreover, retrials are not modeled. See figure 3 for the network of model 2.

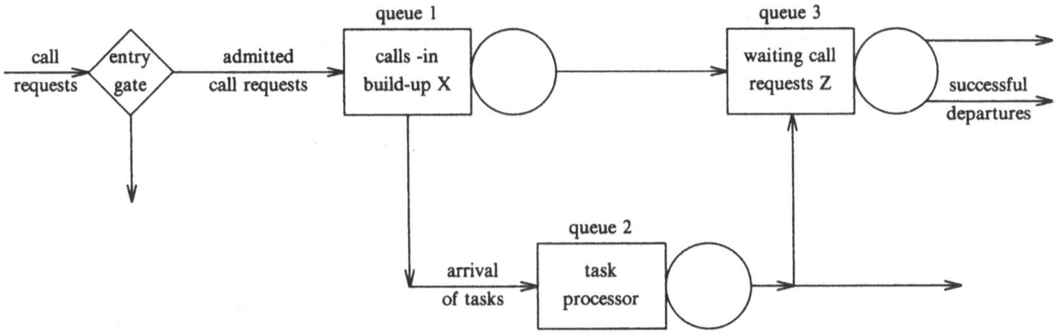

FIGURE 3. Another refined model for overload control (model 2).

Because part of model 2 is identical to model 1, those equations are not duplicated here. This concerns the entry gate, the buffer with calls-in-build-up as modeled by queue 1 and the task processor as modeled by queue 2, with the equations (3.1,3.3,3.4,3.6,3.7,3.8,3.10) with $R=0$.

The process of last tasks. It will be assumed that of every task finished by the processor, thus of $D^Y(t)$, it is a last task of some customer with a certain probability. This is modeled by a random variable $Q(t)$, with $P(\{Q(t)=1\})=c_2$ taken to be the proportion of last tasks over the total number of tasks, here $c_2=\lambda_2/\mu_1$. A disadvantage of this model is that it does not follow the short term fluctuations of the number of calls-in-build-up. The advantage of this model is that it is simple.

$$D^{YL}(t) = \sum_{k=1}^{\infty} Q_1(k) I_{(\tau_k(D^Y) \leq t)}, \tag{3.32}$$

$$dD^{YL}(t) = c_2 \mu_2 I_{(Y(t)>0)} dt + dM_{19}(t), \; D^{YL}(0). \tag{3.33}$$

Call requests waiting for processing of their tasks. The period which a call request has to wait for the processing of its tasks will be represented by queue 3. If there is a departure from queue 1, then there is an arrival at queue 3. The waiting time of each customer at queue 3 is in principle exponentially distributed with mean μ_3^{-1} and assumed to be independent of those of other customers. There is a departure from queue 3 if the patience of a customer runs out or if the last task of a customer is processed.

$$dA^Z(t) = dD^X(t) = \mu_1 X(t)dt + dM_{20}(t), \; A^Z(0), \tag{3.34}$$

$$dD^{ZP}(t) = \mu_3 Z(t)dt + dM_{21}(t), \; D^{ZP}(0), \tag{3.35}$$

$$D^{ZC} = \sum_{s \leq t} I_{(D^{YL}(s-) \geq D^Z(s-))} I_{(Z(s-)>0)} \Delta D^{YL}(s), \tag{3.36}$$

$$Z(t) = Z(0) + A^Z(t) - D^{ZP}(t) - D^{ZC}(t), \tag{3.37}$$

$$dZ(t) = [\mu_1 X(t) - \mu_3 Z(t) - c_2\mu_2 I_{(D^{YL}(t) \geq D^Z(t))} I_{(Y(t)>0)} I_{(Z(t)>0)}]dt + dM_{22}(t), \tag{3.38}$$
$$D^Z(0).$$

Successfully processed call requests. Finally one has to model the process of successfully processed call requests D^S. In the model it is assumed that a processed call request is successful if the number of completed last tasks is larger than or equal to the number of customers that have departed from queue 3,

$$D^S(t) = D^{ZC}(t) = \sum_{s \leq t} I_{(D^{YL}(s-) \geq D^Z(s-))} I_{(Z(s-)>0)} \Delta D^{YL}(s), \tag{3.39}$$

$$dD^S(t) = c_2\mu_2 I_{(D^{YL}(t) \geq D^Z(t))} I_{(Z(t)>0)} I_{(Y(t)>0)} dt + dM_{23}(t), \; D^S(0). \tag{3.40}$$

$$d(D^{YL}(t) - D^Z)(t) = [c_2\mu_2 I_{(Y(t)>0)} - \mu_3 Z(t) \tag{3.41}$$
$$- c_2\mu_2 I_{(D^{YL}(t) \geq D^Z(t))} I_{(Y(t)>0)} I_{(Z(t)>0)}]dt + dM_{24}(t), \; D^{YL}(0) - D^Z(0).$$

The stochastic control system of model 2 consists then of (3.6,3.10,3.38,3.41) with as controlled variable the process D^S of (3.40). Let $\underline{U}$ be the class of admissable control policies that are measurable functions of the past of the processes in the model. This completes the specification of model 2.

4. Stochastic control

In this section the overload control problem is formulated as a stochastic control problem for model 2.

Problem 4.1. *Given the stochastic dynamic system described by model 2 of section 3 with the time index set $T = [t_0, t_1]$, the class of input processes $\underline{U}$ and the cost function*

$$J(u) = -E_u[D^S(t_1) - D^S(t_0)] \tag{4.1}$$
$$= -E_u[\int_{t_0}^{t_1} c_2\mu_2 \, I_{(D^{YL}(t) \geq D^Z(t))} I_{(Z(t)>0)} I_{(Y(t)>0)} \, dt].$$

Determine an optimal control $u^* \in \underline{U}$ such that $J(u^*) \leq J(u)$, for all $u \in \underline{U}$.

Theorem 4.2. *Assume there exists a function $v : T \times N^4 \to R$ satisfying the following differential equation,*

$$v(t_1, k_1, \ldots, k_4) = 0,$$

$$dv(t,k)/dt - c_2\mu_2 I_{(k_4 \geq 0)}(k) I_{(k_3>0)}(k) I_{k_2>0)}(k) \tag{4.2}$$
$$+ [v(t,k_1+1,.) - v(t,k)]\lambda_0 I_{(v(t,k_1+1,.)-v(t,k)<0)}$$

$$+ [v(t,k_1-1,.,k_3+1,.)-v(t,k)]\mu_1 k_1$$
$$+ [v(t,.,k_2+1,.)-v(t,k)]\lambda_1 k_1$$
$$+ [v(t,.,k_2-1,.)-v(t,k)](1-c_2)\mu_2 I_{(k_2>0)}(k)$$
$$+ [v(t,.,k_2-1,k_3-1,.)-v(t,k)]c_2\mu_2 I_{(k_4\geq 0)}(k)I_{(k_2>0)}$$
$$+ [v(t,.,k_2-1,.,k_4+1)-v(t,k)]\mu_2 c_2 I_{(k_4<0)}(k)I_{(k_2>0)}(k)$$
$$+ [v(t,.,k_3-1,k_4-1)-v(t,k)]\mu_3 k_3$$
$$= 0,$$

where

$$\overline{X}(t) = (X(t),Y(t),Z(t),(D^{YL}(t)-D^{Z}(t))), \tag{4.3}$$

denotes the state, where $k^T=(k_1,k_2,k_3,k_4)\in N^4$ *denotes values of the state and a dot denotes components of k that remain unchanged. Then*

$$U^*(t) = I_{R_-}(v(t,X(t-)+1,.)-v(t,X(t-),.)) \tag{4.4}$$

is an optimal control for problem 4.2.

The interpretation of the optimal control law (4.3) is simple. Here $v(t,\overline{X}(t))$ is the estimate of the future cost at time $t\in T$ given the current state $\overline{X}(t)$. Then,

$$v(t,X(t-)+1,.)-v(t,X(t-),.), \tag{4.5}$$

is the change in the estimate of the future cost if a customer is admitted. Thus the control law (4.3) is such that a customer is admitted if in doing so the estimate of the future cost is decreased. The optimal control law is of bang-bang type, it takes only the extremal values 0 or 1. A similar result can be obtained for the stochastic control problem for model 1 although the equivalent of (4.2) is more complex.

The proof of 4.2 is a standard application of dynamic programming and therefore omitted. It is analogous to the proof of theorem 4.1 in [13]. In fact the proof is a special case of the following proposition.

PROPOSITION 4.3. *Assume given a stochastic control system with as state process a pure jump process* $X:\Omega\times T\to R^n$. *The jumps can take only a finite number of values, say* $r_1,\ldots,r_m\in R^n$. *Let* X_i *represent the process that consists of the jumps of X of height* r_i *only. The intensities of these jumps are assumed to be linear in the control process U,*

$$dX_i(t) = [\lambda_{1i}(X(t))+\lambda_{2i}(X(t))U(t)]dt + dM(t), \quad X_i(0). \tag{4.6}$$

Given further a cost function,

$$J(U) = E_U[\int_{t_0}^{t_1} (c_1(X(s)) + c_2(X(s))U(s))\, ds]. \tag{4.7}$$

Then the Bellman-Hamilton-Jacobi equation is linear in the control U,

$$\min_{U(t)\in[0,1]}[\, dv(t,X(t))/dt + c_1(X(t)) + c_2(X(t))U(t) \tag{4.8}$$
$$+ \sum_{i=1}^{i=m}[v(t,X(t)+r_i)-v(t,X(t))][\lambda_{1i}(X(t))+\lambda_{2i}(X(t))U(t)]\,],$$

and the optimal control law is of bang-bang type,

$$U(t) = I_{R^-}(c_2(X(t-) + \sum_{i=1}^{i=m} [v(t,X(t-)+r_i)-v(t,X(t-))] \lambda_{2i}(X(t-))). \qquad (4.9)$$

Comments

1. Instead of the stochastic control problem on a finite horizon one may also consider the infinite horizon problem, either for a discounted cost or for an average cost criterion. As in [13], there exists under certain conditions a time-invariant control law. Although this has not yet been worked out in detail for model 2 it seems that the control law is again of bang-bang type.
2. The stochastic control problem with partial observations still has to be considered. A realistic assumption is that the number of calls-in-build-up, the waiting call requests and the idle time of the processor and can be observed. This partially observed stochastic control problem leads to a stochastic filtering problem for the state of the control system given the observations. This filtering problem has been solved for the hierarchical queueing system of section 2. There it turns out that the resulting stochastic control system with the filter system is again linear in the control. By proposition 4.3 the optimal control law is thus again of bang-bang type.
3. For the application of control algorithms based on the suggested models and stochastic control, more research is necessary. The authors' research program includes an investigation of time-invariant stochastic control laws for average and discounted cost functions, development of algorithms for the numerical approximation of such control laws and of a performance analysis.

REFERENCES

1. B. BENGTSSON (1982). *On some control problems for queues,* Ph.D. thesis, Linköping University, Linköping.
2. R.K. BOEL, P. VARAIYA (1977). Optimal control of jump processes. *SIAM J. Control Optim. 15,* 92-119.
3. R.K. BOEL (to appear). Modelling, estimation and prediction for jump processes. *Advances in statistical signal processing, volume 1,* JAI Press.
4. P. BRÉMAUD (1981). *Point processes and queues - Martingale dynamics,* Springer-Verlag, Berlin.
5. L.J. FORYS (1983). Performance analysis of a new overload strategy. *10th International Teletraffic Congress (ITC).*
6. R.L. FRANKS, R.W. RISHEL (1973). Overload model of telephone network operation. *Bell System Tech. J. 52,* 1589-1615.
7. B. KARLANDER (1973). Control of central processor load in an SPC system. *Ericsson Technics 30,* 221-243.
8. F.C. SCHOUTE (1981). Optimal control and call acceptance in a SPC exchange. *9th International Teletraffic Congres.*
9. F.C. SCHOUTE (1983). *The technical queue: A model for definition and estimation for processor loading,* Report SR2200-83-3743, Philips Telecommunicatie Industrie, Dept. SAS, Hilversum.
10. F.C. SCHOUTE (1983). Adaptive overload control of an SPC exchange. *10th International Teletraffic Congress.*
11. M. SOBEL (1974). Optimal operation of queues. A.B. CLARKE (ed.). *Mathematical methods in queueing theory,* Lecture Notes in Economics and Mathematical Systems, volume 98, Springer-Verlag, Berlin, 231-261.
12. S. STIDHAM JR., N.U. PRABHU (1974). Optimal control of queueing systems. A.B. CLARKE (ed.). *Mathematical methods in queueing theory,* Lecture Notes in Economics and Mathematical Systems, volume 98, Springer-Verlag, Berlin, 263-294.
13. J.H. VAN SCHUPPEN (1984). *Overload control for an SPC telephone exchange - An optimal stochastic control approach,* Report OS-R8404, Centre for Mathematics and Computer Science, Amsterdam.

STOCHASTIC MAXIMUM PRINCIPLE IN THE PROBLEM OF OPTIMAL ABSOLUTELY CONTINUOUS CHANGE OF MEASURE

R.J.Chitashvili

The present paper is an attempt to derive the equation of the maximum principle for adjoint processes in the general optimization problem, which is characterized by a situation when the choice of the control determines the absolutely continuous change of some basic measure. In contrast to the case of diffusion Markov processes ([1],[2]), these equations, generally speaking, are not ordinary differential equations here. We base the derivation on a non-linear equation for the martingale component of the value process ([3]) and an expression for the differential of the maximum of the semi-martingales. The method can be also applied to the case with jump components in the martingales defining measure densities, but we shall restrict ourselves by the continuous case.

1. On the probability space (Ω,F,P) with a flow of σ-algebras $F = (F_t), o\le t\le T$ satisfying the usual conditions let $(M^a, a\in A)$ be a set of square integrable martingales $M^a \in \mathcal{M}^2(P), a\in A$ which define the exponential martingales

$$\rho_t^a = \varepsilon_t(M^a), \quad \rho_t^a > 0 \quad , \quad o\le t\le T \quad , \quad E\,\rho_T^a = 1 \; .$$

It is assumed that the set of decisions A is a compact subset of the metric space, the square characteristics $\langle M^a \rangle$ are dominated by an increasing process K with $K_T \le c < \infty$, the derivative

$$k(a,b) = d\langle M^a, M^b\rangle \,|\, dK$$

is continuous with respect to $a,b\in A$ almost at all points (ω,t) with respect to Dolean's measure μ of the process K $(\mu(\cdot) = E\int_o^T I_{(\cdot)}\,dK)$ and is bounded by a constant on $\Omega \times [o,T]$.

Under these conditions for any control $u\in\mathcal{U}$ representing a predictable process taking values in A the martingale M^u can be defind as a line integral

$$M_t^u = \int_0^t M(ds,u_s) \qquad \text{where} \quad M(\cdot,a) = M^a ,$$

characterized by the property

$$< M^u, m >_t = \int_0^t H_s(m,u_s)\,dK_s ,$$

where $H_t(m,a) = d<M^a,m>_t | dK_t$ for any $m \in \mathcal{M}^2(F,P)$ ([3]) and the exponential martingales

$$\rho^u = \varepsilon(M^u)$$

are densities and define the measures P^u , $u \in \mathcal{U}$.

The mathematical expectation with respect to the measure $P(P^u)$ is denoted by E (E^u) .

The problem is to maximize the mathematical expectation $E^u\eta$ where η is an F_T-measurable random variable which is square integrable. The control quality is estimated by the process $S_t^u = E^u(\eta|F_t)$ (estimator of the control u). The process $\bar{S}_t = \sup_u S_t^u$ is a value process.

The maximum principle is contained in the following assertion.

I. The estimator S^u and the value process $\bar{S}$ are unique solutions of the stochastic equations.

(1) $dS_t^u = -H_t(S^u,u_t)\,dK_t + dm_t^u$, $m^u \in \mathcal{M}^2(P)$, $S_T^u = \eta$,

(2) $d\bar{S}_t = -\max_a H_t(\bar{S},a)\,dK_t + d\bar{m}_t$, $\bar{m} \in \mathcal{M}^2(P)$, $\bar{S}_T = \eta$,

where the Hamiltonian $H_t(\xi,a)$ for the special semimartingale $\xi = M+V$ with $M \in \mathcal{M}^2(P)$ and the predictable process $V \in A$ with bounded variation coincides with $H_t(M,a)$.

The control u is optimal iff

(3) $\max_a H_t(\bar{S},a) = H_t(\bar{S},u_t)$ μ almost everywhere

Linear and non linear equations (1),(2) contain (unknown) martingale summands in the right-hand side, which are uniquely defined by the boundary condition given at the end of the time interval.

These equations can be represented in an equivalent form - as integral equations for the martingale parts ([3])

(1') $m_t^u = E(\eta + \int_0^T H_s(m^u,U_s)\,dK_s | F_t)$,

(2') $\bar{m}_t = E(\eta + \int_0^T \max_a H_s(\bar{m},a)\,dK_s | F_t)$.

2. Passing from equations (1'),(2') to a system of equations for

adjoint processes is motivated by the following.
Let continuous martingales $N^i \in M^{2,c}(P)$, $1 \le i \le n$, exist which are mutually orthogonal with a deterministic characteristics $\langle N^i \rangle$ which, without a loss of generality, are considered to be identical and coincide with the process K, ($\langle N^i \rangle = K$, $i=1,\dots,n$) and let the martingales $M^a, a \in A$ be represented as

(4) $$M_t^a = \sum_i \int_0^t f_i(s,a)\,dN_s^i$$

with bounded processes $f_i(t,a) = f_i(a)$ continuous w.r.t. a.
Naturally, one would also try to search for the desired martingales in the form of the decompositions

$$m_t^u = \sum_i \int_\theta^t \psi_i(s,u)\,dN_s^i + \tilde{m}_t^u ,$$
$$\bar{m}_t = \sum_i \int_0^t \bar{\psi}_i(s)\,dN_s^i + \tilde{m}_t ,$$

where the martingales $\tilde{m}^u, \tilde{m}$ are orthogonal to the system N^i, $1 \le i \le n$.
Now the Hamiltonian $H_t(m^u,a)$ and $H_t(\bar{m},a)$ is expressed in terms of the adjoint processes $\psi_i(u)$ and $\bar{\psi}_i$, $i=1,\dots,n$, so that the following notation

$$H_t(\psi(u),a) = \sum_i \psi_i(t,u) f_i(t,a), \quad H_t(\bar{\psi},a) = \sum_i \bar{\psi}_i(t) f_i(t,a)$$

is natural.
Equations (1'),(2') become now non-explicit equations for the adjoint processes.
For the square integrable F_T-measurable random variables ζ introduce the notation $\phi_i(t,\zeta)$ for the integrand in the representation

$$E(\zeta|F_t) = \sum_i \int_0^t \phi_i(s,\zeta)\,dN_s^i + N_t^\zeta ,$$

where N_t^ζ is a martingale orthogonal to the system $(N^i, 1 \le i \le m)$.
Evidently, $\psi_i(t,u) = \phi_i(t,m_T^u)$, $\bar{\psi}_i(t) = \phi_i(t,\bar{m}_T)$ and equations (1'), (2') can be written as

(5) $$\psi_i(t,u) = \phi_i(t,\eta) + \phi_i\Big(t, \int_0^T H_s(\psi(u),u_s)\,dK_s\Big),$$

(6) $$\bar{\psi}_i(t) = \phi_i(t,\eta) + \phi_i\Big(t, \int_0^T \max_a H_s(\bar{\psi},a)\,dK_s\Big),$$

and now our aim is to reduce equations (5),(6) to the differential form.

3. First we shall deal with linear equation (5) and, for simplicity, the

index u is omitted. So we consider a linear equation of the following form

(7) $\psi_i(t) = \phi_i(t,\eta) + \phi_i(t,\int_o^T H_s dK_s)$

where $H_t = \sum_i \psi_i(t) f_i(t)$.

Let $Y = Y_{[o,T]}$ denote a class of the processes ξ which admit the following representation

(8) $\xi_t = \xi_o + V_t^\xi + \sum_i \int_o^t \beta_i(s) dN_s^i + m_t^\xi$,

where β_i, V^ξ, m^ξ are continuous processes, the martingale m^ξ is orthogonal to the elements of the set $(N^i, i=1,\ldots,n)$, $V^\xi \in A^c$. We shall use the following notation $\xi^{(i)}$, $1\leq i\leq n$, and $\xi_t^{(o)}$ for the processes $\beta_i(t), 1\leq i\leq n$ and V_t^ξ entering (8), so $\xi_t^{(i)} = \beta_i(t), 1\leq i\leq n$, $\xi_t^{(o)} = V_t^\xi$. The subclass $Y_{[o,T]}$ of the processes ξ such that the expressions $\phi_i(t,\xi_s)$ also belong to $Y_{[o,s]}$ for all $0\leq t\leq s\leq T$ is denoted as $Y^2 = Y^2_{[o,T]}$.

Let $\xi \in Y^2$. For $t\leq s\leq T$ we have

$$\phi_i(t,\xi_s) = \phi_i(t,\xi_o+\xi_s^{(o)}+\sum_j \int_o^s \xi_s^{(j)} dN_s^j + m_s^\xi) = \xi_t^{(i)} + \phi_i(t,\xi_s^{(o)})$$

In View of the obvious property

$$\phi_i(t,\xi_s) = 0 \ , \quad s < t \ ,$$

and the linearity of the operator ϕ_i the last relation can be rewritten as

$$\phi_i(t,\xi_s) = \xi_t^{(i)} + \phi_i(t,\xi_s^{(o)} - \xi_t^{(o)}).$$

This implies that $\phi_i(t,\xi_t) = \xi_t^{(i)}$.

Thus, for $\xi \in Y^2$ we can write a representation for $t\leq s$

(9) $\Phi_i(t,\xi_s) = \xi_s^{(i)} - \sum_j \int_t^s \phi_i^{(j)} (z,\xi(s)) dN_s^j -$
$\qquad - (\phi_i^{(a)}(s,\xi_s) - \phi_i^{(o)}(t,\xi_s)) + \tilde{N}_s^i - \tilde{N}_t^i$,

where $\tilde{N}^i$ is a martingale orthogonal to the set $(N^i, i=1,\ldots,n)$.

If we assume that the processes $f_i(t)$, $\psi_i(t)$ from (7) belong to the class Y^2 equation (7) will take the following form

(10) $\psi_i(t) = \phi_i(t,\eta) + \int_t^T H_s^{(i)} dK_s - \int_t^T (\phi_i^{(o)}(s,H_s) - \phi_i^{(o)}(t,H_s)) dK_s -$

$$- \sum_j \int_t^T (\int_t^s \phi_i^{(j)} (z,H_s) dN_z^i) dK_s + \int_t^T (\tilde{N}_s^i - \tilde{N}_t^i) dK_s \ .$$

Assume that $\phi_i(t,\eta)\in Y, 1\leq i\leq n$ and introduce the following notation

(11) $\Gamma_i(t,H) = -\phi_i^{(o)}(t,\eta) - \int_t^T \phi_i^{(o)}(t,H_s)dK_s$.

Now the system of equations for $\psi = (\psi_i, 1\leq i\leq n)$ can be written in the differential form

(12) $d\psi_i(t) = -\Gamma_i(dt,H) - H_t^{(i)}dK_t + dm_t^i$, $1\leq i\leq n$,

with boundary condition at the end of the time interval $\psi_i(T) = \phi_i(T,\eta)$ where $m^i\in M^{2,c}(P)$ is some martingale.

Further, by Ito's formula, for the processes $\psi_j, f_j\in Y$ we have

(13) $(\psi_j(t)f_j(t))^{(i)} = \psi_j^{(i)}(t)f_j(t) + \psi_j(t)f_j^{(i)}(t)$.

It is also evident that

$$\psi_i^{(j)}(t)f_j(t)dK_t = d < \psi_i, \int_o^{\cdot} f_j(s)dN_s^j >_t$$

and, hence

(14) $\sum_j \psi_i^{(j)}(t)f_j(t)dK_t = d < \psi_i, M >_t$

where $M_t = \sum_j \int_o^t f_j(s)dN_s^j$.

In view of (13) and (14) the expression $H_t^{(i)}$ can be written as

$$H_t^{(i)} = \sum_j \psi_i(t)f_j^{(i)}(t) + \frac{d}{dK_t} < \psi_i, M >_t + \sum_j (\psi_j^{(i)}(t) - \psi_i^{(j)}(t))f_j(t).$$

Consider the last summand in (13). It follows from (10) that

$$\psi_i^{(j)}(t) = \phi_i^{(j)}(t,\eta) + \int_t^T \phi_i^{(j)}(t,H_s)dK_s .$$

Introduce the following notation

(15) $\delta_i(t,H) = \sum_j (\phi_j^{(i)}(t,\eta) - \phi_i^{(j)}(t,\eta))f_j(t) + \sum_j (\int_t^T (\phi_i^{(j)}(t,H_s) - \phi_j^{(i)}(t,H_s))dK_s)f_j(t)$.

The final form of the equation for the process ψ is

(16) $d\psi_i(t) = -\Gamma_i(dt,H) - \delta_i(t,H)dK_t - \sum_j \psi_j(t)f_j^{(i)}(t)dK_t + d\tilde{m}_t^i$

with the boundary condition $\psi_i(T) = \phi_i(T,\eta)$, where

$$\tilde{m}_t^i = m_t^i - < \psi_i, M >_t = m_t^i - < m^i, M >_t$$

and m^i is a martingale component in the decomposition of the process ψ_i, $m^i\in M^{2,c}(P)$.

The obtained result is statet as an assertion.

II. Let for a fixed $u\in U$ the processes $f_i(t,u_t), H_t(\psi(u),U_t)$ belong to the class Y^2 , then the adjoint process $(\psi_i(t,u), 1\leq i\leq n)$ is a solution of (16) where it is implied that

$$H_t = H_t(\psi(u),U_t),\ f_j(t) = f_j(t,U_t),\ M_t = M_t^u$$

with $\tilde{m}^i \in M^{2,c}(P^u)$.

The last assertion follows from Girsanov's theorem.

4. As a corollary an analogue of Haussmann's formula ([4]) can be given with the solution of (16) represented in the form of the conditional mathematical expectation.

Consider the matrix-valued process B_{ts} , $0 \leq t \leq s \leq T$, which is a solution of the equation

$$d_t B_{ts} = -G_t B_{ts} dK_t\ ,\ B_{tt} = I\ ,$$

where G is a matrix-valued process with the elements $G_t(i,j)=f_j^{(i)}(t)$, I is a unit matrix.

Applying the vector form for the processes $\psi(t) = (\psi_i(t),\ 1 \leq i \leq n)$, $\Gamma(t,H)=(\Gamma_i(t,H), 1 \leq i \leq n), \delta(t,H)=(\delta_i(t,H),\ 1 \leq i \leq n)$, $\phi(T,\eta) = (\phi_i(T,\eta),\ 1 \leq i \leq n)$ in terms of the fundamental solution B_{ts} we can express the vector $\psi(t)$ as

$$(17)\quad \psi(t) = E(\varepsilon_{tT}(M) B_{tT} \phi(T,\eta) | F_t) + E(\int_t^T \varepsilon_{ts}(B) B_{ts}(\Gamma(ds,H) + \delta(s,H) dK_s) | F_t)\ ,$$

where $\varepsilon_{tT}(M) = \varepsilon_T(M) \varepsilon_t^{-1}(M)$.

Formula (17) is obtained by means of the standard technique for the solution of equations with a boundary value at the righthand end of the time interval (for such equations see also [5]) - simple calculation of the difference shows that the process

$$\varepsilon_t(M) B_{ot} \psi(t) + \int_o^t \varepsilon_s(M) B_{os}(\Gamma(ds,H) + \delta(s,H) dK_s)$$

is a martingale w.r.t. the measure with the corresponding density $\varepsilon(M)$ and, taking into consideration the boundary value, we can obtain (17).

To compare (17) with Haussmann's formula it is convenient to restrict oneself by the one-dimensional case.

Let the basic measure P be a distribution of the Wiener process on $\Omega = C_{[o,T]}$, the martingale M defining the absolutely continuous transformation be given as

$$M_t = \int_0^t f(s,z_s)dz_s \ , \quad z \in C_{[o,T]}$$

where $f(t,x), x \in R^{(1)}$ is some differentiable function with respect to x, η is a random variable expressed as a differentiable on $C_{[o,T]}$ functional then the adjoint process ψ can be written as ([4])

$$(18) \quad \psi(t) = E(\int_t^T \varepsilon_{tT}(M)\exp(-\int_t^\tau \frac{\partial}{\partial x} f(s,z_s)ds)R(d\tau)|F_t)$$

where $R(t) = R(t,z)$, $z \in C_{[o,T]}$ are functions of bounded variation entering the expression for the derivative of the functional η.

In order to compare (18) with (17) the process R nonadapted to the flow F should be replaced by its filter. To be more exact, if $\hat{R}$ denotes a dual predictable projection of the process R, i.e. a predictable process with bounded variation in the representation

$$E(\varepsilon_{tT}(M)R(t)|F_t) = \hat{R}(t) + \hat{m}_t \ ,$$

where $\hat{m}$ is a martingale part then (18) can be written as

$$(19) \quad \psi(t) = E(\int_t^T \varepsilon_{ts}(M)\exp(-\int_t^s \frac{\partial}{\partial x} f(v,z_v)dv)\hat{R}(ds)|F_t) \ .$$

Now by the equality

$$f_1^{(1)}(t) = \frac{\partial}{\partial x} f(t,z_t)$$

which is evident from the definition of $f_i^{(j)}$, the relation between formulas (17) and (18) is expressed as

$$(20) \quad B_{ts} = \exp(-\int_t^s \frac{\partial}{\partial x} f(v,z_v)dv) \ ,$$

$$\hat{R}(T)-\hat{R}(t) = \phi(T,\eta) + (\Gamma(T,H)-\Gamma(t,H)) + \int_t^T \delta(s,H)dK_s \ .$$

5. The expression of $\hat{R}$ is by no means defined only by the random variable η, but also by the martingale M (the function f) by means of which the desired process ψ in (16) enters the integral expressions Γ and δ (by means of H).

But in the case of a simple functional η the expressions Γ and δ become zero. Indeed, let $w = (w^1,\dots,w^n)$, w^i be independent Wiener Processes, $\eta = \varphi(w_T^1,\dots,w_T^n)$ where φ is a differentiable function on $R^{(n)}$

$$M_t = \sum_{i=1}^{n} \int_o^t f_i(s,w_s^1,\dots,w_s^n)\,dw_s^i ,$$

where f_i, $1\le i\le n$ are differentiable. Under such assumptions the expression

$$s_t = E(\varepsilon_{tT}(M)\eta|F_t)$$

is represented by a solution of the differential equation

$$\frac{\partial}{\partial t} g + \sum_{i=1}^{n} \frac{\partial}{\partial x^i} g + \frac{1}{2}\sum_{i=1}^{n} \frac{\partial^2}{(\partial x^i)^2} g = o \ , \ g(T,x)=\varphi(x), \ x\in R^{(n)}$$

in the following form

$$s_t = g(t,w_t^1,\dots,w_t^n) \ .$$

The adjoint process (t) is expressed in terms of derivatives of

$$\psi_i(t) = \frac{\partial}{\partial x^i} g(t,w_t^1,\dots,w_t^n) \ .$$

Further, for any twice differentiable function $h(x)$ on $R^{(n)}$ by Clark's formula ([6]), for $t<s$ we have

$$\phi_i(t,h(w_s)) = E(\frac{\partial}{\partial x^i} h(w_s)\ F_t\ ) \ .$$

Consequently, it follows: first, that $\phi_i(t,h(w_s))$ is a martingale and, hence,

$$\phi_i^{(o)}(t,h(w_s)) = O \ .$$

On the other hand, again applying Clark's formula, we have

$$\phi_i^{(j)}(t,h(w_s)) = E(\frac{\partial^2}{\partial x_i\partial x_j} h(w_s)\ F_t) \ ,$$

and, evidently, $\phi_i^{(j)}(t,h(w_s)) = \phi_j^{(i)}(t,h(w_s))$. Now we must apply this result to the processes $\phi_i^{(o)}(t,\eta), \phi_i^{(j)}(t,\eta)$, $\phi_i^{(o)}(t,H_s)$, $\phi_i^{(j)}(t,H_s), 1\le i,j\le n$, in the expressions Γ_i, δ_i to verify that $\Gamma_i=\delta_i=O$, $1\le i\le n$. Under these conditions (16) takes the form of an ordinary differential equations.

When we derive the differential equations for adjoint processes for non-linear equations (2) or (6) we have only to calculate the "derivatives" of $(\max_a H_t(\bar\psi,a))^{(i)}$, $1\le i\le n$.

In particular, an important consequence of the maximum principle is the assertion that for any optimal control u

$$(\max_a \sum_j \bar\psi_j(t)f_j(t,a))^i = \sum_j \psi_j^{(i)}(t)f_j(t,U_t) + \sum_j \psi_j(t)f_j^{(i)}(t,U_t) . \quad (21)$$

Note that for an arbitrary control u and relation (21) is incorrect, e.g. for $A \subset R^{(1)}$ if $f_j(a)$ is differentiable with respect to the parameter a and the process u belongs to the class Y then

$$(f_j(t,U_t))^{(i)} = f_j^{(i)}(t,U_t) + \frac{\partial}{\partial a} f_j(t,U_t)U_t^{(i)} .$$

We shall need a formula of the differential for the maximum of the semimartingales.

Lemma. Let $\xi^K, K=1,\dots l$ be continuous semimartingles. Then

$$\bar{\xi}^l = \max_K \xi_t^K = \bar{\xi}_o^l + \sum_J \frac{1}{N(J)} \int_o^t I_{\left[\xi_s^K = \bar{\xi}_s^l, K\in J, \xi_s^K < \bar{\xi}_s^l, K\notin J\right]} d\xi_s^K + \bar{V}_t^l, \qquad (22)$$

where J is a subset of the set of indices $(1,\dots,l)$, $\bar{V}^l$ is an increasing process (continuous), $N(J)$ is the number of elements of the set J .

For the case when $l=2$ the process $\bar{\xi}_t^2$ can be represented as $\bar{\xi}_t^2 = (\xi_t^1 - \xi_t^2)^+ + \xi_t^2$ and we can apply the formula of the differential of the positive part of the semimartingale ([7])

$$\xi_t^+ = \xi_o^+ + \int_o^t I_{[\xi_s > o]} d\xi_s + \frac{1}{2}\int_o^t I_{[\xi_s = o]} d\xi_s + \frac{1}{2} L_t(\xi)$$

where $L_t(\xi)$ is the local time (spent at zero) of the process ξ .

This directly implies (22) with $\bar{V}_t^2 = L_t(\xi^1 - \xi^2)$.

For $l > 2$ the formula is derived by induction using the symmetrized recurrence relation

$$\bar{\xi}_t^{l+1} = \frac{1}{l+1} \sum_{j=1}^{l+1} \max(\max_{K\neq j} \xi_t^K , \xi_t^j) .$$

Corollary. If $m^K, 1\le K\le l$, $\bar{m}^l$ are martingale components of the processes $\xi^K, 1\le K\le l$, and $\bar{\xi}^l$ then for any F-adapted process u taking values in $(1,\dots,l)$ such that $\bar{\xi}_t^l = \xi_t^{u_t}$ a.s.

$$\bar{m}_t^l = \bar{m}_o^l + \sum_{K=1}^{l} \int_o^t I_{[u_s=K]} dm_s^K . \qquad (23)$$

The assertion follows from (22) if we apply the fact that (see [7]) for (any continuous) semimartingales ξ^i, ξ^j

$$\int_o^T I_{[\xi_s^i = \xi_s^j]} d\langle m^i - m^j\rangle_s = 0 \qquad \text{a.s.}$$

The following assertion based on the concept of finite approximation is given without a proof.

III. Let $\xi_t(a)$ be a process continuous w.r.t. a belonging to Y for all a , $\xi_t^{(o)}(a)$ be dominated by some increasing process $\bar{K}$, $\xi_t^o(a)=\int_o^t \varphi_s^o(a)\,d\bar{K}_s$ with $\xi_t^{(i)}(a)$, $1\leq i\leq n$, $\varphi_t^o(a)$ continuous w.r.t. a,

$$\int_o^T \sup_a [\xi_t^i(a)]^2 dK_t<\infty \ , \ \int_o^T \sup_a |\varphi_s^o(a)|\,d\bar{K}_s<\infty \ ,$$

and the martingale m^ξ in representation (8) is independent of a.

Then $\bar{\xi}_t = \sup_a \xi_t(a)$ belongs to Y and for any F-adapted measurable process u with $\bar{\xi}_t = \xi_t(u_t)$

$$\bar{\xi}_t^{(i)} = (\xi_t(u_t))^{(i)} = \xi_t^{(i)}(u_t), \ 1\leq i\leq n, \quad \mu \ \text{a.s.}$$

is satisfied.

Corollary. Let the processes $f_i(t,a)$, $\bar{H}_t = \max_a H_t(\bar{\psi},a)$ belong to the class Y^2 and, besides, $f_i(t,a)$, $1\leq i\leq n$, satisfy the conditions of assertion III. Then for any optimal control u the adjoint process $\bar{\psi}(t) = (\bar{\psi}_i(t),\ 1\leq i\leq n)$ is a solution of the equation

$d\bar{\psi}_i=-(\Gamma_i(dt,\bar{H})-\delta_i(t,\bar{H})\,dK_t)-\sum_j \bar{\psi}_j(t)f_j^{(i)}(t,u_t)\,dK_t+d\tilde{m}_t^i$, with the boundary condition $\bar{\psi}_i(T)=\phi_i(T,\eta)$ with $\tilde{m}^i\in M^{2,c}(P^u)$, $1\leq i\leq n$.

REFERENCES

1. Kushner, H.J., Necessary conditions for continuous parameter stochastic optimization problems, SIAM J.Control, 2, 1973, pp.587-594.
2. Haussmann U.G., On the stochastic maximum principle, SIAM J.Control optim., 16,no.2, 1978, 236-251.
3. Chitashvili R.J., Martingale ideology in the theory of controlled stochastic processes, Lect.Notes in Math., 1021, 1983, 73-92.
4. Haussmann, U.G., Functionals of Ito processes as stochastic integrals, SIAM J.Control, 16, no.2, 1978, 252-269.
5. Bismut J.M., An introduction to duality in random mechanics, Lect. Notes in Contr. and Inf.Sciences, 16, Stoch.Contr.Theory and Stoch. Diff.Systems, Bad Honnef, 1979, 42-60.
6. Clark J.M., The representation of functionals of Brownian motion by stochastic integrals, AMS 41, 4, 1970, 1282-1295.
7. Jacod J., Lect.Notes in Math., Calcul Stochastique et Problèmes de Martingales , Springer-Verlag, 714, 1979.

Asymptotic Properties of Least-Squares Estimators in Semimartingale Regression Models

N. Christopeit
Institut für Ökonometrie und Operations Research
Universität Bonn
Adenauerallee 24-42
D-5300 Bonn 1

1. Introduction and main results

We shall be concerned with the following class of linear regression models:

(1.1) $$y_t = y_o + B \int_o^t x_s d\theta_s + M_t , \quad t \geq 0 .$$

The m-dimensional process (y_t) is a semimartingale which is generated as the sum of the k-dimensional input process (x_t), integrated with respect to some increasing process (θ_t) and weighted with some $(n\times k)$-parameter matrix B, plus an unobservable disturbance (M_t), which we shall assume to be an m-dimensional martingale.

The model (1.1) covers a variety of special models, in particular the usual discrete time model with stochastic regressors:

(1.2) $$y_t = B x_t + \varepsilon_t, \quad t = 1,2,...,$$

the continuous time Ito equation

(1.3) $$dy_t = B x_t d_t + C dw_t, \quad t \geq 0 ,$$

where (w_t) is Brownian motion, and the Skorokhod equation

$$y_t = y_o + B \int_o^t x_s ds + \int_o^t \sigma_s dw_s + \int_o^t \int_Z c_s(z) q(ds,dz) ,$$

where $q(\omega,ds,dz) = p(\omega,ds,dz) - ds \otimes \alpha(dz)$, p being a Poisson point process with Lévy measure α.

We shall consider the following estimate for B:

(1.4) $$\hat{B}_T' = \left(\int_0^T x_s x_s' d\theta_s\right)^{-1} \int_0^T x_s dy_s'$$

based on the observations up to time T and assuming that $\int_0^T x_s x_s' d\theta_s$ is nonsingular. For the discrete time model (1.2), (1.4) is just the ordinary least squares estimate; for the system (1.3), it is the maximum likelihood estimate (if C is nonsingular). In the general case, $\hat{B}_T$ may be formally obtained by minimizing

$$\int_0^T \left\| \frac{dy_s}{d\theta_s} - B x_s \right\|^2 d\theta_s \ .$$

We shall present here some results on *strong consistency* and *asymptotic normality* of $\hat{B}_T$. For the proofs, the reader is referred to [1]. Moreover, it is shown how these general results work in the setting of *autoregressive Ito processes*.

2. Assumptions and main results.

Throughout we shall assume that the processes to be considered live on some probability space $(\Omega, \mathcal{F}, P)$ with filtration $(\mathcal{F}_t)$ satisfying the "usual hypotheses". A process will be called regular if it is adapted and all its paths have left and right hand limits at each point of time.

We make the following assumptions.

(A) (i) (M_t) is a local L^2-martingale, $M_o = 0$.

(ii) (x_t) is a regular left continuous process.

(iii) (θ_t) is a regular right continuous increasing process.

(B) Let $M_t = M_t^c + M_t^d$ be the decomposition of (M_t) into its continuous and purely discontinuous parts, and let $\langle X \rangle$ and $[X]$ denote the first and second increasing process, resp., of a local L^2-martingale X.

For all $\delta > 0$

(i) $$\int_0^t \phi_s d\langle M^c \rangle_s = O\left(\left[\int_0^t \phi_s d\theta_s^c\right]^{1+\delta}\right)$$

and

(ii) $$\int_0^t \phi_s d\langle M^d\rangle_s = O([\int_0^t \phi_s \, d\theta_s]^{1+\delta})$$

holds with probability one for every nonnegative predictable process (ϕ_t).

We shall also consider the following strengthened form of (B):

(B') (i) and (ii) hold for $\delta = 0$. Moreover,

(iii) $$\int_0^t \phi_s d[M^d]_s = O(\int_0^t \phi_s \, d\theta_s) \quad \text{a.e.}$$

Denote

$$Z_t = \int_0^t x_s x_s' d\theta_s ,$$

and let $\lambda_{max}(t)$ and $\lambda_{min}(t)$ denote the maximal and minimal eigenvalue of (Z_t), resp.

The *crucial assumption* is

(C) $\lambda_{min}(T) \to \infty$ a.e. and

$$(\log \lambda_{max}(T))^{1+\delta} = O(\lambda_{min}(T)) \quad \text{a.e.}$$

for some $\delta > 0$.

Or, in its weaker form:

(C') $\lambda_{min}(T) \to \infty$ a.e. and $(\log(\lambda_{max}(T)) = o(\lambda_{min}(T)$ a.e.

We then have the following result.

Theorem 1. Assume (A) - (C). Then the least-squares estimate (1.4) converges a.e. to the true parameter matrix B. If, instead of (B) the stronger assumption (B') holds, then (C) may be weakened to (C').

Sketch of proof.

$$\begin{aligned} ||\hat{B}_T - B||^2 &= ||Z_T^{-1} \int_0^T x_s dM_s'||^2 \\ &\le ||Z_T^{-1/2}||^2 ||Z_T^{-1/2} \int_0^T x_s dM_s'||^2 \\ &\le \sqrt{k}\, \lambda_{min}^{-1}(T) \cdot Q_T , \end{aligned}$$

where

$$Q_T = ||Z_T^{-1/2} \int_0^T x_s dM_s'||^2 .$$

The main problem now consists in showing

Lemma 1. Suppose that, for some $c > 0$, $\tau = \inf\{t : \lambda_{min}(t) \geq c\} < \infty$ a.e.. Then, under assumptions (A) - (B),

(2.1) $$Q_T = o((\log \lambda_{max}(T))^{1+\delta}) + O(1) \quad \text{a.e.}$$

for every $\delta > 0$. If (B) is replaced by the stronger assumption (B'), then (2.1) can be strengthened to

(2.3) $$Q_T = O(\log \lambda_{max}(T)) \quad \text{a.e.}$$

The proof can be found in [1].

For the discrete time case (1.2), the crucial conditions (C) are just those obtained in [2]. The example given there shows that (C) cannot be improved.

Asymptotic normality can be obtained if existence of suitably normalizing matrices is assumed and certain restrictions on the jumps of (M_t) are imposed. For details cf. [1].

3. Linear stochastic differential equations.

We consider systems

(3.1) $$dy_t = Ay_t dt + Cdw_t, \quad 0 \leq t \leq T ,$$

where (w_t) is multidimensional Brownian motion, and the unknown parameter matrix A is to be estimated by least squares. If C is non-singular, this is also the ML-estimator obtained by maximizing the density $d\mu_y/d\mu_\xi$, where μ_X denotes the law of the process X in the space of continuous functions and $d\xi_t = Cdw_t$.

We shall assume throughout: *(A,C) is a controllable pair.*

In order to verify assumption (C) in section 2, we have to investigate the asymptotic behavior of the largest and smallest eigenvalue of the matrix $Z_T = \int_0^T y_t y_t' dt$.

The proofs of the following assertions (3.2) - (3.5) can be found in a forthcoming paper by the author together with K. Helmes.

a) *Purely explosive case*, i.e. all eigenvalues of A have real parts > 0. Then

$$(3.2) \qquad \lim_{T\to\infty} \frac{1}{T} \log \lambda_{max}(T) = 2M \quad \text{a.e.},$$

$$(3.3) \qquad \lim_{T\to\infty} \frac{1}{T} \log \lambda_{min}(T) = 2m \quad \text{a.e.},$$

where M and m denote the largest resp. smallest real part of the eigenvalues of A.

b) *Purely nonexplosive case*, i.e. all eigenvalues of A have real parts $\leq$ 0. Then

$$(3.4) \qquad \liminf_{T\to\infty} \frac{1}{T} \lambda_{min}(T) > 0 \quad \text{a.e.}$$

Moreover, if ρ denotes the largest multiplicity of all distinct eigenvalues of A with zero real part (and $\rho = 0$ if all eigenvalues lie in the open left halfplane), then

$$(3.5) \qquad \lambda_{max}(T) = O(T) \qquad \text{a.e. if } \rho = 0,$$

$$= O(T^{2\rho+\varepsilon}) \quad \text{a.e. if } \rho \geq 1$$

for every $\varepsilon > 0$.

Compare the results in the discrete time case [3].

Hence Theorem 1 applies to system (3.1). It remains an open task to carry over the results obtained for the *mixed case* (eigenvalues in the left and in the right halfplane) in the discrete time setting (cf. [3]) to the model (3.1).

As an example of special interest, consider the r-th order linear differential equation

$$(3.6) \qquad \eta_t^{(r)} = \alpha_1 \eta_t^{(r-1)} + \dots + \alpha_r \eta_t + \xi_t ,$$

where $\eta^{(i)}$ denotes the i-th derivate and ξ is white noise. As is well known, (3.6) may be written in the form (3.1) by introducing the augmented process $y(t) = (\eta_t,\ldots,\eta_t^{(r-1)})'$ and putting

$$A = \begin{pmatrix} 0 & 1 & 0 & \cdots & 0 \\ 0 & 0 & 1 & \cdots & 0 \\ \vdots & & \ddots & \ddots & \vdots \\ 0 & \cdots & \cdots & 0 & 1 \\ \alpha_r & \cdots & \cdots & \cdots & \alpha_1 \end{pmatrix} , \quad C = (0,\ldots,0,\sigma)' .$$

The pair (A,C) is controllable. Moreover, it is easily seen that the last row of $\hat{A}$ is the ML-estimator of the parameter vector $\alpha' = (\alpha_1,\ldots,\alpha_r)$ obtained by maximizing the density of the law of $dy_r = \alpha' dyt + \sigma dw$ with respect to Wiener measure.

References

[1] N. Christopeit, Quasi-Least-Squares Estimation in Semimartingale Regression Models. Technical Report, Bonn 1984.

[2] T.L. Lai and C.Z. Wei, Least Squares Estimates in Stochastic Regression Models with Applications to Identification and Control of Dynamic Systems, Ann. Statist. 10 (1982) 154-166.

[3] T.L. Lai and C.Z. Wei, Asymptotic Properties of General Autoregressive Models and Strong Consistency of Least Squares Estimates of Their Parameters, Journal of Multivariate Analysis 13 (1983) 1-23.

A SOLUTION to the PARTIALLY OBSERVED CONTROL PROBLEM of LINEAR SYSTEMS , with NON-QUADRATIC COST.

Robert COHEN

Centre National d'Etudes des Télécommunications, PAA/TIM/DRI
38-40, rue du général Leclerc.
92131 Issy-les-Moulineaux. FRANCE

INTRODUCTION

In this paper, we deal with the continuous stochastic control problem of linear diffusion processes, with initial gaussian distribution, only known through a linear noisy observation. We solve this partially observed linear control problem with a non-quadratic cost, over the set of the admissible controls defined to be the collection of all the processes that are adapted to the observation's filtration. In addition we prove that the optimal control can be chosen to be a markovian function of the filter.

We can give a first formulation of the studied problem: The unknown state process X, which evolves according to the stochastic differential equation:

$$dX_t = F_t.X_t.dt + b(t,u_t).dt + G_t.dB_t$$

with an initial gaussian distributed r.v., must be controlled, until time T, through its drift term by choosing at each time t the parameter u_t.

This choice is based on the observation process Y solution of:

$$dY_t = H_t.X_t.dt + C_t.dW_t \quad , \; Y_0=0$$

The control problem is to find u* in $\mathcal{U}$, the set of the admissible controls such that the mean cost function J, associated to X, achieves its infimum over $\mathcal{U}$:

$$J(u^*) = \operatorname*{Inf}_{\mathcal{U}} J(.)$$

The process u* is called an optimal policy to the problem. A separation theorem is obtained if u* can be expressed in term of the filter of the system (X,Y).

The result is obtained along various contributions by introducing several additionnal hypothesis concerning either the cost function, or the set of the admissible controls, or the existence of a suitable formulation. Recent or less recent results show that the weak formulation of control problem in complete observation is a powerful approach. Dealing with the partial observation system (X,Y), we are tempted to consider the so-called separated problem associated with the filter. Then we face to a fresh difficulty that the weak formulation we consider for the filter must correspond to a real filtered system (X,Y). For the subject we are interested in, the known contributions can be classified in three groups:

- Using the fixed given space $(\Omega,\underline{\underline{A}},P)$: for example [16],[07]
- Changing the probability on $(\Omega,\underline{\underline{A}})$: for example [05]
- Changing the space (weak formulation): for example [04],[02]

We propose here a purely probabilistic method to solve the problem over the set $\mathfrak{U}$. This is done by attracting attention on an auxiliary problem in complete information which can be solved by control density method [01],[05],[06], and by using control functions. Then, using results for strong solution of stochastic differential equation [15], we are able to carry the optimal control on the initial space.
All the propositions are extensively proved in [03], consequently we will only give sketches of proofs.

I- THE CONTROL PROBLEM FORMULATION

This section is devoted to the controlled system (X,Y) constructed by the reference probability method. This part is classical, however we give, for further use, a more detailled construction.

On the <u>initial space</u> $(\Omega,\underline{\underline{A}},P_0)$ the system (X,Y) is described as the solution of the s.d.e.:

$$(I.1)\qquad \begin{cases} dX_t = F_t.X_t.dt + G_t.dB_t\ , & X_0 \simeq N(\hat{X}_0,Q_0) \\ dY_t = C_t.dW_t^0 & , Y_0=0 \end{cases}$$

for t running into [0,T], N(x,q) is the notation for gaussian distribution with mean value x and covariance matrix q; F,G,C are measurables bounded functions from [0,T] into $R^n xR^n$; X_0 is independent of the Brownian motion (B,W^0); for every t, $\underline{\underline{F}}_t$ is the σ-field generated by X_0,B_s,W_s^0 for s in [0,t]; $\underline{\underline{F}}=(\underline{\underline{F}}_t,0\le t\le T)$ is the filtration. In the same way $\underline{\underline{Y}}$ denote the filtration of the observation where $\underline{\underline{Y}}_t$ is the σ-field generated by Y_s for s in [0,t], equivalently by W^0_s.

The <u>reference probability space</u> $(\Omega,\underline{\underline{F}},\underline{\underline{Y}},P)$ is obtained by a Girsanov transform and the so-called free system, (X,Y) is solution of:

$$(I.2)\qquad \begin{cases} dX_t = F_t.X_t.dt + G_t.dB_t \\ dY_t = H_t.X_t.dt + C_t.dW_t \end{cases}$$

for a given measurable bounded function H .

<u>Definition:</u> Let U be a fixed compact set of R^p.
An admissible control is any U-valued process $u=(u_t,t \in [0,T])$ which is progressively measurable with respect to the filtration $\underline{\underline{Y}}$, generated by the observation Y. The set of all the admissible controls is denoted by $\mathcal{U}$.

The construction of the filtering model by the reference probability method avoid any vicious circle in the inter-dependence between the filtration and the controls.

Let b be a measurable bounded function from [0,T]xU into R^n, continuous in u. For G invertible with bounded inverse, we define by Girsanov transform for each admissible u a probability P^u equivalent to P. On the space $(\Omega,\underline{\underline{F}},\underline{\underline{Y}},P^u)$ the system (X,Y) evolves according to the s.d.e.:

$$(I.3)\qquad \begin{cases} dX_t = F_t.X_t.dt + b(t,u_t).dt + G_t.dB^u_t \\ dY_t = H_t.X_t.dt + C_t.dW_t \end{cases}$$

Since X_0 is $\underline{\underline{F}}_0$-measurable and P^u coincides with P on $\underline{\underline{F}}_0$, $E^u(X_0)$ does not depend on u, E^u denotes the mean under P^u.

The control problem consists in choosing u in $\mathcal{U}$ which minimizes the mean cost function:

$$J(u) = E^u\left(\int_0^T l(s,X_s,u_s).ds + m(X_T) \right)$$

the running cost l is supposed measurable bounded and continuous in u, the terminal cost m is supposed measurable bounded and a.s. continuous in t. We denote this problem by $((X,Y),J,\mathcal{U})$.

II- THE SEPARATED PROBLEM

The separated problem is a filter associated problem obtained from the previous one through the filtering theory. The first part of this section shows that such a general problem still suffers ambiguities in spite of the results of [02], [04], [05], [07]. In second part, we prove by impulse control techniques that the same infimum is achieved by J over $\mathcal{U}$ and by the filter's associated cost $\hat{J}$ over the set $\hat{\mathcal{U}}$ of (defined below) separated controls.

The filtering theory [11] asserts us that the conditionnal distribution of X under P^u given $\underline{\underline{Y}}$ is gaussian with mean $\hat{X}^u$ and error covariance matrix Q^u, solution of:

$$d\hat{X}^u_t = F_t.\hat{X}^u_t.dt + b(t,u_t).dt + K_t.C_t.d\hat{W}^u_t$$

(II.1)

$$K_t = Q_t.H'_t.(CC')^{-1}_t$$

with the innovation process $\hat{W}^u$ defined by:

(II.2) $$d\hat{W}^u_t = C^{-1}_t.(\ dY_t \ -H_t.\hat{X}^u_t.dt\ )$$

Because the coefficients of the Riccati's equation and the initial value do not depend on u neither do the matrix Q^u; we denote it simply Q. In this gaussian finite dimensionnal control problem all the information is contained in the filter. Let $\hat{g}$ denotes the convolution between g and the density N, then the filter associated cost $\hat{J}$ is defined by conditionning in J by $\underline{\underline{Y}}$, and we get:

$$J(u) = E^u \left(\int_0^T \hat{l}(s,\hat{X}^u_s,u_s).ds + \hat{m}(\hat{X}^u_T) \right) \triangleq \hat{J}(u)$$

We are tempted to define a full observation control problem for the filter. The new problem $(\hat{X}^u, \hat{J}(u), \mathcal{U})$ is still not a complete one because the Brownian motion $\hat{W}^u$ is specified to be (II.2), and the admissibles controls are still $\underline{\underline{Y}}$-adapted and not $\underline{\underline{\hat{X}}}^u$-adapted.

Do the results on filtering theory hold when u is no more in $\mathcal{U}$ but is a function of the filter? This is the first difficulty which concerns the solution of non-trivial s.d.e..
Even in case we are able to define a full observation separated problem for controls (functions or processes) it is not possible to solve it by the control of density method: if we try to cancel the control drift term b(.,.) in (II.1) by a change of

probability, the Brownian motion that arises from this operation will depend on u, and so will be the equivalent probability; if we come back on the initial space, we get Y as a Brownian motion but b(.,.) will not be cancelled. There is no process Z on a space $(\Omega,\underline{\underline{F}},Q)$, such that Z and Q are independent of u, and such that $\hat{X}^u$ under P^u is obtained from both Z and Q by a Girsanov transform. We either have to strengthen the hypothesis either try another way, that is done in part III and IV.

Let us make precise two definitions.

Definitions:

1. The set of impulse control processes is defined to be:
$\mathcal{V} = \{ v_t = \sum_{k\geq 0} I_k.1_{\{T_k<t\leq T_{k+1}\}}$; where $\{T_k\}$ is a sequence of $\underline{\underline{Y}}$-stopping times such that $T_k \nearrow \infty$, I_k is, for each k, an $\underline{\underline{Y}}_{T_k}$-r.v.$\}$
2. Let $\mathcal{S}$ be the set of all the functions f:

$f: [0,T]\mathrm{x}C([0,T];R^n) \to C([0,T];R^n)$, progressively measurable with respect to the Borel filtration, such that :

(II.3) $dZ_t = F_t.Z_t.dt + b(t,f(t,\{Z_s,0\leq s\leq t\})).dt + K_t.(dY_t - H_t.Z_t.dt)$
admits a unique strong solution .

Then it is easy to check the two following points:
1. Each function f of $\mathcal{S}$ generates an admissible control in sense that the process $u. = f(.,\{Z_s,0\leq s\leq .\})$ belongs to $\mathcal{U}$.
2. For such an admissible control u, the filter $\hat{X}^u$ is solution of

(II.1) $d\hat{X}^u_t = F_t.\hat{X}^u_t.dt + b(t,u_t).dt + K_t.(dY_t - H_t.X^u_t.dt)$

which is, compare with (II.3), exactly Z : $Z=\hat{X}^u$. That is any strong solution to the equation (II.1) is a filter process.
This is the meaning of our notation: $\hat{X}^u = \hat{X}^f$. The set of processes $\hat{\mathcal{U}} = \{u.=f(.,\{Z_s,0\leq s\leq .\}) = f(.,\{\hat{X}^u_s,0\leq s\leq .\}) = f(.,\{\hat{X}^f_s,0\leq s\leq .\}),\ f \in \mathcal{S}\}$ is called the set of separated controls.

The following results concerning the separated control problem will be useful in the last part of this paper. In our setting we prove:

1.Theorem.

1a. $\inf_{\mathcal{U}} J(.) = \inf_{\mathcal{V}} J(.)$

1b. $\inf_{\mathcal{V}} J(.) = \inf_{\mathcal{V}} \hat{J}(.)$

2a. Define for any impulse control v:

$$\hat{J}_n(v) = E^v\left(\int_0^T \hat{l}(s,\hat{X}^v_s,v_s).ds + \hat{m}(\hat{X}^v_T) + 1/n.\sum_{k\geq 0} 1_{\{T_k \leq T\}} \right)$$

then : $c^*_n = \operatorname{Inf}_{\mathcal{V}} \hat{J}_n(.)$ is a bounded, decreasing sequence,

2b. limit of which, c^*, can be identified to $\operatorname{Inf}_{\mathcal{V}} \hat{J}$.

3. For each n, $c^*_n \geq \operatorname{Inf}_{\mathcal{S}} \hat{J}(f(.,\{\hat{X}^f_s, 0\leq s\leq .\}))$

2. Corollary. $\operatorname{Inf}_{\mathcal{U}} J(.) \geq \operatorname{Inf}_{\mathcal{S}} \hat{J}(f(.,\{\hat{X}^f_s, 0\leq s\leq .\}))$

Sketch of proof. 1a) For each admissible control u we can find a sequence $\{v^n, n\geq 0\}$ of $\mathcal{V}$ which converges to u dtxdP-a.s.. Using a technical computation, we show that:

$$|J(u)-J(v)| \leq k.[\{E^u(\int_0^T |l(s,X_s,u_s)-l(s,X_s,v_s)|^2.ds)\}^{\frac{1}{2}} +$$

$$+ \{E(\int_0^T [L^u_s.(b(s,u_s) - b(s,v_s)]^2.ds)\}^{\frac{1}{2}}]$$

the result follows from continuity of l, b in u.
1b) Each impulse control is an admissible one thus, $J(v)=\hat{J}(v)$.
2a) The cost $\hat{J}_n$ is seen to be associated to an impulse control problem with impulse cost function constant 1/n. For n going to infinity, we obtain an impulse control problem with a vanishing cost. Such a problem have been studied in [12],[09;thIII.1.7], [06], or [13] in various situations and under several other hypothesis. The fixed real time T induces here a non-homogeneous treatment of the impulse control problem. Using results in the complete observation case [10], we prove the result (see [03]).
2b) It is easy to verify that the limit c^* is equal to $\operatorname{Inf}_{\mathcal{V}} \hat{J}$.

3) From the separation theorem for impulse control, as in [14], in our setting of non-homogeneous impulse control problem [03], for each n there exists an optimal separated control v^* for $\hat{J}_n$:

$$c^*_n \geq \operatorname{Inf}_{\mathcal{U}} \hat{J}(u) = \operatorname{Inf}_{\mathcal{S}} \hat{J}(f(.,\{\hat{X}^f_s, 0\leq s\leq .\}))$$

The corollary follows by passing to the limit in the last inequality. Let us emphasize that an optimal separated control exists for the cost $\hat{J}_n$, for each n, but it may not exists for (the null impulse cost) $\hat{J}$: what is proved here is an equality about the minimal cost.

III- THE AUXILIARY PROBLEM

The separated problem was obtained first by a change of probability and then by a filtering procedure. We reverse the order of these two operations: we first compute the filter $\hat{X}$ of the free system (that is (X,Y) under P) and then change the probability space [03]. This apparently disconnects us from the previous problem. The problem is said to be auxiliary in sense that it gives a space on which the Hamilton-Bellman-Jacobi's equation can be easily solved, and provides a markovian function f*, expected to generates the solution to the given problem. The space $(\Omega,\underline{\underline{F}},P)$ is for the second time a reference space.

3.Proposition. The conditionnal distribution of X under P given Y is gaussian with mean $\hat{X}$, and error covariance matrix $\hat{Q}$, solution of:

(III.1) $$d\hat{X}_t = F_t.\hat{X}_t.dt + K_t.C_t.d\hat{W}_t$$

with the innovation process $\hat{W}$ defined by:

(III.2) $$d\hat{W}_t = C^{-1}.(dY_t - H_t.\hat{X}_t.dt)$$

We verify that the error covariance matrix $\hat{Q}$ is solution to the same Riccati's equation with the same initial value as the error covariance of the filter,then it follows that for any t in [0,T] and any admissible u : $\hat{Q}_t=Q_t=Q_t^u$, and the gain matrix K is the same for both $\hat{X}^u$ and $\hat{X}$, for any u. For a control problem in complete observation, the admissible controls have to be $\underline{\underline{\hat{X}}}$-adapted, where $\underline{\underline{\hat{X}}}$ is the filtration generated by $\hat{X}$. Let $\hat{\mathfrak{U}}$ the set of admissible controls for this new problem.

If KC is invertible, we can define a probability $\hat{P}^u$ equivalent to P , for each admissible u,having the exponential martingale $\hat{L}^u$ for Radon-Nikodym derivative with respect to P.

4.Proposition. There exists $\hat{W}^u$ an $\underline{\underline{Y}}/\hat{P}^u$-brownian motion such that:

(III.3) $$d\hat{X}_t = F_t.\hat{X}_t.dt + b(t,u_t).dt + K_t.C_t.d\hat{W}_t^u$$

$\hat{X}_0$ does not depend on u because X_0 is $\underline{\underline{F}}_0$ measurable and $\hat{P}^u$ coincides with P on $\underline{\underline{F}}_0$.

The processes $\hat{X}$ is the filter of (X,Y)under P but is no more the filter of (X,Y) under $\hat{P}^u$, neither the mean cost $\hat{J}$, associated to $\hat{X}$, can be obtained from J by conditionning in by $\underline{\underline{Y}}$, as was done for $\hat{J}$. We fix $\hat{J}$ to be:

$$\tilde{J}(u) = \hat{E}^u\left(\int_0^T \hat{l}(s,\hat{X}_s,u_s).ds + \hat{m}(\hat{X}_T) \right)$$

where the functions $\hat{l}$ and $\hat{m}$ are the same as in $\hat{J}$.

Following the result 1a) of theorem 1 we get:

5.Theorem. $\quad \inf_{\tilde{\mathcal{U}}} \tilde{J} = \inf_{\tilde{\mathcal{V}}} \tilde{J}$

The set $\tilde{\mathcal{V}}$ of impulse controls is defined in the same way as the set $\mathcal{V}$, the filtration $\underline{\underline{\hat{X}}}$ being substituted to the filtration $\underline{\underline{Y}}$.
As $\{\hat{L}_T^u\ , u \in \tilde{\mathcal{U}}\}$ is a uniformly integrable family of r.v., it is possible, using [06] to solve the full observation problem $(\hat{X},\tilde{J},u \in \tilde{\mathcal{U}})$ using the Doob-Meyer decomposition theorem of supermartingale for the value function:

$$\text{(III.4)} \quad \tilde{V}(S) = \inf_{\tilde{\mathcal{U}}} \hat{E}^u\left(\int_{S\wedge T}^T \hat{l}(s,\hat{X}_s,u_s).ds + \hat{m}(\hat{X}_T) \;/\; \underline{\underline{\hat{X}}}_{S\wedge T}\right)$$

6.Theorem.
The value function is markovian, and there exists a markovian function f* such that:

$$\inf_{\tilde{\mathcal{V}}} \tilde{J}(.) = \tilde{J}(f^*(.,\hat{X}.))$$

To be able to link this problem with the separated one, we show that because the process $f(.,\{\hat{X}_s, 0\le s\le .\})$ belongs to $\tilde{\mathcal{U}}$ for any f of $\mathcal{S}$:

7.Proposition.

$$\inf_{\mathcal{S}} \tilde{J}(f(.,\{\hat{X}_s,0\le s\le .\})) \ge \tilde{J}(f^*(.,\hat{X}.))$$

We are going to make use of the function f* in order to solve the partially observed control problem.

IV- THE SOLUTION TO THE INITIAL PROBLEM

We prove that starting from the free system the drift transformation and the filtering operations commute in sense of the uniqueness of the law of the obtained process for all functions of the set $\mathcal{S}$. This is the link between the separated problem obtained by drift transform and filtering, and the auxiliary problem obtained by filtering and drift transformation.

In part II we saw that:

$$(*)\quad \inf_{\mathcal{U}} J(.) = \inf_{\mathcal{V}} J(.) = \inf_{\mathcal{V}} \mathfrak{J}(.) \geq \inf_{\mathcal{S}} \mathfrak{J}(f(.;\{\hat{X}^f_s, 0\leq s\leq .\}))$$

in part III we saw that:

$$(**)\qquad \inf_{\mathcal{S}} \mathfrak{J}(f(.,\{\hat{X}_s, 0\leq s\leq .\})) \geq \mathfrak{J}(f^*(.,\hat{X}.))$$

<u>8.Proposition.</u> For any f in $\mathcal{S}$:

1a. $$\mathfrak{J}(f(.,\{\hat{X}^f_s, 0\leq s\leq .\})) = \mathfrak{J}(f(.,\{\hat{X}_s, 0\leq s\leq .\}))$$

1b. $$\inf_{\mathcal{S}} \mathfrak{J}(f(.,\{\hat{X}^f_s, 0\leq s\leq .\})) = \inf_{\mathcal{S}} J(f(.,\{\hat{X}_s, 0\leq s\leq .\}))$$

2. Let $\hat{X}^*$ be the filter controlled by f^*, that is : $\hat{X}^* = \hat{X}^{f^*}$ thus

$$\mathfrak{J}(f^*(.,\hat{X}.)) = \mathfrak{J}(f^*(.,\hat{X}^*_.))$$

Sketch of proof. For any f of $\mathcal{S}$, the filter equation admits a unique strong solution, and by the uniqueness law theorem [08;th.IV.4.2], the law of $\hat{X}$ and the law of $\hat{X}^f$ are the same. As the mean cost take into account only the law of the concerned processes this proves 1a). The equality 1b) follows from 1a) by taking the infimum over the set $\mathcal{S}$. By a straightforward extension [03] of Veretennikov's result [15], we see that f^* is indeed in the set $\mathcal{S}$; using equality 1a) we get 2).

We are able to conclude :

<u>9.Separation theorem.</u>[03] The optimal separated policy to the partially observed control problem $((X,Y),J,\mathcal{U})$ is $f^*(.,\hat{X}^*_.)$, where f^* is the markovian optimal function of the auxiliary problem $(\hat{X},\mathfrak{J},\mathcal{U})$.

Because $f^*(.,\hat{X}^*_.)$ is an admissible control, the proof follows by combining (*),(**), and proposition 8:

$$\inf_{\mathcal{U}} J \geq \mathfrak{J}(f^*(.,\hat{X}^*_.)) = J(f^*(.,\hat{X}^*_.))$$

<u>Remark.</u> To solve the case of the partially observation problem for dim Y < dim X, one need to solve the auxiliary (full observation) control problem with a degenerated matrix diffusion K.C. Our approach can be extended to any finite dimensionnal control problem.

<u>CONCLUSION</u>

Introducing an auxiliary complete observation control problem, using the separated controls and a Veretennikov's theorem extension, we have proved, in a probabilistic manner -by density control method- that the partially observed control problem has a separated optimal policy amongst the admissible controls defined to be all the processes adapted to the observation's filtration.

Acknowledgement
The author is grateful to G. Mazziotto and J. Szpirglas for valuable discussions and useful comments.

References

[01] Benes V.E.: Existence of optimal stochastic control law. SIAM J. of Control, 1971, vol.9, p.446-472.

[02] Christopeit N./Helmes K.: Separation principle for partially observed linear control problem. Lect. Notes in cont. & inf. sciences, vol.61, p.36-60, Springer-Verlag 1983.

[03] Cohen R.: Théorème de séparation pour le contrôle des diffusions linéaires en observation partielle. Thèse de 3eme cycle, Université Paris VI, 1985.

[04] Davis M.H.A.: Separation principle in stochastic control via Girsanov solution. SIAM J. of Cont., 1976, vol.14, p.176-188.

[05] Davis M.H.A./Varaiya P.: Dynamic programming conditions for partially observable systems. SIAM J. of Cont., 1973, vol.11, p.226-261.

[06] El Karoui N.: Aspects probabilistes du contrôle stochastique. Ecole d'été de probabilité de Saint-Flour 1979. Lect. Notes in Math., vol.876, p.73-238, Springer-Verlag 1981.

[07] Haussmann U.G.: Optimal control of partially observed diffusions via separation principle. Lect. Notes in cont. & inf. sciences, vol.43, p.302-311, Springer-Verlag 1982.

[08] Ikeda N./Watanabe S.: Stochastic differential equations and diffusion processes. North-Holland/Kodansha 1981.

[09] Krylov N.V.: Controlled diffusion processes. Application of Math. vol.14, Springer-Verlag 1980.

[10] Lepeltier J.P./Marchal B.: Techniques probabilistes dans le controle impulsionnel. Stochastics, 1979, vol.2, p.243-286.

[11] Liptser R.S./Shiryaev A.N.: Statistics of random processes. Applications of mathematics,vol.15.Springer-Verlag,1977.

[12] Ménaldi L./Robin M.: On some cheap control problems for diffusions processes. Trans. Am. Math. Soc., 1983.

[13] G. Mazziotto : Approximate impulse control for partially observed systems. Proceed. of the conf. on modelling and filtering, Rome 12.1984.

[14] G. Mazziotto,J. Szpirglas : Separation principle for impulse control with partial information. Stochastics, 1983, vol.10, p.47-73.

[15] A.J. Veretennikov: On strong solutions and explicit formulas for solutions of stochastic integral equations. Math. USSR Sbornik,1981,vol.39,p.387-403.

[16] W.M. Wonham: On the separation theorem of stochastic control. SIAM J. of cont.,1968,vol.6,p.312-326.

STATIONARY CONTROL OF BROWNIAN MOTION IN SEVERAL DIMENSIONS

R. MITCHELL COX & IOANNIS KARATZAS*
Department of Statistics, Columbia University
New York, N.Y. 10027/USA

ABSTRACT

We envision a Brownian particle which is undergoing a controlled motion in R^d, $d \geq 2$. The element of control is the drift of this motion, and can take values in the unit ball $\Gamma = \{z \in R^d;\ |z| \leq 1\}$. The decision-maker cannot anticipate the future, so his choice of drift $u(t,.)$ at time t can only depend upon the history of the path $\{\xi_s;\ 0 \leq s \leq t\}$ up to time t, which is assumed to be completely observable. He attempts to keep the particle as close to the origin as possible, since he pays a cost $\varphi(|\xi_t|)$ per unit time for missing the origin; the function $\varphi(.)$ can be quite general but may be viewed here, for simplicity, as quadratic. Keeping a desired course can be accomplished by "pushing" the particle in the right direction, i.e., by selecting an appropriate vector in the control set Γ; however, such an action entails a cost (per unit time) which is proportional to the length of the drift vector. The <u>stationary control problem</u> is to find, in the class $\mathcal{C}$ of nonanticipative functionals $u(t,\xi)$: $[0,\infty) \times C^d[0,\infty) \to \Gamma$ (henceforth called "control laws"), a law u^* which minimizes long-run average expected cost per unit time. In particular, we seek a constant $\lambda > 0$ such that

(1) $$\lambda = \lim_{T\to\infty} \frac{1}{T} E_x^* \int_0^T \{\varphi(|\xi_t^*|) + |u^*(t,\xi_t^*)|\}dt; \qquad \forall\ x \in R^d$$

(2) $$\lambda \leq \underline{\lim}_{T\to\infty} \frac{1}{T} E_x^u \int_0^T \{\varphi(|\xi_t|) + |u(t,\xi_t)|\}dt; \qquad \forall\ x \in R^d,\ u \in \mathcal{C}.$$

It is clear that the problem at hand amounts to a careful balancing between cost of deviation from the origin and cost of action. The former can be reduced by applying continually a drift of maximal (unit)

---

*Research supported in part by the U.S. National Science Foundation under grant NSF MCS-81-03435-A01.

length along the ray towards the origin; if the latter cost were absent, such a pattern of action would in fact be optimal (cf. [1], [5]). On the other hand, one can avoid being taxed for action by never doing anything and leaving the situation uncontrolled, but only with disastrous effects on performance. Evidently, the decision-maker needs to find a compromise between doing too much and doing too little or nothing.

It turns out that the proper balance has to be drawn thus: to exert full force towards the origin if the particle is outside a sphere of radius $b>0$ centred at the origin, and to remain idle otherwise; i.e.,

$$(3)\qquad u^*(t,\xi) = -\frac{\xi_t}{|\xi_t|}\,1_{\{|\xi_t|\geq b\}};\quad t\geq 0,\ \xi\in C^d[0,\infty).$$

Under very mild conditions on the cost function $\varphi(.)$, it is shown that b can be so selected that the control law in (3) is optimal, at least against laws which give rise to a process ξ admitting an invariant measure. In particular, a detailed analysis of the Bellman equation of dynamic programming shows that the positive constants λ, b in (1), (3) are uniquely determined by the pair of equations

$$(4)\qquad \lambda = H_1(b) = H_2(b)$$

where the functions $H_1(.)$, $H_2(.)$ are <u>weighted averages</u> of the same function $\psi(s) \triangleq \varphi(s) + \frac{d-1}{2s}$; $s>0$ over the disjoint intervals $(0,r)$ and (r,∞), with weights s^{d-1} and $f(s) \triangleq e^{-2s}\,s^{d-1}$, respectively:

$$(5)\qquad H_1(r) = \frac{d}{r^d}\int_0^r s^{d-1}\psi(s)ds,\quad H_2(r) = \frac{\int_r^\infty f(s)\psi(s)ds}{\int_r^\infty f(s)ds};\quad r>0.$$

To prove optimality of u^* in (3) against all laws in $\mathcal{G}$, we undertake a thorough study of the control problem on a finite time-horizon $[0,T]$, with value function $V(T,x)$. The crux of the argument is the relation

$$(6)\qquad V(T,x) = \lambda T + o(T),\quad \text{as}\quad T\to\infty,$$

valid for every initial position $x\in\mathbb{R}^d$, which is established with the help of the maximum principle for parabolic operators and the comparison theorem in [5].

Detailed results and proofs will appear in [4], which extends to the multidimensional case and refines the results of article [2]. A fairly comprehensive theory for the stationary control problem, with general control sets Γ and cost structures, has been recently developed by Cox in [3] by exploiting the connection of the stationary with the discounted stochastic control problem.

REFERENCES:

[1] BENEŠ, V.E. (1975) Composition and invariance methods for solving some stochastic control problems. Adv. Appl. Probability 7, 299-329.

[2] BENEŠ, V.E. & KARATZAS, I. (1981) Optimal stationary linear control of the Wiener process. J. Optim. Theory Appl. 35, 611-633.

[3] COX, R.M. (1984) Stationary and Discounted Stochastic Control. Doctoral Dissertation, Columbia University, New York.

[4] COX, R.M. & KARATZAS, I. (1985) Stationary control of Brownian motion in several dimensions. Adv. Appl. Probability 18 (3), to appear.

[5] IKEDA, N. & WATANABE, S. (1977) A comparison theorem for solutions of stochastic differential equations and its applications. Osaka J. Math. 14, 619-633.

CONTROL OF PIECEWISE-DETERMINISTIC PROCESSES VIA DISCRETE-TIME DYNAMIC PROGRAMMING

M.H.A. Davis
Department of Electrical Engineering,
Imperial College, London SW7 2BT, England

Abstract

Controlled piecewise-deterministic Markov processes have deterministic trajectories punctuated by random jumps, at which the sample path is right-continuous. By considering the sequence of states visited by the process at its jump times, it is shown that a discounted infinite horizon control problem can be reformulated as a discrete-time Markov decision problem (the 'positive' case). Under certain continuity assumptions it is shown that an optimal stationary policy exists in relaxed controls.

1. Throughout the paper we shall consider piecewise deterministic processes(PDP) taking values in an open subset E of R^d defined by $E = \{x \in R^d : \psi(x) < 0\}$ where ψ is some C^1 function such that $|\nabla\psi(x)| \geq 1$ for $x \in \partial E = \{x : \psi(x) = 0\}$. E may or may not be connected and/or have compact closure. $\underline{\underline{E}}$ denotes the Borel sets of E.

A PDP is determined by its three "local characteristics"

(i) A Lipschitz-continuous vector field f; this determines a flow $\phi(t,x)$ in E satisfying

$$\frac{\partial}{\partial t}\phi(t,x) = f(\phi(t,x))$$

$$\phi(0,x) = x .$$

We denote $t^*(x) = \inf\{t>0 : \phi(t,x) \in \partial E\}$.

(ii) A jump rate $\lambda : E \to R_+$; throughout the paper we shall assume that λ is bounded, i.e. $\lambda \in L_\infty(E)$.

(iii) A transition measure $Q : \underline{\underline{E}} \times (E \cup \partial E) \to [0,1]$.

Construction of the sample path x_t of the PDP starting from $x \in E$ is simple: we define

$$x_t = \phi(t,x) \qquad \text{for } t < T_1,$$

where T_1 is the first jump time, whose distribution is given by

$$P_x[T_1 > t] = \begin{cases} \exp\left(-\int_0^t \lambda(\phi(s,x))ds\right) & t < t^*(x) \\ 0 & , \ t \geq t^*(x). \end{cases}$$

The conditional distribution of x_{T_1} given T_1 is specified by the transition measure Q:

$$P_x[x_{T_1} \in A | T_1] = Q(A; \phi(T_1,x))$$

[The trajectory is right-continuous and $x_{T_1 -} = \phi(T_1,x)$]. The process now restarts at x_{T_1}. We assume that the sequence of jump times $T_1, T_2, \ldots$ satisfies $\lim_n T_n = \infty$ P_x-a.s. for all $x \in E$. [It is possible to violate this condition, even with λ bounded]. The reader is referred to Davis [2] for further details, general properties and examples. To the author's knowledge the complete literature on PDPs under that name consists at the present time of the following papers in addition to [2]: Vermes [8] and Soner [6] (optimal control); Lenhart-Liao [5] and Gugerli [4] (optimal stopping). PDPs are however closely related to the so-called Markov Decision Drift Processes (van der Duyn Schouten [7]; Yushkevich [10]). Also,PDPs are right processes (though not Feller processes) so the general theory of El Karoui [3] applies to them. However, by using the special structure of PDPs one can get more specific results.

2. Control problems arise when the local characteristics f, λ, Q depends on a control parameter $u \in U \subset R^m$. In the case of Q it is necessary to distinguish between control in the state space ($x \in E$) and on the boundary ($x \in \partial E$). In the present paper we shall suppose there is no control on the boundary so that there is a fixed transition measure $Q_\partial(dy;x)$ for $x \in \partial E$ describing jumps from ∂E and a parametrized family $Q_0(dy;x,u)$ for $x \in E$ describing jumps from interior points. The following assumptions will be in force throughout.

(A1) U is compact

(A2) $f: E \times U \to R^d$ is continuous, and Lipschitz continuous in $x \in E$ uniformly in $u \in U$

(A3) $\lambda: E \times U \to R_+$ is bounded and continuous

(A4) $Q_0: E \times U \to P(E)$ and $Q_\partial: \partial E \to P(E)$ are continuous.[Here $P(E)$ denotes the set of probability measures on E with the topology of weak convergence].

The "usual" class of admissible controls in Markovian problems is that of state feedback controls $u_t = u(x_t)$. This is however not the appropriate choice here since the deterministic parts of the trajectory should then satisfy the ordinary differential equation $\dot{x}_t = f(x_t, u(x_t))$, whereas this equation only has a unique solution for "smooth" $u(\cdot)$. Following Vermes [8], we obtain a less restrictive class ofcontrols by adjoining to the process two extra states z_t, the state at the last jump time, and τ_t, the time since the last jump. Admissible controls are now functions $u(z_t, \tau_t)$ so that the deterministic dynamics on the interval $t \in [T_i, T_{i+1}[$ become

$$\dot{x}_t = f(x_t, u(z_t, \tau_t))$$

$$(1) \qquad \dot{z}_t = 0, \qquad z_{T_i} = x_{T_i}$$

$$\dot{\tau}_t = 1, \qquad \tau_{T_i} = 0.$$

Note that (a) this system of equations has a unique solution for any measurable function $u: E \times R_+ \to U$, by the Carathéodory theorem (we denote by $\underline{\underline{U}}$ the set of such functions) (b) the augmented process $\tilde{x}_t = (x_t, z_t, \tau_t)$ is still a PDP, and (c) (z_t, τ_t) determines (x_t), so controls of the above form still represent "complete observations"; see [2,8] for further details. Given $u \in \underline{\underline{U}}$, the sample path of the process $\tilde{x}_t$ is constructed as in §1, using (1) above. As before we make the following assumption

(A5) For any $u \in \underline{\underline{U}}$ we have $P_x[\lim_n T_n = \infty] = 1$ for all $x \in E$.

Let $\ell_0: E \times U \to R_+$ and $\ell_\partial: \partial E \to R_+$ be bounded continuous functions and δ be a positive constant. The <u>cost</u> of control $u \in \underline{\underline{U}}$ is given by

$$J_x(u) = E_x\left[\int_0^\infty e^{-\delta t} \ell_0(x_t, u)(z_t, \tau_t)dt + \sum_i \exp(-\delta T_i)\, \ell_\partial(x_{T_i}) I_{(x_{T_i -} \in \partial E)} \right].$$

There would be no gain in generality in allowing time-dependent coefficients, since "time" can always be included as one component of x_t. Similarly, <u>finite-horizon</u> problems are included in the above formulation.

3. We now reformulate the control problem as an infinite-horizon discrete-time dynamic programming problem by considering the sequence of states $\{x(T_k), k=0,1,\ldots\}$ (by convention, $T_0 = 0$).

First note that the admissible controls are "piecewise open loop" in that at each time T_k we choose a measurable time function $u: R_+ \to U$ (depending on $z(T_k)$) such that our control will be $u_t = u(t - T_k)$ on the interval $[T_k, T_{k+1}[$. Denote by $\underline{\underline{C}}$ the set of such functions. Secondly, explicit mention of the discount factor can be avoided by introducing "killing". We adjoin an isolated point Δ to the state space and define a new process $\hat{x}_t$ on $E_\Delta = E \cup \{\Delta\}$ by

$$\hat{x}_t = \begin{cases} \tilde{x}_t & t < T \\ \Delta & t \geq T \end{cases}$$

where T is independent exponentially distributed time ($P_x[T>t] = e^{-\delta t}$ for all $x \in E$). All functions $\chi: E \times U \to R$ are extended to $E_\Delta \times U$ by setting $\chi(\Delta, u) = 0$ for all $u \in U$. Let

$\underline{F}_t$ denote the natural filtration of $\hat{x}_t$. Then in view of the strong Markov property the cost $J_x(u)$ can be written in terms of the killed process $\hat{x}_t$ as

$$J_x(u) = E_x\left[\int_0^\infty \ell_0(\hat{x}_t,u(\hat{z}_t,\hat{\tau}_t))dt + \sum_k \ell_\partial(\hat{x}_{T_k}-)I_{(\hat{x}(T_k-)\in\partial E}\right]$$

$$=E_x \sum_{k=0}^\infty E_x\left[\int_{T_k}^{T_{k+1}} \ell_0(\hat{x}_t,u(\hat{z}_t,\hat{\tau}_t))dt + \ell_\partial(\hat{x}_{T_{k+1}}-)I_{(\hat{x}_{T_{k+1}}-\in\partial E)}\middle|\underline{F}_{T_k}\right]$$

(2) $$= E_x \sum_{k=0}^\infty g(\hat{x}_{T_k},\hat{u}_{T_k})$$

where $\hat{u}_{T_k} := u(\hat{x}_{T_k},\cdot)$ and, for $x\in E$ and $u\in\underline{C}$,

$$g(x,u):=E_x\left[\int_0^{T_1} \ell_0(\hat{x}_t,u(t))dt + \ell_\partial(\hat{x}_{T_1}-)I_{(\hat{x}_{T_1}-\in\partial E)}\right]$$

Note that $g(\Delta,u)=0$ and hence that the sum in (2) has only finitely many non-zero terms, P_x-a.s. For $(x,u)\in \underline{E}\times C$ and $A\in\underline{E}_\Delta$, define

$$q(A;x,u) := P_x[\hat{x}_{T_1}\in A]$$

where $\hat{x}_t$ is the (killed) process with control $u(t)$ for $t<T_1$.

For notational simplicity we shall suppose henceforth that killing is included in the original specification of the process, so that we can drop both the "hats" (^) and explicit inclusion of the cemetary state Δ. Note that this implies $\lambda(x,u)\geq\delta$ for all (x,u). For $(x,u)\in E\times\underline{C}$ let $\phi_t^u(x)$ be the deterministic flow given by

$$\frac{\partial}{\partial t}\phi_t^u(u) = f(\phi_t^u(x),u(t))$$

$$\phi_0^u(x) = x.$$

Then by the construction of the PDP given in §1,

(3) $$q(A;x,u) = \int_0^{t^*_u(x)} Q(A;\phi_t^u(x),u(t))\lambda(\phi_t^u(x),u(t))n^u(t)dt$$

$$+ Q_\partial(A;\phi^u_{t^*_u(x)}(x))\eta^u(t^*_u(x))$$

where $t^*_u(x) = \inf\{t:\phi^u_t(x) \in \partial E\}$ and

$$\eta^u(t) = \exp(-\int_0^t \lambda(\phi^u_s(x),u(s))ds)), \qquad t \leq t^*_u(x).$$

Similarly, we find that

$$(4) \qquad g(x,u) = \int_0^{t^*_u(x)} \ell_0(\phi^u_t(x),u(t))\eta^u(t)dt + \ell_\partial(\phi^{u*}_{t^*_u(x)}(x))\eta^u(t^*_u(x)).$$

Now equations (2) - (4) constitute a discrete-time Markov decision model (DTMDM) for a process $z_k(=x_{T_k})$, k=0,1,.. governed by the stochastic kernel q, i.e. such that

$$P^u[z_{k+1} \in A|z_k] = q(A;z_k,u_k)$$

and having cost

$$J_x(u) = E_x\sum_k g(z_k,u_k)$$

corresponding to a control policy u. This is a positive programming problem in terminology of Bertsekas and Shreve [1]. The main objective now is to state conditions under which there exists a stationary optimal policy, i.e map $\mu^0:E\to\underline{\underline{C}}$ such that the policy $u_k = \mu^0(z_k)$ minimizes $J_x(\mu)$ over all policies μ. Note that stationary policies μ coincide with admissible controls $u(z_t,\tau_t)=\mu(z_t)(\tau_t)$ for the original PDP problem. Thus such a result would imply the existence of an optimal policy $u^0\in\underline{\underline{U}}$. To obtain this result however we have to enlarge the admissible controls to include relaxed controls.

4. A relaxed control is a measurable function $v:R_+\to\underline{\underline{P}}(U)$. We denote by $\underline{\underline{R}}$ the set of relaxed controls. Using a relaxed control amounts to randomizing at each time t over the set of possible control values U rather than choosing a specific value $u\in U$. Of course ordinary controls $\underline{\underline{C}}$ are imbedded in $\underline{\underline{R}}$ by taking $v_t(dy) = \delta_{u(t)}(dy)$ for $u\in\underline{\underline{C}}$ where δ_ξ is the Dirac measure at $\xi\in U$.

Fix $v\in\underline{\underline{R}}$, $\theta\in C(E)$ and $x\in E$ and define functions ϕ_t, η_t, I^θ_t and C_t by the following equations

(5) $$\dot{\phi}_t = \int_U f(\phi_t,u)v_t(du), \qquad \phi_0 = x$$

(6) $$\dot{\eta}_t = -\int_U \lambda(\phi_t,u)\eta_t v_t(du), \qquad \eta_0 = 1$$

(7) $$\dot{I}^\theta_t = \int_U \theta(y)Q(dy;\phi_t,u)\lambda(\phi_t,u)\eta_t\, v_t(du), \qquad I^\theta_0 = 0$$

(8) $$\dot{C}_t = \int_U \ell_0(\phi_t,u)\eta_t\, v_t(du), \qquad C_0 = 0\ .$$

Then it is clear from (2) - (4) above that the DTMDM with relaxed controls is determined by a transition function $q:E\times\underline{\underline{R}}\to\underline{\underline{P}}(E)$ given by

(9) $$\int_E \theta(y)q(dy;x,v) := I^\theta_{t^*} + \int_U \theta(y)Q(dy;\phi_{t^*})\eta_{t^*}$$

and a one-stage cost function $g:E\times\underline{\underline{R}}\to R_+$ given by

(10) $$g(x,v) := C_{t^*} + \ell_\partial(\phi_{t^*})\eta_{t^*}$$

where, as usual, $t^* = \inf\{t:\phi_t\in\partial E\}$. A <u>stationary policy</u> is a map $\mu:E\to\underline{\underline{R}}$ and the cost corresponding to μ is

(11) $$J_x(\mu) = E_x \sum_k g(z_k,\mu(z_k))$$

where z_k is the discrete-time Markov process with transition kernel $q(A;x,\mu(x))$.

5. The purpose of this section is to derive conditions under which the functions q g of (9),(10) are continuous with respect to L.C. Young's topology on $\underline{\underline{R}}$. This is defined as follows, following Warga [9]. $X=L_1(R_+;C(U))$ is a Banach space whose dual space is $X^* = L_\infty(R_+;C^*(U))$ with the pairing

$$(h,v) = \int_0^\infty \int_U h(t,u)\, v_t(du)dt$$

for $h\in X$ and $v\in X^*$. Let $\|\cdot\|$ denote the total variation norm in $C^*(U)$. The <u>unit ball</u> in X^* is

$$B_1 = \{v \in L_\infty(R_+;C^*(U)) : \operatorname*{ess\,sup}_{t\in R_+} \|v_t(\cdot)\| \leq 1\}$$

and this is w^*-compact (Alaoglu's theorem). The set of relaxed controls $\underline{\underline{R}}$ is a subset of B_1 which is easily seen to be w^*-closed. Thus $\underline{\underline{R}}$ _is a_ w^*-_compact subset of_ X^*. The _Young topology_ on $\underline{\underline{R}}$ is the relative topology, and this is the only topology on $\underline{\underline{R}}$ considered henceforth. The following theorem is not hard to prove.

Theorem 1 [9,p 325]. Suppose $f : R^d x U \to R^d$ is continuous in $(x,u) \in R^d x U$, Lipschitz continuous in $x \in R^d$, and ϕ_t is the unique solution of (5) for some $v \in R$, $x \in R^d$. Then the map $(x,v) \to \phi(\cdot)$ is continuous from $R^d x \underline{\underline{R}}$ to $C([0,T];R^d)$ for any $T \in R_+$ (with respect to the uniform norm on $C([0,T];R^d)$.)

We now come to the main technical result of this paper.

Theorem 2 Suppose conditions (A1) - (A7) hold, where (A6) and (A7) are given below. Then $q : E x \underline{\underline{R}} \to \underline{\underline{P}}(E)$ and $g : E x \underline{\underline{R}} \to R_+$ are continuous functions.

(A6) $\lambda(x,u) \geq \delta > 0$

(A7) For $(x,v) \in E x \underline{\underline{R}}$ either (a) or (b) holds:

(a) $t^* = \infty$

(b) $t^* < \infty$ and $\liminf_{t\uparrow t^*} \dfrac{\langle x_{t^*} - x_t, \nabla\psi(x_{t^*})\rangle}{t^* - t} > 0.$

($\langle\cdot,\cdot\rangle$ is the inner product in R^d)

Proof To show that q is continuous we need to show that if $(x_n,v_n) \to (x,v)$ then

(12) $$\int_E \theta(y)q(dy;x_n,v_n) \to \int_E \theta(y)q(dy;x,v)$$

for all $\theta \in C(E)$. Let ϕ_t be the solution of (5) and ϕ^n_t the solution with (x_n,v_n) replacing (x,v), and let η_t, η^n_t and I_t, I^n_t be the corresponding solutions of (6) and (7) for $\theta \in C(E)$ fixed. Then as in (9)

(13) $$\int_E \theta(y)q(dy;x_n,v_n) = I^n_{t^*_n} + \int_E \theta(y)Q(dy;\phi^n_{t^*_n})\eta^n_{t^*_n}.$$

It follows from (A6) that $\eta^n_t \leq e^{-\delta t}$ for all n and that $\eta_t \leq e^{-\delta t}$.

(a) $(t^*=\infty)$ Let $T_\varepsilon(t)$ denote the ε-tube around ϕ, i.e.

$$T_\varepsilon(t) = \{y \in R^d : \min_{0\leq s\leq t} |\phi_s - y| < \varepsilon\}$$

Theorem 1 then implies that $\{\phi^n_s, 0\leq s\leq t\} \subset T_\varepsilon(t)$ for n sufficiently large and hence, since $T_\varepsilon(t) \subset E$ for small ε, that $t^*_n \to \infty$ as $n \to \infty$. Thus the last term in (13) converges to zero as $n\to\infty$ and it remains to show that $I^n_{t^*_n} \to I_\infty$. Now (A6) implies that $I^n_t \to I^n_\infty$ as $t \to \infty$ uniformly in n and hence that $I^n_\infty \to I_\infty$ as $n \to \infty$: for any $T>0$ we can write

$$|I_\infty - I^n_\infty| \le |I_\infty - I_T| + |I_T - I^n_T| + |I^n_T - I^n_\infty|.$$

We can now choose T (independently of n) and then n to secure $|I^n_\infty - I_\infty| < \varepsilon$ for any $\varepsilon > 0$. Finally, we have

$$|I_\infty - I^n_{t^*_n}| \le |I_\infty - I^n_\infty| + |I^n_{t^*_n} - I^n_\infty| .$$

The result follows, using again the uniform convergence of I^n_t to I^n_∞ and the fact that $t^*_n \to \infty$ as $n \to \infty$.

(b) $(t^*<\infty)$ Let $\xi_t = (\phi_t, \eta_t, I^\theta_t, C_t)$ denote the solution of equations (6)- (9) and ξ^n_t denote this solution with (x_n, v_n) replacing (x,v). These evolve in the space $\hat{E} = E\times R^3 = \{(x,y):\Phi(x,y) < 0\}$ where $\Phi(x,y)=\psi(x)$ for $x\in R^d$, $y\in R^3$. It follows from the assumptions that if x_t satisfies (A7)(b) then ξ_t satisfies a similar condition, i.e.

(14) $$\liminf_{t\to t^*} \frac{\langle \xi_{t^*} - \xi_t, \nu\rangle}{t^*-t} > 0$$

where $\nu = \nabla\Phi(\xi_{t^*})[\|\nabla\Phi(\xi^{-1}_{t^*})\|]^{-1}$ is the outward normal to $\partial\hat{E}$ at ξ_{t^*}. Now the right hand side of (13) takes the form $h(\xi^n_{t^*_n})$ for some continuous function $h(\cdot)$, so to establish (12) it suffices to prove that $\xi^n_{t^*_n} \to \xi_{t^*}$ as $n \to \infty$. We do this by showing that

(15) $$\operatorname{diam}(\hat{T}_\varepsilon(t^*) \cap \partial\hat{E}) \to 0 \quad \text{as } \varepsilon \downarrow 0.$$

where $\hat{T}_\varepsilon(t)$ is the ε-tube around $\{\xi_s, 0\le s\le t\}$. Since for arbitrary t and ε, $\{\xi^n_s, 0\le s\le t\}$ $\hat{T}_\varepsilon(t)$ for sufficiently large n, (15) will imply that eventually t^*_n is finite and that then $|\xi^n_{t^*_n} - \xi_t| < \operatorname{diam}(T_\varepsilon(t^*) \cap \partial E)$. From the hypothesis that $\hat{E}$ has C^1 boundary we see that (15) holds if

(16) $$\operatorname{diam}(\hat{T}_\varepsilon(t^*) \cap \Gamma) \to 0 \quad \text{as } \varepsilon \downarrow 0,$$

where Γ is the tangent hyperplane to $\partial\hat{E}$ at ξ_{t^*}. We thus have to establish (16).

Take $t_1 \in [0,t^*[$ and $\gamma > 0$ such that for all $t \in [t_1,t^*[$

(17) $$\langle \xi_{t^*} - \xi_t, \nu\rangle > \gamma(t^*-t) .$$

Such t_1,γ exist, by (14). Since $\{\xi_s, 0 \le s \le t_1\} \subset \hat{\hat{E}}$, there exists $\varepsilon_1 > 0$ such that $\hat{T}_{\varepsilon_1}(t_1) \subset \hat{E}$. Thus for $\varepsilon \in]0,\varepsilon_1]$, $\hat{T}_\varepsilon(t^*) \cap \Gamma = [\hat{T}_\varepsilon(t^*)-\hat{T}_\varepsilon(t_1)] \cap \Gamma$, i.e. we can discard the trajectory prior to t_1. From (A2) the velocity of ξ_t is locally bounded, i.e. there exists a constant α such that for $t \in [t_1,t^*[$

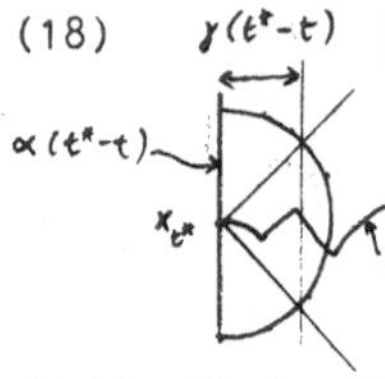

(18) $$|\xi_t - \xi_{t^*}| < \alpha(t^*-t).$$

Now (17) and (18) imply that the trajectory segment $\Xi = \{\xi_t, 0 \le t \le t^*\}$ is contained in a cone of internal angle $2\cos^{-1}(\gamma/\alpha)$ (see figure). Hence if $\varepsilon < \varepsilon_1$, and $y \in \Gamma$ is such that $\mathrm{dist}(y,\Xi) < \varepsilon$ then $|y-\xi_{t^*}| < \alpha\varepsilon/\gamma$, i.e. diam $(\hat{T}_\varepsilon(t^*) \cap \Gamma) < 2\alpha\varepsilon/\gamma$. This establishes (16) and completes the proof.

Continuity of g is proved in an exactly similar manner.

6. With Theorem 2 established the remaining results follow from the theory of discrete-time stochastic control as developed by Bertsekas and Shreve [1]. The DTMDM (9)-(11) satisfies the conditions of the lower-semi-continuous (l.s.c.) model [1,p 208]. The appropriate class of controls is that of Borel policies. Let $\underline{\underline{B}}$ denote the set of Borel-measurable maps $\mu : E \to \underline{\underline{R}}$. A Borel policy is a sequence $\pi = \{\mu_0,\mu_1,\ldots\}$ where each μ_i is in $\underline{\underline{B}}$. π is stationary if $\mu_k = \mu_0$ for all k. For $v \in \underline{\underline{R}}$, $\mu \in \underline{\underline{B}}$, $h: E \to R_+$ l.s.c. and $x \in E$ we define

$$H(x,v,h) = g(x,v) + \int_E (h(y)q(dy;x,v)$$

$$T_\mu h(x) = H(x,\mu(x),h)$$

and

$$Th(x) = \inf_{\mu \in \underline{\underline{B}}} T_\mu h(x)$$

It follows from [Proposition 8.6] that for any integer N there exists a policy $\pi_N^0 = \{\mu_0^0,\ldots,\mu_{N-1}^0\}$ which minimizes the N-horizon cost (sum from k=0 to N-1 in (11)), that the minimal cost is $T^N h_0(x)$ where $h_0(x) := 0$ for all x and that $T^N h_0(\cdot)$ is lower semi-continuous. Here, of course, we are interested in the infinite horizon case. By appealing to the theory involving monotonicity assumptions we obtain the following theorem, which is the main result of this paper.

<u>Theorem 3</u> There exists a stationary optimal Borel policy for the DTMDM (9)-(11).
<u>Proof</u> We refer throughout to [1, Chapter 5]. $H(x,v,h)$ clearly satisfies Assumption I:

$$h_0(x) \leq H(x,v,h_0), \quad x \in E, \quad v \in \underline{\underline{R}}$$

Now H has the stochastic interpretation

$$H(x,v,h) = E_x\Big[\int_0^{T_1}\int_U \ell_0(x_t,u)v_t(du)dt + \ell_\partial(x_{T_1}-)I_{(x_{T_1}-\in\partial E)} + h(x_{T_1})\Big]$$

where x_t is the PDP corresponding to control v on $[0,T_1[$. By the monotone convergence theorem, if $h_0 \leq h_1 \leq h_2 \ldots$ then

$$\lim_{k\to\infty} H(x,u,h_k) = H(x,n,\lim_{k\to\infty} h_k);$$

and if r is any positive number, then

$$H(x,v,h) \leq H(x,v,h+r) = H(x,v,h) + r.$$

Thus, Assumption I.1 and I.2 are satisfied. By Proposition 5.10 of [1] a stationary policy exists if the set

$$\{v \in \underline{\underline{R}} : H(x,v,T^k h_0) \leq \lambda\} \tag{19}$$

is compact for each $x \in E$, $\lambda \in R$ and integer k. Now $T^k h$ is ℓ.s.c. and the map $v \to H(x,v,h)$ is ℓ.s.c. for any ℓ.s.c. function h: indeed, if $h_n \uparrow h$ is an approximating sequence of continuous functions then by monotone convergence

$$\int_E h_n(y)q(dy;x,v) \uparrow \int_E h(y)q(dy;x,v).$$

Since the functions on the left are continuous in v this shows that $v \to \int_E h(y)q(dy;x,v)$ is ℓ.s.c.; and we know that $v \to g(x,v)$ is continuous. Hence $v \to H(x,v,h)$ is ℓ.s.c., and therefore the set at (19) is a closed subset of the compact space $\underline{\underline{R}}$. This completes the proof.

7. We conclude with some remarks on Theorem 3. This result was established under broadly similar conditions but using completely different methods by Vermes [8]. Boundary controls are included in [8]. The key assumption is the non-tangency condition (A7)(b). Vermes used the stronger condition

$$\inf_{u \in U} \langle f(x,u), \nabla\psi(x)\rangle > 0 , \qquad x \in \partial E$$

but even in the form given here (A7)(b) is unpleasantly strong and easily violated in applications (for example, uniform retardation to zero velocity on the boundary does not satisfy it). The condition actually required is (15); (A7)(b) implies (15) but weaker conditions of the form

$$\liminf_{t \uparrow t^*} \frac{\langle \xi_{t^*} - \xi_t, \nu\rangle}{b(t^*-t)} > 0,$$

where b(·) is some strictly increasing function such as $b(t) = t^2$, would probably suffice.

Finally we remark that stationary Borel policies for the discrete time problem are equivalent to stationary randomized controls in $\underline{\underline{U}}$ for the original PDP (see §2).

Acknowledgements I am indebted to Richard Vinter and to João Do Val for helpful discussions

References

1. D. Bertsekas and S.E. Shreve, Stochastic Optimal Control: the Discrete-time Case, Academic Press, New York, 1978.

2. M.H.A. Davis, Piecewise-deterministic Markov processes:a general class of non diffusion stochastic models, J. Royal Statist. Soc(B)46(1984)

3. N. El Karoui, Les aspectes probabilistes du contrôle stochastique, Lecture Notes in Mathematics, Springer-Verlag, Berlin.

4. U. Gugerli, Optimal stopping of piecewise deterministic Markov processes, sub mitted to Stochastics.

5. S.M. Lenhart and Y.C. Liao, Integro-differential equations associated with optimal stopping of a piecewise-deterministic process, Stochastics, to appear 1985.

6. M. Soner, PhD Thesis, Brown University, 1984

7. F.A. van der Duyn Schouten, Markov Decision Drift Processes, CWI Tract, Amsterdam 1983.

8. D. Vermes, Optimal control of piecewise-deterministic Markov processes, Stochastics, 14(1985)165-208.

9. J. Warga, Optimal Control of differential and functional equations, Academic Press New York 1972

10 A.A. Yushkevich, Continuous-time Markov decision processes with interventions, Stochastics 9(1983)235-274.

REVERSE TIME SMOOTHING FOR POINT PROCESS OBSERVATIONS

Robert J. Elliott
Department of Statistics and Applied Probability
University of Alberta, Canada

1. INTRODUCTION.

Consider a signal process ξ_t, $t \geq 0$, and an observed point process N_t with intensity $h(t,\xi_t) \geq 0$. Suppose $\underline{\underline{G}} = \{\underline{G}_t\}$ is the filtration generated by N; then the filtering problem is concerned with obtaining a recursive expression for $E[\xi_t | \underline{G}_t]$. The filtering problem for point process observations has been discussed by many authors; see, for example, Brémaud [2], and Liptser and Shiryayev Vol. 2 [5]. A treatment using optional and predictable projections and special semimartingales is outlined in section 2.

Given a terminal time $\tau > 0$ the smoothing problem discusses the estimate $E[\xi_s | \underline{G}_t]$ for $s < \tau$. Write $B\underline{\underline{G}} = \{B\underline{G}_t\}$ for the reverse time filtration generated by $\sigma\{N_\tau, N_{u-} : t \leq u \leq \tau\}$. Then it is shown in Anderson and Rhodes [1] how the smoothing problem involves estimates of the form $E[\xi_{s-} | B\underline{G}_s]$. In section 3 we obtain a reverse time semimartingale decomposition of the observation point process, such that $N_{s-} - \int_s^\tau \kappa_u du$ is a reverse time martingale in s, where $\kappa_u = -h(u,\xi_u)/\int_0^u h(v,\xi_v)dv$. This is used to obtain a reverse time expression for $E[\xi_{s-} | B\underline{G}_s]$. If $T > 0$ is a random time and $\xi_t = I_{t \geq T}$ it is proved that $I_{t>T} + \int_t^T (f_u \xi_u / F_u) du$ is a reverse time martingale in t, where $F_t = \int_0^t f_u du = P(T \leq t)$, and a reverse time expression is obtained for $E[\xi_{s-} | B\underline{G}_s]$.

Finally, when filtering and smoothing problems are discussed using the reference probability method a stochastic exponential $\Lambda_{s,t}(z)$ is introduced. The differentiability of $\Lambda_{s,t}(z)$ in s is treated in section 4.

2. POINT PROCESS FILTERING.

CONVENTION. All filtrations will be supposed right continuous and complete.

Consider a SIGNAL PROCESS which is a real semimartingale given by

$$\xi_t = \xi_0 + \int_0^t \beta_u du + M_t, \quad \text{for} \quad t \geq 0.$$

Let $\underline{\underline{F}} = \{\underline{F}_t\}$ be the filtration generated by $\{\xi_s : 0 \leq s \leq t\}$. The OBSERVATION PROCESS is a point process N_t. Write $\underline{\underline{G}} = \{\underline{G}_t\}$ for the filtration generated by $\sigma\{N_s : 0 \leq s \leq t\}$. Then $\underline{\underline{H}}$ will denote the filtration generated by $\underline{H}_t = \underline{G}_t \vee \underline{F}_\infty$. We suppose the observation process N is a conditional $\underline{\underline{H}}$ Poisson process with intensity $h(t,\xi_t) \geq 0$. Then the process $Q_t = N_t - \int_0^t h(u,\xi_u)du$ is an $\underline{\underline{H}}$ martingale.

NOTATION 2.1. For any (non-negative) $\underline{H}_\infty$ measurable process α write $\hat{\alpha}$ for its $\underline{\underline{G}}$ optional projection $\Pi_0(\alpha)$ and $\hat{\hat{\alpha}}$ for its $\underline{\underline{G}}$ predictable projection $\Pi_p(\alpha)$. Then, for any $t \geq 0$,

$$\hat{\alpha}_t = E[\alpha_t | \underline{G}_t] \quad \text{and}$$
$$\hat{\hat{\alpha}}_t = E[\alpha_t | \underline{G}_{t-}];$$

(however, because care must be taken with measurability in t these equations cannot be used to define the projections.)

LEMMA 2.2. For any non-negative $\underline{H}_\infty$ measurable locally bounded process α

$$\int_0^t \hat{\alpha}_u du = \int_0^t \hat{\hat{\alpha}}_u du \text{ a.s.} \quad \text{for all} \quad t \geq 0.$$

PROOF For any $\underline{\underline{G}}$ stopping time T the process $X = I_{]\!]0,T\wedge t]\!]}$ is $\underline{\underline{G}}$ predictable. Therefore, from Theorems 6.39 and 7.25 of [3]

$$E[\int_0^\infty \Pi_p(X\hat{\alpha})_u du] = E[\int_0^\infty X_u \hat{\hat{\alpha}}_u du]$$
$$= E[\int_0^{T\wedge t} \hat{\alpha}_u du] = E[\int_0^{T\wedge t} \Pi_p(\alpha)_u du]$$
$$= E[\int_0^{T\wedge t} \hat{\hat{\alpha}}_u du].$$

Applying Theorem 4.18 of [3] the process $\phi_s = \int_0^{s\wedge t} \hat{\alpha}_u du - \int_0^{s\wedge t} \hat{\hat{\alpha}}_u du$ is a $\underline{\underline{G}}$ martingale; however, it is predictable and of bounded variation and so zero. The result follows because the above t is arbitrary.

REMARKS 2.3. Note that, (writing $h_u = h(u,\xi_u)$),

$$\tilde{Q}_t = N_t - \int_0^t \hat{\hat{h}}_u du$$

is a $\underline{\underline{G}}$ martingale. Define

$$\tilde{M}_t = \hat{\xi}_t - \hat{\xi}_0 - \int_0^t \hat{\beta}_u du.$$

(From the above lemma $\int_0^t \hat{\beta}_u du = \int_0^t \hat{\hat{\beta}}_u du$). For each $t \geq 0$ $\hat{\xi}_t = E[\xi_t | \underline{G}_t]$; we wish to obtain a recursive equation for $\hat{\xi}_t$. Now it is easily checked that $\tilde{M}_t$ is a $\underline{\underline{G}}$ martingale and so, (see Chapter 3, Theorem 9 of [2]), it has a representation $\tilde{M}_t = \int_0^t \gamma_u d\tilde{Q}_u$, where γ is a $\underline{\underline{G}}$ predictable process. The filtering problem is to find an explicit expression for γ.

PROPOSITION 2.4. $\gamma_u = (\hat{\hat{h}}_u)^{-1}(\widehat{\widehat{\xi_{u-}h_u}} - \hat{\xi}_{u-}\hat{\hat{h}}_u + \langle M,N\rangle_t)$ if $\hat{\hat{h}}_u \neq 0$, and $\gamma_u = 0$ otherwise.

PROOF. Applying the product rule,

$$\xi_t N_t = \int_0^t \xi_{u-}(dQ_u + h_u du) + \int_0^t N_{u-}(\beta_u du + dM_u) + \sum_{0<u\leq t} \Delta\xi_u \Delta N_u \tag{1}$$

Here the sum $\sum_{0<u\leq t} \Delta\xi_u \Delta N_u = \sum_{0<u\leq t} \Delta M_u \Delta N_u$ is the process $[M,N]_t$. Suppose its dual $\underline{\underline{G}}$ predictable projection $\langle M,N\rangle_t$ exists. Taking the $\underline{\underline{G}}$ optional projection of the processes in (1), the process H^1 is a $\underline{\underline{G}}$ martingale where

$$H_t^1 = \widehat{\xi_t N_t} - \int_0^t \widehat{\widehat{\xi_{u-}h_u}} du - \int_0^t \widehat{N_{u-}\beta_u} du - \langle M,N\rangle_t .$$

Note, therefore, that $\widehat{\xi_t N_t} = \hat{\xi}_t N_t = N_t E[\xi_t | \underline{G}_t]$ is a special $\underline{\underline{G}}$ semimartingale, and so its decomposition as the sum of a martingale and a predictable bounded variation process is unique. However, we also have

$$\hat{\xi}_t N_t = \int_0^t \hat{\xi}_{u-}(d\tilde{Q}_u + \hat{\hat{h}}_u du) + \int_0^t N_{u-}(\hat{\beta}_u du + \gamma_u d\tilde{Q}_u) + \sum_{0<u\leq t} \Delta\hat{\xi}_u \Delta N_u .$$

Now $\Delta\hat{\xi} = \gamma_u \Delta N_u$, so the last sum is

$$\sum_{0<u\leq t} \gamma_u (\Delta N_u)^2 = \sum_{0<u\leq t} \gamma_u \Delta N_u = \int_0^t \gamma_u dN_u = \int_0^t \gamma_u d\tilde{Q}_u + \int_0^t \gamma_u \hat{\hat{h}}_u du .$$

Then H^2 is a $\underline{\underline{G}}$ martingale, where

$$H_t^2 = \hat{\xi}_t N_t - \int_0^2 \hat{\xi}_{u-}\hat{\hat{h}}_u du - \int_0^t N_{u-}\hat{\hat{\beta}}_u du - \int_0^t \gamma_u \hat{\hat{h}}_u du . \tag{2}$$

Because $\hat{\xi}N$ is a special semimartingale with a unique decomposition, equating the bounded variation processes in (1) and (2):

$$\gamma_u \hat{\hat{h}}_u = \widehat{\widehat{\xi_{u-}h_u}} - \hat{\xi}_{u-}\hat{\hat{h}}_u + \langle M,N\rangle_t$$

giving $\gamma_u = (\hat{\hat{h}}_u)^{-1}(\widehat{\widehat{\xi_{u-}h_u}} - \hat{\xi}_{u-}\hat{\hat{h}}_u + \langle M,N\rangle_t)$ if $h_u \neq 0$. For any set $A \in \underline{G}_s$

$$E[I_A \int_s^t dN_u] = E[I_A \int_s^t \hat{\hat{h}}_u du]$$

so γ_u can be taken to be 0 on any set where $\hat{\hat{h}}_u = 0$.

REMARKS 2.5. The random variables γ_u are $\underline{G}_{u-}$ measurable. If M and N have no jumps in common $[M,N] = 0$, so $\langle M,N\rangle = 0$.

DISORDER PROBLEM 2.6. (See [2]). Suppose $T > 0$ is failure time with an absolutely continuous distribution function $F_t = P(T \leq t) = \int_0^t f(u)du$. Consider the process $\xi_t = I_{t \geq T}$. With respect to the filtration $\underline{\underline{F}}$ generated by ξ the process $M_t = I_{t \geq T} - \int_0^t (1-\xi_u)f_u(1-F_u)^{-1}du$ is a martingale, so the Doob-Meyer decomposition of ξ is

$$\xi_t = \int_0^t f_u(1-\xi_u)(1-F_u)^{-1}du + M_t.$$

Suppose at the failure time T the intensity of a point process N changes from a constant a to a constant b. Then the intensity of N is

$$h_u = a + (b-a)\xi_{u-}$$

so $\hat{\hat{h}}_u = a + (b-a)\hat{\xi}_{u-}$

The jump time T is a.s. not a jump of N so Proposition 2.4 give

$$\hat{\xi}_t = \int_0^t f_u(1-\hat{\xi}_u)(1-F_u)^{-1}du + \int_0^t (b-a)\big(\hat{\xi}_{u-}-(\hat{\xi}_{u-})^2\big)\big(a+(b-a)\hat{\xi}_{u-}\big)^{-1}d\tilde{Q}_u.$$

3. REVERSE TIME EQUATIONS.

Consider a signal process ξ which is, as above, a semimartingale with respect to its own filtration. Suppose t runs backwards from some terminal time τ. Write $B\underline{\underline{F}} = \{B\underline{F}_t\}$, $0 \leq t \leq \tau$, for the, (left continuous - because time is decreasing), filtration generated by $G\{\xi_{s-},\xi_\tau : t \leq s \leq \tau\}$. (The 'B' here and below will denote 'backward'.) Suppose $\xi_t = \xi_{t-}$ is a reverse time $B\underline{\underline{F}}$ semimartingale of the form $\overline{\xi}_t = \overline{\xi}_\tau + \int_t^\tau \overline{\beta}_u du + \overline{M}_t$, $0 \leq t \leq \tau$, where $\overline{M}$ is a reverse time martingale, (that is $E[M_{t-\delta}|B\underline{F}_t] = M_t$ a.s.). Write $\tau - t = s$, $t - u = v$ and $\zeta_s = \xi_{\tau-s} = \xi_t$, so the signal process is given by

$$\zeta_s = \zeta_0 + \int_0^s \overline{\beta}_{\tau-v}dv + \mu_s$$

where $\mu_s = \overline{M}_{\tau-s}$ is a forward martingale in s.

The observation process will again be a point process N such that, with respect to the filtration $\underline{\underline{H}}$, N is a forward conditional Poisson process with bounded intensity $h_t = h(t,\xi_t)$. However, we shall consider the process N run in reverse time from terminal time τ. We shall assume that ξ and N have no common jumps, so both the forward and backward variation processes $[\xi,N]$ are 0. Consider the left continuous process $\overline{N}_t = N_{t-}$. $\underline{\underline{BG}}$ will denote the reverse filtration $\{\underline{BG}_t\}$ generated by $G\{N_s, N_\tau : t \le s \le \tau\}$, and write $\underline{\underline{BH}}$ for the reverse filtration $\underline{BH}_t = \underline{BG}_t \vee \underline{BF}_0$. Then, for example by Baye's rule and the continuity of the probabilities, for $s \le t$

$$P[\overline{N}_s = k|\overline{N}_t = m \text{ and } \underline{BF}_0] = \binom{m}{k}\left(\int_s^t h_u du\right)^{m-k}\left(\int_0^s h_u du\right)^k\left(\int_0^t h_u du\right)^{-m},$$

and $E[\overline{N}_s|\overline{N}_t$ and $\underline{BF}_0] = \left(\int_0^s h_u du\right)\left(\int_0^t h_u du\right)^{-1}\overline{N}_t$. By the Markov property and conditionally independent increments

$$E[\overline{N}_{s-\delta} - \overline{N}_s|\underline{BH}_s] = E[\overline{N}_{s-\delta} - N_s|\overline{N}_s \text{ and } \underline{BF}_0] = -\left(\int_{s-\delta}^s h_u du\right)\left(\int_0^s h_u du\right)^{-1}\overline{N}_s .$$

Consequently $E|E[\overline{N}_{s-\delta} - \overline{N}_s|\underline{BH}_s]| = \int_{s-\delta}^s h_u du$ and so

$$\int_0^t E|E[\overline{N}_{s-\delta} - \overline{N}_s|\underline{BH}_s]|ds \text{ is } O(h).$$

By Stricker's theorem, [7], $\overline{N}_s$ is a reverse time $\underline{\underline{BH}}$ quasimartingale and it has a decomposition

$$\overline{N}_t = \overline{N}_\tau + \int_t^\tau \kappa_u du + \overline{Q}_t \tag{3}$$

where, from the above calculation,

$$\kappa_u = -h_u / \int_0^u h_v dv. \tag{4}$$

Equation (3) for $\overline{N}$ defines the reverse time observation process. As for the signal write $\tau - t = s$, $\tau - u = v$ and $\eta_s = \overline{N}_{\tau-s}$, so

$$\eta_s = \eta_0 + \int_0^s \kappa_{\tau-v} dv + \nu_s$$

where ν is a forward martingale in s. With $\wedge$ and $\hat{\hat{}}$ denoting the $\underline{\underline{BG}}$ optional and predictable projection, applying Proposition 2.4

$$\hat{\zeta}_s = \hat{\zeta}_0 + \int_0^s \hat{\hat{\beta}}_{\tau-v} dv + \int_0^s \delta_v (d\eta_v - \hat{\hat{\kappa}}_{\tau-v} dv)$$

where $\delta_v = (\widehat{\widehat{\kappa}}_{\tau-v})^{-1}(\widehat{\widehat{\xi_v\kappa_{\tau-v}}}-\hat{\xi}_v\widehat{\widehat{\kappa}}_{\tau-v})$ if $\widehat{\widehat{\kappa}}_{\tau-v} \neq 0$, and $\delta_v = 0$ otherwise.

That is, $\hat{\xi}_t = \hat{\xi}_\tau + \int_t^\tau \hat{\bar{\beta}}_u du + \int_t^\tau \delta_{\tau-u}(d\bar{N}_u - \widehat{\widehat{\kappa}}_u du)$ where

$$\delta_{\tau-u} = (\widehat{\widehat{\kappa}}_u)^{-1}(\widehat{\widehat{\xi_u\kappa_u}}-\hat{\xi}_u\widehat{\widehat{\kappa}}_u)I(\kappa_u \neq 0) \tag{5}$$

and κ_u is given by (4).

REVERSE DISORDER PROBLEM. Consider the single jump process $\xi_t = I_{t \geq T}$ for t running backwards from τ. Then $E[\xi_s|\xi_t] = \xi_t F_s/F_t$, for $s \leq t$. Write $\bar{\xi} = \xi_{t-}$, so by continuity

$$E[\bar{\xi}_s|\bar{\xi}_t] = \bar{\xi}_t F_s/F_t .$$

If, as above, $B\underline{\underline{F}}$ denotes the (left continuous) filtration generated by $\bar{\xi}$ then

$$E[\bar{\xi}_{s-\delta}-\bar{\xi}_s|B\underline{F}_s] = E[\bar{\xi}_{s-\delta}-\bar{\xi}_s|\bar{\xi}_s] = \bar{\xi}_s(F_{s-\delta}-F_s)F_s^{-1} = -\bar{\xi}_s \int_{s-\delta}^{s} f_u du/F_s .$$

Therefore, Stricker's result, [7], can again be applied to show that $\bar{\xi}_t$ is a reverse time $B\underline{\underline{F}}$ quasimartingale with a decomposition

$$\bar{\xi}_t = \bar{\xi}_\tau + \int_t^\tau \bar{\beta}_u du + \bar{M}_t$$

where $\bar{M}_t$ is a reverse time $B\underline{\underline{F}}$ martingale and $\bar{\beta}_u = -\bar{\xi}_u f_u/F_u$. If the intensity of the consitional Poisson process N is given by

$$h_u = a + (b-a)\xi_u$$

then the reverse time formula (5) gives

$$\hat{\bar{\xi}}_t = \hat{\bar{\xi}}_\tau + \int_t^\tau \hat{\bar{\beta}}_u du + \int_t^\tau \delta_{\tau-u}(d\bar{N}_u - \widehat{\widehat{\kappa}}_u du)$$

where $\delta_{\tau-u}$ is given by (5) and

$$\kappa_u = -(a+(b-a)\xi_u)/\int_0^u (a+(b-a)\xi_v)dv.$$

4. REFERENCE PROBABILITY SMOOTHING.

In the reference probability method, (see [2],[3]), a new probability measure P_h related to the signal process through the intensity $h(u,\xi_u)$, and absolutely continuous with respect to a 'reference probability' P, governs the statistics of the observation process N. It is supposed that, under P,N is a standard Poisson process of intensity 1 on $(\Omega,\underline{\underline{H}})$, where as in section 2 $\underline{\underline{H}} = \{\underline{H}_t\}$ and $\underline{H}_t = \underline{G}_t \vee \underline{F}_\infty$. Suppose the signal process is the single jump process $\xi_t = I_{t>T}$ and write

$$Z_{s,t}(z) = z \mp I_{(z=0)}\xi_t,$$

where $z = \xi_s$. For $0 \leq s \leq t$ consider the family of martingale exponentials

$$\Lambda_{s,t}(z) = \Big(\prod_{s<u\le t} h(u,Z_{s,u}(z))\Delta N_u\Big)\exp\Big(\int_s^t (1-h(u,Z_{s,u}(z)))du\Big).$$

Suppose P_h is defined by

$$E\Big[\frac{dP_h}{dP}\,\Big|\,\underline{H}_t\Big] = \Lambda_{0,t}(z).$$

Under P_h the process N is an $\underline{\underline{H}}$ conditional Poisson process with intensity $h(u,Z_{0,u}(z))$. Now

$$E_{P_h}[\xi_t|\underline{G}_t] = \frac{E_P[\xi_t\Lambda_{0,t}(z)|\underline{G}_t]}{E_P[\Lambda_{0,t}(z)|\underline{G}_t]} .$$

Write $\underline{D}_s^t$ for the σ-field generated by $N_v - N_u$, $s \le u \le v \le t$. As in Pardoux [6] and Elliott [4], for the smoothing problem it is of interest to investigate the reverse time differentiability in s of $\hat{\Lambda}_{s,t}(z) = E_P[\Lambda_{s,t}(z)|\underline{D}_s^t]$. The following result is proved similarly to one for Markov chains in [4].

THEOREM 4.1. Write $c = \inf\{t: F_t = 1\}$ and suppose $t < c$. Then

$$\hat{\Lambda}_{s,t}(z) - 1 = \int_s^t \hat{\Lambda}_{u,t}(z)((h(u,z)-1)(d\overline{N}_u-du) + \int_s^t (1-z)\hat{\Lambda}_{u,t}(z)f_u(1-F_u)^{-1}du.$$

SKETCH PROOF. Consider a partition $\Delta = \{s = t_0<t_1<\ldots<t_n = t\}$, with $|\Delta| = \max_k(t_{k+1}-t_k)$. Then

$$\Lambda_{s,t}(z) - 1 = \sum_{k=0}^{n-1}(\Lambda_{t_k,t}(z) - \Lambda_{t_{k+1},t}(z)).$$

Recall $\Lambda_{s,t}(z) = 1 + \int_s^t \Lambda_{s,u-}(z)(h(u,Z_{s,u-}(z))-1)(dN_u-du)$. Now

$$\Lambda_{t_k,t}(z) - \Lambda_{t_{k+1},t}(z) = \Lambda_{t_k,t_{k+1}}(z)(\Lambda_{t_{k+1},t}(Z_{t_k,t_{k+1}}(z)) - \Lambda_{t_{k+1},t}(z))$$
$$+ \Lambda_{t_{k+1},t}(z)(\Lambda_{t_k,t_{k+1}}(z)-1) = J_k + K_k \text{ , say.}$$

Then
$$J_k = (\Lambda_{t_k,t_{k+1}}(z)-1)(\Lambda_{t_{k+1},t}(Z_{t_k,t_{k+1}}(z)) - \Lambda_{t_{k+1},t}(z))$$
$$+ (\Lambda_{t_{k+1},t}(Z_{t_k,t_{k+1}}(z)) - \Lambda_{t_{k+1},t}(z)) = J_k^{(1)} + J_k^{(2)}, \text{ say.}$$

However, $E|J_k^{(1)}| \le \text{Const.}|t_{k+1}-t_k|\dfrac{(F_{t_{k+1}}-F_{t_k})}{(1-F_{t_{k+1}})}$ so

$E|\sum_{k=0}^{n-1} \hat{J}_k^{(1)}| \le \text{Const.}|\Delta|(F_t-F_s)/(1-F_t)$ and this has limit 0 as $|\Delta| \to 0$. On the other hand, $J_k^{(2)} \ne 0$ only if $Z_{t_k,t_{k+1}}(z) \ne z$, which is so if $z = 0$ and $Z_{t_k,t_{k+1}}(z) = 1$, that is if the Z process starting at 0 at t_k has a jump before t_{k+1}; therefore, it can be shown that as $|\Delta| \to 0$

$$\sum_{k=0}^{n-1} \hat{J}_k^{(2)} \to \int_s^t (1-z)\hat{\Lambda}_{u,t}(z) f_u (1-F_u)^{-1} du.$$

Similarly,

$$K_k = \Lambda_{t_{k+1},t}(z)\Big(\int_{t_k}^{t_{k+1}} \Lambda_{t_{k,u-}}(z)\big(h(u,Z_{t_k,u}(z))-1\big)(dN_u - du\Big)$$

and as $|\Delta| \to 0$

$$\sum_{k=0}^{n-1} \hat{K}_k \to \int_s^t \hat{\Lambda}_{u,t}(z)\big(h(u,z)-1\big)(d\bar{N}_u - du\big)$$

so the result follows.

REFERENCES

1. B.D.O. Anderson and I.B. Rhodes, Smoothing algorithms for nonlinear finite dimensional systems. Stochastics 9(1983), 139-165.

2. P. Brémaud, Point Processes and Queues. Springer-Verlag, New York. 1981.

3. R.J. Elliott, Stochastic Calculus and Applications. Applications of Mathematics Vol. 18. Springer-Verlag. New York. 1982.

4. R.J. Elliott, Reverse time differentiation and smoothing formulae for a finite state Markov process. Annals of Probability.

5. R.S. Liptser and A.N. Shiryayev, Statistics of Random Processes. Vol. 2. Applications of Mathematics Vol. 6. Springer-Verlag. New York. 1978.

6. E. Pardoux, Equations du filtrage non-linéaire, de la prediction et du lissage. Stochastics 6(1982), 193-231.

7. C. Stricker, Une characterization des quasimartingales. Séminaire de Probabilitiés IX, Lecture Notes in Math. 465. Springer-Verlag. Berlin 1975.

A FINITELY ADDITIVE VERSION OF POINCARE'S RECURRENCE THEOREM

H. Francke
Inst. f. Theor. Physik
Domagkstr. 71
D-4400 Münster

D. Plachky and W. Thomsen
Inst. f. Math. Statistik
Einsteinstr. 62
D-4400 Münster

In this paper we give a finitely additive version of Poincaré's recurrence theorem, which is a general statement on a motion in a measure space (dynamical system), cf. [1] and [4]. This part is related to a paper of Barone and Bhaskara Rao [2]. An application of this result proves that a probability charge ν on all subsets of a group G is strongly continuous and purely finitely additive, if $\nu^g = \nu$ for some $g \in G$ of infinite order. In the second section we compute the mean and the variance of the recurrence time in an invariant measure space.

1. A finitely additive version of Poincaré's recurrence theorem

We adhere to the notation of [3]. $\mathbb{A}$ stands for an algebra of subsets of a set Ω, ν is a probability charge on $\mathbb{A}$, i.e. a positive normed finitely additive set function, and ν_* is the inner charge of ν, i.e. $\nu_*(B) = \sup\{\nu(A) \mid A \subset B,\ A \in \mathbb{A}\}$. $T:\Omega \to \Omega$ denotes a mapping such that $T^{-1}(\mathbb{A}) \subset \mathbb{A}$ and $\nu^T = \nu$. Given $A \in \mathbb{A}$ and $n \in \mathbb{N}$, we put

$$S(n,A) = \{\omega \in A \mid \sum_{i=1}^{\infty} 1_{T^{-i}(A)}(\omega) \geq n\}.$$

This is the subset of all $\omega \in A$, such that the orbit $(T^i(\omega),\ i \in \mathbb{N})$ is at least n times in A. Observe that the set $S(1,A) = A \cap \bigcup_{i=1}^{\infty} T^{-i}(A)$ is not in $\mathbb{A}$ in general, cf. [2].

Theorem 1. The equality

$$\nu_*(S(n,A)) = \nu(A)$$

holds for all $A \in \mathbb{A}$ and $n \in \mathbb{N}$.

Proof: Given $A \in \mathbb{A}$ and $n \in \mathbb{N}$, we have $\sum_{k=1}^{n} \nu(\bigcap_{i=1}^{k} T^{-i}(A^c)) =$
$= \sum_{k=1}^{n} \nu(\bigcap_{i=0}^{k} T^{-i}(A^c)) + \sum_{k=1}^{n} \nu(A \setminus \bigcup_{i=1}^{k} T^{-i}(A))$ which implies
$\sum_{k=1}^{n} \nu(A \setminus \bigcup_{i=1}^{k} T^{-i}(A)) = \nu(A^c) - \nu(\bigcap_{i=0}^{n} T^{-i}(A^c))$. Therefore,

$\lim_{n\to\infty} \nu(A \setminus \bigcup_{i=1}^{n} T^{-i}(A)) = 0$ and hence $\nu_*(S(1,A)) = \nu(A)$. Now let $S(1,A,i_1) = A \cap \bigcup_{i=1}^{i_1} T^{-i}(A) \in \mathbb{A}$ be the set of all $\omega \in A$ with $T^i(\omega) \in A$ for some $1 \leq i \leq i_1$. By induction, we put

$$S(n+1,\ A,i_1,\dots,i_{n+1}) = S(1,S(n,A,i_1,\dots,i_n),\ i_{n+1}) \in \mathbb{A}.$$

As easily seen, $S(n,A,i_1,\dots,i_n) \subset S(n,A) \subset A$. Furthermore, $\lim_{i\to\infty} \nu(S(1,B,i)) = \nu(B)$ holds for all $B \in \mathbb{A}$. Thus we find $i_\nu \in \mathbb{N}$, $\nu = 1,\dots,n$ with $\nu(S(n,A,i_1,\dots,i_n)) \geq \nu(S(n-1,A,i_1,\dots,i_{n-1})) - \frac{\varepsilon}{2^n}$, and hence $\nu(A) \geq \nu_*(S(n,A)) \geq \nu(S(n,A,i_1,\dots,i_n)) \geq \nu(A) - \varepsilon$ which implies the assertion.

Remarks.

1. Let us note that by virtue of the Markov-Kakutani fixed point theorem there is a T-invariant probability charge μ on $\sigma(\mathbb{A})$ extending ν. Moreover, if $T(\Omega) = \Omega$ and $T^{-1}(\mathbb{A}) = \mathbb{A}$, then $\nu^* = (\nu^*)^T$. Thus, given $A_o \in \sigma(\mathbb{A})$, there exists even a T-invariant extension of ν on $\sigma(\mathbb{A})$ with $\mu(A_o) = \nu^*(A_o)$. This result remains true, if T is replaced by a family $\mathfrak{T}$ of pairwise commuting transformations such that $\nu^T = \nu$, $T(\Omega) = \Omega$, $T^{-1}(\mathbb{A}) = \mathbb{A}$ holds for all $T \in \mathfrak{T}$.

2. If $\mathfrak{T}$ is an arbitrary family of transformations $T: \Omega \to \Omega$ with $T^{-1}(\mathbb{A}) \subset \mathbb{A}$, then a statement due to J.v. Neumann (cf. [5]) shows that a necessary and sufficient condition for the existence of a $\mathfrak{T}$-invariant probability charge ν on $\mathbb{A}$ is

$$\inf_{\omega\in\Omega} \sum_{i=1}^{n} (f_i(\omega) - f_i \circ T_i(\omega)) \leq 0$$

for all finite subsets $\{T_1,\dots,T_n\} \subset \mathfrak{T}$ and all finite subsets $\{f_1,\dots,f_n\}$ of bounded functions $f_i: \Omega \to \mathbb{R}$ which are in the sup-norm closed linear hull of the indicators 1_A, $A \in \mathbb{A}$.

3. Let us point out that either there exists a σ-additive $\mathfrak{T}$-invariant probability charge on $\mathbb{A}$ or each $\mathfrak{T}$-invariant probability charge on $\mathbb{A}$ is purely finitely additive. This follows easily from the representation of the σ-additive part of a probability charge ([3], p. 242), since the invariance property carries over to the σ-additive part.

Application. Let G be a group and $g_o \in G$ of infinite order. According to Remark 1 above there exists a probability charge ν on the σ-algebra of all subsets of G with $\nu^{g_o} = \nu$. It will be shown that ν is strongly continuous (in the sense of [3]) and purely finitely additive. To this end let G_o denote the cyclic infinite subgroup $\{g_o^n: n \in \mathbb{Z}\}$ and let S be a complete system of representatives for the left cosets of G_o. Obviously, $gS \cap g'S = \emptyset$ for $g, g' \in G_o$ with $g \neq g'$. Therefore, the finitely additive version of Poincaré's recurrence theorem yields $\nu(S) = \nu(S \cap \bigcup_{n=1}^{\infty} g_o^{-n} S) = \nu(\emptyset) = 0$. Hence, λ defined by $\lambda(A) = \nu(\bigcup_{n\in A} g_o^{-n} S)$, $A \subset \mathbb{Z}$, is a translation-invariant probability charge on the set of all subsets of $\mathbb{Z}$ vanishing on the singletons. Since λ is not σ-additive, Remark 3 above shows that ν is a pure charge. Finally, $\lambda(n\mathbb{Z}+r) = \lambda(n\mathbb{Z})$ implies $\lambda(n\mathbb{Z}+r) = \frac{1}{n}$ for $r = 0,\dots,n-1$ and $n \in \mathbb{N}$. This proves ν to be strongly continuous.

2. Mean and variance of the recurrence time

In this section we assume additionally that $\mathcal{A}$ is a σ-Algebra and P is a T-invariant probability measure on $\mathcal{A}$. Given $A \in \mathcal{A}$ with $P(A) > 0$, the random variable $X_A: \Omega \to \mathbb{N} \cup \{\infty\}$, $X_A(\omega) = \inf\{n \in \mathbb{N} \mid T^n(\omega) \in A\}$, is the first entrance time into A, $E_P(X_A|A) = \frac{1}{P(A)} \int X_A \, dP$ is the mean (conditional) recurrence time and $V_P(X_A|A)$ is the (conditional) variance of the recurrence time. Subsequently we shall compute these values.

<u>Theorem</u> 2. Suppose $A \in \mathcal{A}$ with $0 < P(A)$. Then

a) $E_P(X_A)$ is finite if and only if $\sum_{n=1}^{\infty} P(\bigcap_{i=1}^{n} T^{-i}(A^c)) < \infty$ and if $E_P(X_A)$ is finite, then $E_P(X_A) = 1 + \sum_{n=1}^{\infty} P(\bigcap_{i=1}^{n} T^{-i}(A^c))$.

b) $E_P(X_A|A) = \dfrac{P(X_A < \infty)}{P(A)}$.

<u>Proof:</u> The first part follows by virtue of $P(X_A > n) = P(\bigcap_{i=1}^{n} T^{-i}(A^c))$ and $E_P X_A = \infty \cdot P(X_A = \infty) + \sum_{n=1}^{\infty} nP(X_A = n) = \infty \, P(X_A = \infty) + \sum_{n=1}^{\infty} P(X_A > n) + P(X_A < \infty)$.
In order to prove the second part, observe that Theorem 1 yields $P(A \cap \{X_A = \infty\}) = 0$. Therefore, $E_P(X_A|A) = \frac{1}{P(A)} \left(\sum_{n=1}^{\infty} P(A \cap \{X_A \geq n\}\right) =$

$$= \frac{1}{P(A)} \{P(A) + \sum_{n=1}^{\infty} P(A \cap \bigcap_{i=1}^{n} T^{-i}(A^c))\} = \frac{1}{P(A)} (1 - P(\bigcap_{i=0}^{\infty} T^{-i}(A^c)),$$

where the last equality follows from the first part in the proof of Theorem 1. This yields $E_P(X_A|A) = \frac{1}{P(A)} P(\bigcup_{i=0}^{\infty} T^{-i}(A))$.

<u>Remark:</u> If additionally P is ergodic, then $E_P(X_A|A) = \frac{1}{P(A)}$ holds

(cf. [7]) by virtue of $P(X_A < \infty) = P(\bigcup_{i=1}^{\infty} T^{-i}(A)) = P(\limsup_{i\to\infty} T^{-i}(A)) = 1$. Complementary, observe that $E_P(X_A|A) = \frac{1}{P(A)}$ is true, if $E_P(X_A)$ is finite.

Theorem 3. Suppose $A \in \mathcal{A}$ with $0 < P(A) < 1$. Then $V_P(X_A|A) < \infty$ if and only if $E_P(X_A) < \infty$, and if $V_P(X_A|A) < \infty$, then $V_P(X_A|A) = \frac{P(A^c)}{P(A)} \{2\, E_P(X_A|A^c) - E_P(X_A|A)\}$.

Proof: Observe that $\sum_{k=1}^{n} k\, P(A \cap \bigcap_{i=1}^{k} T^{-i}(A^c)) = \sum_{k=0}^{n-1} P(\bigcap_{i=0}^{k} T^{-i}(A^c)) - n\, P(\bigcap_{i=0}^{n} T^{-i}(A^c))$ holds for all $n \in \mathbb{N}$. Suppose that $E_P(X_A) < \infty$. This implies $\lim_{n\to\infty} n\, P(\bigcap_{i=1}^{n} T^{-i}(A^c)) = 0$, and hence $E_P(X_A) = 1 + \sum_{k=1}^{\infty} k\, P(A \cap \bigcap_{i=1}^{k} T^{-i}(A^c))$. Thus, $E_P(X_A^2|A) = \frac{1}{P(A)} \sum_{n=1}^{\infty} n^2\, P(A \cap \{X_A = n\}) = \frac{1}{P(A)} \sum_{n=1}^{\infty} (2n-1) P(A \cap \{X_A \geq n\}) = \frac{2}{P(A)} \sum_{n=1}^{\infty} n\, P(A \cap \bigcap_{i=1}^{n} T^{-i}(A^c)) + E_P(X_A|A)$. This implies $V_P(X_A|A) = E_P(X_A^2|A) - (E_P(X_A|A))^2 = \frac{2}{P(A)} (E_P(X_A) - 1) + E_P(X_A|A) - (E_P(X_A|A))^2 = \frac{P(A^c)}{P(A)} (2\, E_P(X_A|A^c) - E_P(X_A|A))$, since $E(X_A|A) = \frac{1}{P(A)}$. Converserly, if $E_P(X_A) = \infty$, then $\sum_{k=1}^{\infty} k\, P(A \cap \bigcap_{i=1}^{k} T^{-i}(A^c)) = \infty$ and hence $V_P(X_A|A) = \infty$.

Remark (due to U. Krengel): Part b) of Theorem 2 is Kac's recurrence theorem (cf. [6], p. 19); [6] contains also references concerning the computation of the corresponding variance.

References

[1] Arnold, V. I.: Mathematical Methods in Classical Mechanics. Springer, New York, 1978.

[2] Barone, E. and K. P. S. Bhaskara Rao: Poincaré's Recurrence Theorem for Finitely Additive Measures. Rendiconti di Mathematica 1 (1981), 521 - 526.

[3] Bhaskara Rao, K. P. S. and M. Bhaskara Rao: Theory of Charges. A Study of Finitely Additive Measures. Academic Press, New York, 1983.

[4] Cornfield, J. P., Fomin, S. V., and Ya. G. Sinai: Ergodic Theory. Springer, New York, 1984.

[5] Greenleaf, F. P.: Invariant Means on Topological Groups. Van Nostrand, New York, 1969.

[6] Krengel, U.: Ergodic Theorems. De Gruyter, Studies in Mathematics 6, 1985.

[7] Ornstein, D. S., Rudolph, D. J., and B. Weiss: Equivalence of measure preserving transformations. Memoirs of the American Math. Soc, 37 (1982).

GIRSANOV AND FEYNMANN-KAC FORMULAS IN THE DISCRETE STOCHASTIC MECHANICS

G. Del Grosso and R. Marra
Dipartimento di Matematica Università di Roma "La Sapienza".

1. INTRODUCTION.

In the probabilistic approach to Quantum Mechanics and Euclidean Quantum Field Theory a powerful tool is given by the functional integration based on path integral; in this context a crucial role is played by the Girsanov and Feynmann-Kac formulas (1,2). In the Nelson's Stochastic Mechanics (3), which provides a quantization procedure for classical dynamical systems based on the theory of stochastic differential equations,these formulas are very useful to study the diffusion processes involved. The discrete Stochastic Mechanics, proposed by Guerra and Marra (4), provides an extension of the Nelson's scheme to a general quantum system based on the use of the time continuous Markov chains. The main purpose of this paper is to formulate a Girsanov formula in this discrete context and analyze the relation between the last one and the Feynmann-Kac formula.
In section 3 a Girsanov formula for Markov chains is given. In section 4 we show that in certain cases the stochastic process associated to the imaginary time Schrodinger equation is the same as the process associated to the ground state by the discrete Stochastic Mechanics,in the sense that the corresponding semigroups are unitary equivalent. This connection was pointed out for the euclidean scalar field in (7) and exploited in the case of scalar Schrodinger particles in (8). Finally we investigate as the equivalence between the semigroups induces the equivalence between the Girsanov and Feynmann-Kac formulas.

2. THE DISCRETE STOCHASTIC MECHANICS.

In the paper (4) the mathods of stochastic quantization, in the frame of Nelson Stochastic Mechanics, have been extended to a general quantum system. The approach is based on the stochastic variational principle (9) for controlled Markov random processes taking values on a discrete (finite or denumerable) configuration space. Inthis discrete formulation we consider a quantum system on a separable Hilbert space
If φ_i , $i = 1,2,...$ is an ortonormal basis, let us introduce for a generic wave function ψ , the components $\psi_i=\langle\varphi_i,\psi\rangle$ and consider the Schrödinger equation in the abstract form

$$i\hbar \quad \partial_t\psi = H\psi \qquad 1$$

or for the components

$$i\hbar \; \dot{\psi}_i = \sum_j h_{ij} \exp(i\alpha_{ij}/\hbar) \psi_j \qquad 2$$

In the right hand side of formula 2 the matrix elements of the Hamiltonian H have been written so that $h_{ij} \geq 0$. Self-adjointness of H implies

$$h_{ij} = h_{ji} , \quad \alpha_{ij} = -\alpha_{ji} , \quad \alpha_{ii} = 0 \qquad 3$$

It is convenient to introduce real variables $(\rho_i , S_i), \rho_i \geq 0$ such that

$$\psi_i = \rho_i^{1/2} \exp(iS_i / \hbar) \qquad 4$$

Then the equation 2 splits into two real ones:

$$\dot{\rho}_i = \sum_j (2h_{ij} /\hbar) (\rho_i \rho_j)^{1/2} \sin\beta_{ij} \qquad 5$$

$$\dot{S}_i = - \sum_j h_{ij} (\rho_i / \rho_i)^{1/2} \cos\beta_{ij} \qquad 6$$

where $\hbar \beta_{ij} = \alpha_{ij} + S_j - S_i , \beta_{ij} = -\beta_{ji} , \beta_{ii} = 0.$

Note that 5 is a continuity equation and implies the conservation of the total probability so that $\sum_i \rho_i (t) = 1 \quad \forall t$

For a suitable time-reversal invariant choice of the stochastic action, the variational principle provides the following expression for the generator A of the process associated to each solution ψ of the equation 1

$$A = ((a_{ij})) , \quad a_{ij} = (h_{ij}/\hbar)(\rho_j/\rho_i) (1 + \sin\beta_{ij}) \qquad 7$$

Of course, the occupation density $\rho_i(t)$ is given $\forall t$ by $|\psi_i(t)|^2$. Since it is actually difficult to obtain an explicit expressions of the process generated by the matrix A should be interesting to give a Girsanov formula in order to get a rather explicit expression for the semigroup exp tA.

3. SEMIMARTINGALE REPRESENTATION OF A MARKOV CHAIN: GIRSANOV FORMULA.

It is a fact that, altough point-process systems and Wiener driven systems are qualitatively different , they present striking similarities when the martingale point of view is adopted (10,11). It seems therefore natural, looking for the Girsanov and Feynmann-Kac formulas for a Markov chain, to try to express it in terms of point processes, taking into account that the Poisson process plays, in this context, a particular role, likewise in the continuous case where the driving process is usually a Wiener one.
Let E the one point compactification of the one dimensional lattice $\mathbb{Z}$ i.e. $E= \mathbb{Z} \cup\{\infty\}$ and let X(t) be a stable and conservative Markov chain with state space E. Let us denote by $A = ((a_{ij}))$ the generator of X(t) so that for a function f(k), $k\in\mathbb{Z}$ vanishing off a finite subset of $\mathbb{Z}$

$$\lim_{t\to o} (\sum_{j\in Z} p_{ij}f(j) - f(i)) t^{-1} = (Af)(i) . \qquad 8$$

Let $D_E [O,T)$ denote the right continuous nondecreasing E-valued function y s.t. y(O) = O and y(t) - y(t) is O or 1, $t\in[O,T)$. We endow $D_E [O,T)$ by the Skorohod topology when a topology is needed. Let $N_{ij}(t)$ $i,j\in\mathbb{Z}$ be the indipendent Poisson processes counting each one the transition $i \to j$ of $X(t)$ and $G_t = \bigvee_{i,j} \mathcal{F}_t^{N_{i,j}}$ denote the internal history of these processes. Let us put $\mathcal{F}_t = G_t \vee \mathcal{F}_\infty^X$; then the $\mathcal{F}_t$-intensity of the process N_{ij} is given by

$$\lambda_{ij} = 1_{\{x(s) = i\}} a_{ij} \quad . \qquad 9$$

Let τ be the explosion time defined by

$$\tau = \inf \{t \in [0,T) : X(t) \text{ or } X(t^-) = \infty\} .$$

Then for the chain $X(t)$ the following representation holds:

$$X(t) = \begin{cases} X(0) + \sum_{\substack{i,j \in Z \\ i \neq j}} (j-i) N_{ij}(t) & t < \tau_\infty \\ 0 & t \geq \tau_\infty \end{cases} \qquad 10$$

The process given in 10 is actually a semimartingale solving a local martingale problem for the operator

$$(Af)(i) = \begin{cases} \sum_{\substack{i,j \in Z \\ i \neq j}} \lambda_{ij} [f(i+j) - f(i)] & i \neq \infty \\ 0 & i = \infty \end{cases} \qquad 11$$

(see 12, 13, 14). Let $\mathbb{P}$ denote the probability measure on the trajectories space $D_E[0,T)$ under which the processes $N_{ij}(t)$ have intensity λ_{ij} and let $\tilde{\mathbb{P}}$ be the measure on the same space under which the processes $N_{ij}(t)$ have intensity

$$\tilde{\lambda}_{ij} = 1_{\{x(s^-) = i\}} \tilde{a}_{ij} \quad . \qquad 12$$

For each $t \in [0,T)$ let $\mathbb{P}_t$ and $\tilde{\mathbb{P}}_t$ be the restriction of the measure $\mathbb{P}$ and $\tilde{\mathbb{P}}$ respectively on $(D_E[0,T), \mathcal{F}_t)$. Let us suppose λ_{ij} $\mathcal{F}_t$-predictable and consider the process L_t, $t \in [0,T)$ given by

$$L_t = \prod_{i \neq j} L_t^{ij} \qquad 13$$

where

$$L_t^{ij} = \prod_{0 \leq s \leq t} ((\lambda_{ij}/\tilde{\lambda}_{ij})^{\Delta N_{ij}(s)}) \exp\{\int_0^t (\tilde{\lambda}_{ij} - \lambda_{ij}) ds\} \quad . \qquad 14$$

It is possible to show (14) that if $E[L_t] = 1$, then L_t is a $(\mathbb{P}, \mathcal{F}_t)$ martingale and $d\mathbb{P}/d\tilde{\mathbb{P}} = L_t$. The proof given in (14) is a generalization of the proof given in (10) to a denumerable number of counting processes. Let us rewrite equation 13 in the form

$$L_t = \exp\{\int_0^t \sum_{\substack{i,j \\ i \neq j}} \log(\lambda_{ij}/\tilde{\lambda}_{ij}) dN_{ij}(s) + \sum_{i \neq j} \int_0^t (\tilde{\lambda}_{ij} - \lambda_{ij}) ds\} . \qquad 15$$

Now, taking into account the expressions 9 and 12, equation 15 becomes

$$L_t = \exp\{\int_0^t \sum_{\substack{i,j \\ i \neq j}} \log(a_{ij}/\tilde{a}_{ij}) dN_{ij}(s) + \int_0^t (-\tilde{a}_{ii} + a_{ii}) ds\} , \qquad 16$$

where in the second integral on the r.h.s. the terms $\tilde{a}_{ii}$ and a_{ii} have to be read as evaluated for i s.t. $X(s^-) = i$.

A good utilization of 16 depends strictly from the choice of the parameters $\tilde{\lambda}_{ij}$. In other words, a good choice of $\tilde{\lambda}_{ij}$ should be such that the associated process be completely known and the stochastic integral appearing in 16 become easy to avaluate. To give an example, it is well known in the continuous case (1) that if the process associated to the ground state of the equation 1 is known, then the Girsanov formula gives information on the behaviour of the processes associated to the ex.cited states.
Let us remark that the stochastic integral in 16 disappears if there exists a function $f(x) \neq 0$, $x \in Z$ s.t.

$$a_{ij} = \tilde{a}_{ij} \, (f_j \, / \, f_i) \, . \qquad 17$$

In this case in fact we have

$$\sum_{i,j \in Z \; i \neq j} \log(a_{ij}/\tilde{a}_{ij}) \; dN_{ij}(s) = dF(s)$$

where

$$F(x) = \log f(x) \, .$$

This is actually what happens for some particular processes in Stochastic Mechanics.

4. RELATION BETWEEN GIRSANOV AND FEYNMANN-KAC FORMULAS.

If one is interested in spectral properties of the Hamiltonian H the Feynmann-Kac and the Feynmann-Kac-Ito formulas are powerfull tools (2). They provide a probabilistic expressions of the solution of the imaginary time Schrödinger equation. In the discrete case the expression for the semigroup generated by an Hamiltonian H in terms of path integrals of a finite Markov chain has been given in (5,6). More precisely in (5) is given the "analogous" of the Feynmann-Kac formula for Hamiltonian of the form $H = -(1/2)\Delta + V$, and in (6) the "analogous" of Feynmann-Kac-Ito formula for the quantum Hamiltonian for a particle in a magnetic field.
Our specific interest is to investigate the relation between Feynmann-Kac and Girsanov formulas in the discrete framework and to show their equivalence in some particular cases.
Let us recall briefly what happens in the continuous case with an example (1,2;15). Consider the following Hamiltonian

$$H = - \, (1/2)(d^2 \, / \, dx^2) + V \qquad 18$$

operating in the Hilbert space $\mathcal{H} = L^2(R,dx)$, where V is a nice real function. Let E be defined by

$$E = \inf \operatorname{spec} (H) \qquad 19$$

and Ω be the corresponding eigenfunction (which is positive).
Let us define the unitary transformation U between $L^2(R,dx)$ and $L^2(R,\Omega^2 dx)$ by

$$Uf = f\Omega^{-1} \qquad \forall\ f\varepsilon\ L^2\ (R,dx) \tag{20}$$

and consider the operator $\tilde{H}$

$$\tilde{H} = U\ (H - E)\ U^{-1} \quad .$$

The explicit form of $\tilde{H}$ is given by

$$\tilde{H} = -(1/2)(d^2/dx^2\ ) + b\nabla - E \tag{21}$$

where

$$b = \nabla \log \Omega \quad . \tag{22}$$

It is easy now to conclude that $\tilde{H}$ is the generator of a diffusion process with drift given by 22, so that using the Girsanov formula it is possible to express the process generated by $\tilde{H}$ in terms of a Wiener process. We have in fact for $f\varepsilon\ \mathcal{D}(\exp\text{-}t\tilde{H})$

$$(\exp(-t\tilde{H})f)(x) = E_x[f(x(t))\exp\{\int_0^t b(x(s))dx(s) - (1/2)\ b^2(x(s))ds\}] \tag{23}$$

where the expectation is taken w.r. to a Wiener measure.
Now since the following equality holds

$$b(x(s))\ dx(s)\ -(1/2)\ b^2(x(s))ds = d(\log\Omega)\ -(V-E)ds \tag{24}$$

we have

$$(\exp(-tH)f)(x) = (\exp(tE))E_x\{[\Omega(x(t))/\Omega(x(0))]f(x(t))\ \exp - \int_0^t V(x(s))ds\ \} \tag{25}$$

Finally performing back the unitary transformation U we obtain

$$(\exp(-tH)f)(x) = E_x\{\ f(x(t))\ \exp - \int_0^t V(x(s))ds\ \} \tag{26}$$

which is the Feynmann-Kac formula for the operator H.
Let us now look at the discrete space and consider a discrete Hamiltonian of the form

$$H = -B + V \tag{27}$$

where $B = ((b_{ij}))$ is the generator of a stable and conservative Markov chain and V is a lower bounded function on $\mathbb{Z}$ (in particular the operator B can be the discrete Laplacian: this is one of the most interesting cases). Let X(t) denote the process associated via Stochastic Mechanics to the ground state of 27. Since exp(-tH) is positivity preserving the lowest eigenvalue E of the operator H is simple and the corresponding eigenfunction ψ is positive (17). It follows that taking into account the expressions 3, 4, 7, we have for the matrix elements of H and for ψ

$$h_{ij} = b_{ij} , \quad \alpha_{ij} = \pi , \quad s_i = o \Rightarrow \quad \psi_i = (\rho_i)^{1/2} \qquad 28$$

then for the generator $A = ((a_{ij}))$ of $X(t)$ we have

$$a_{ij} = b_{ij} \, (\rho_j/\rho_i)^{1/2} \, . \qquad 29$$

The following theorem shows that also in the discrete case the process associated to the imaginary time Schrödinger equation is equavilent to the one associated to the ground state by Stochastic Mechanics.

THEOREM

Let $\mathcal{H}$ be the space $\rho_2(\mathbb{Z})$ and let us indicate with $\mathcal{H}_\rho$ the space $\rho_2(\mathbb{Z})$ with the norm given by

$$||f|| = \sum_{i\in Z} |f_i|^2 \, \rho_i \quad . \qquad 30$$

Consider the unitary transformation U

$$U : \quad \mathcal{H} \to \mathcal{H}_\rho$$
$$f \to Uf = f \, \rho^{-1/2} \quad . \qquad 31$$

Then the generator A of the process X(t) is given by

$$A = \; - \rho^{-1/2} \; (H - E) \; \rho^{1/2}. \qquad 32$$

Proof:

The result follows easily from the explicit calculation if the 6, expressed in the form $-E = \sum_{i,j \atop i\neq j} h_{ij}(\rho_j/\rho_i) \cos \beta_{ij}$, is taken into account.

Remark:

Consider a Hamiltonian H satisfying the following condition: $\alpha_{ij} = \varphi_j - \varphi_i \; \forall \; i,j\in\mathbb{Z}$ for some function (gradient condition) and assume that the corresponding semigroup exp(-tH) exists. Let us suppose that a solution ψ of the Schrödinger equation exists s.t. $\rho_i \neq 0 \quad \forall \; i,$ $\hbar \, \beta_{ij} = \alpha_{ij} - S_j + S_i = \pi$; denote by p the corresponding eigenvalue. Then the theorem extends also to this case. More precisely, the process associated with ψ is generated by $A = - \psi^{-1} \; (H - p)\psi$.

Therefore our results cover for example the case of the Pauli Hamiltonian for 1/2 spin for the ground state, as already remarked in (16). Now if we want to express the process X(t) in terms of the process y(t) generated by B, we can use the representation 10 and apply the Girsanov formula 16. Therefore we have

$$L_t = \exp \{ \int_o^t \sum_{i,j \atop i\neq j} \log \frac{a_{ij}}{b_{ij}} \, dN_{ij}(s) + \int_o^t (a_{ii} - b_{ii}) ds \} \qquad 33$$

Then it follows from formula 29

$$L_t = (\rho(y(t))/\rho(y(0)))^{1/2} \exp \int_0^t (a_{ii} - b_{ii})ds \qquad 34$$

Using the equations 5, and 6 for the ground state we have

$$- E = \sum_{j\ j\neq i} b_{ij}(\rho_j/\rho_i)^{1/2} + b_{ii} - V \qquad 35$$

which implies

$$a_{ii} - b_{ii} = E - V$$

where

$$H\ \rho^{1/2} = E\ \rho^{1/2}$$

Then we can write the equation 34 in the form

$$L_t = (\ \rho(y(t))/\ \rho(y(0)))^{1/2} \exp \int_0^t (E - V)ds \qquad 36$$

so for the semigroup we have finally

$$(\exp\ (tA)\ f)(y) = E_y \{f(y(t))\ L_t\} \ .$$

Using now the theorem, if we look for the expression of the semigroup exp(-tH) it is enough to perform back the transformation 32 to obtain

$$(\exp(-tH)\ f)(y) = E_y \{\exp - \int_0^t V(y(s))\ ds\} \ . \qquad 37$$

The expectations are taken w.r. the measure associated to the process y. 37 is the Feynmann-Kac formula for the Hamiltonian H.
To conclude in the discrete case every thing goes in perfect analogy with the continuous one.
Finally let us remark that the equivalence between the Feynmann-Kac and Girsanov formulas induced by the transformation which uses as mapping U the multiplication by the ground state function, works just for Hamiltonians of the form considered in the theorem. This equivalence seems in fact to fail for other Hamiltonians as in the continuous case it does not work for a scalar particle in an electromagnetic field.
In the reference (14) the results exposed in this paper are used to investigate the relation between the processes associated in Stochastic Mechanics to coordinates and momenta respectively for the quantistic pendulum system.

REFERENCES.

(1) F.Guerra. "Structural aspects of Stochastic Mechanics and Stochastic Field Theory". Phys. Rep. 77 , 1981 .

(2) B. Simon. "Functional integration and quantum physics". Academic Press, 1979

(3) E. Nelson. "Quantum fluctuation". Princeton, N.J., Princeton University Press, 1985

(4) F. Guerra, R. Marra. "Discrete stochastic variational principle and quantum mechanics". Phys. Rev., D 29 (3) 1984

(5) M. Kac. "Integration in function spaces and some of its applications". Scuola Normale Superiore di Pisa. Lezioni Fermiane, 1980

(6) G. F. De Angelis, G. Jona Lasinio, M. Sirugue. "Probabilistic solution of Pauli type equation". J. Phys., A: Math. Gen. 16, 1983

(7) F. Guerra, P. Ruggiero. Phys. Rev. Lett. 31, 1022, 1973

(8) S. Albeverio, R. Köegh-Kronn, L. Streit. J. Math. Phys. 18, 907, 1974

(9) F. Guerra, L. Morato. Phys. Rev., D 27, 1774, 1983

(10) P. Bremaud. "Point processes and queues: martingale dynamics". Springer, 1981

(11) C. Dellacherie, P. A. Meyer. "Probabilities and potential". North Holland Math. Studies, vol. 29 (1978), vol. 72 (1982)

(12) T. Kurtz. "Representation and approximation of counting processes" Lec. Not. Control Inf. Sci. 42, 1982

(13) S. N. Ethier, T. Kurtz. "Markov processes: characterization and convergence" to appear

(14) G. Del Grosso, R. Marra. to appear

(15) G. Del Grosso, A. Gerardi, F. Marchetti. "Lateral diffusion and eigenvalue estimates". Bull. Math. Biol. 45 (4) 1983

(16) G. F. De Angelis, G. Jona Lasinio. J. Phys., A: Math. Gen. 15, 2053, 1982

(17) M. C. Reed, B. Simon. "Methods of modern mathematical physics, vol. IV, Analysis of operator. Academic Press, 1978

EXISTENCE OF OPTIMAL MARKOVIAN CONTROLS FOR DEGENERATE DIFFUSIONS

U. G. Haussmann
Mathematics Department
University of British Columbia
Vancouver, Canada

1. Introduction

We consider the question of existence of an optimal control for the problem

$$\inf\{Ef(X_\rho): u \in \mathcal{U}\} \tag{1.1}$$

where X_t is a weak solution of

$$dX_t = b(X_t,u_t)dt + \sigma(X_t,u_t)dw_t, \qquad X_0 = x, \tag{1.2}$$

and ρ is the first exit time of X from an open set D. Here $\mathcal{U}$ denotes the set of all U-valued (U is a compact, metrizable set) progressively measurable processes u on some space $(\Omega,\mathcal{F},\{\mathcal{F}_t\},P)$ for which (1.2) has a solution for some standard Brownian motion w, all of which may vary with u. Under some regularity and convexity conditions we shall show that an optimal control exists, which is moreover Markovian. Since σ may be degenerate, then this problem includes the case where f, b, and σ depend on time as well as the case where the payoff also contains an integral term. The proof is based on a compactness argument for measures, Krylov's Markovian selection theorem and a measurable selection theorem. S.R.S. Varadhan first suggested the underlying idea, we are providing the technical arguments here.

As far as the literature is concerned beyond some early results proving existence of optimal relaxed controls, (Becker and Mandrekar, 1969), or of optimal adapted controls when b is linear in u (Fleming and Nisio, 1966), the work of Beneš (1971), Davis (1973), and Kushner (1975) must be cited. Kushner used the method of Tonelli based on compactness which requires some convexity, whereas Davis based his proof on the Hamilton-Jacobi theory and Girsanov's theorem, thus requiring σ to be independent of u and to be non-degenerate. The result of Beneš is weaker in that it requires both sets of hypotheses. These results only establish the existence of an optimal adapted (to X_t) control - it need not be Markovian. Yamada (1973) under hypotheses similar to Beneš' proved the existence of an optimal Markovian control using compactness arguments. Subsequently Fleming and Rishel (1975), Bismut (1976) and El Karoui (1981) extended the method of Hamilton-Jacobi to establish the existence of an optimal Markovian control under Davis' hypotheses. Krylov (1980) and Lions (1983) were able to weaken this condition to

$$\sup_{u\in U} \lambda^*\sigma(x,u)\ \sigma(x,u)^*\lambda > 0 \qquad \forall\ D, \quad \forall\lambda \neq 0$$

with * denoting transpose. In this article we extend the other approach, i.e. that of Tonelli, to the Markovian case, thus requiring convexity but no non-degeneracy. El Karoui (to appear) has independently derived a similar result using the same methods.

We state the problem and the result in the next section and devote section three to the proof.

I would like to thank Professor P.L. Lions for some discussions concerning this problem.

2. Formulation and Result.

We are given a compact metrisable set U, an open set D in $\mathbf{R}^n$ with closure $\bar{D}$, bounded continuous functions

$$\sigma\colon \bar{D} \times U \to \mathbf{R}^d \otimes \mathbf{R}^d,$$

$$b\colon \bar{D} \times U \to \mathbf{R}^d,$$

and a measurable function

$$f\colon \bar{D} \to \mathbf{R} \cup \{\infty\}$$

bounded below. For x in D let $\mathcal{U}_{sx}$ be the set of elements

$$\alpha = (\Omega^\alpha,\ F^\alpha,\ \{F_t^\alpha\},\ P^\alpha,\ \{w_t^\alpha\}_{t\geq s},\ \{X_t^\alpha\}_{t\geq s},\ \{u_t^\alpha\}_{t\geq s})$$

where $(\Omega^\alpha, F^\alpha, \{F_t^\alpha\}, P^\alpha)$ is an arbitrary filtered probability space, $\{w_t^\alpha\}$ is a standard Brownian motion on this space, $\{u_t^\alpha\}$ is a progressively measurable U-valued process on this space, and $\{X_t^\alpha\}$ is a right continuous, almost surely continuous, $\mathbf{R}^n$-valued, progressively measurable process which satisfies (1.2) for $t \geq s$ as well as $X_s^\alpha = x$. Let E^α denote expectation with respect to P^α and let ρ be the first exit time of $\{X_t^\alpha\}$ from D. The problem is

$$\min\{E^\alpha f(X_\rho^\alpha):\quad \alpha \in \mathcal{U}_{0x}\}, \tag{2.1}$$

We mention that it can always be arranged that σ be a square matrix as assumed above.

We require some convexity of b, σ. Specifically let $a(x,u) = \sigma(x,u)\sigma(x,u)^*$ and set

$$\tilde{K}(x) = \begin{cases} \{(a(x,u),\ b(x,u)):\ u \in U\} & \text{if } x \in D \\ \{0\} & \text{otherwise.} \end{cases}$$

Now the requirement is

$$\tilde{K}(x) \text{ is convex for each } x. \tag{2.2}$$

Finally we also require some continuity of f. If ω is an element of $C(0,\infty;\mathbf{R}^d)$, the space of continuous functions $[0,\infty) \to \mathbf{R}^d$ under uniform convergence on compact subsets, then $\rho(\omega)$ is well defined, possibly $+\infty$. We require:

$$\omega \to f(\omega_{\rho(\omega)}) \text{ is lower semicontinuous a.s.} \tag{2.3}$$

Here the a.s. refers to the law of X^α for any and <u>all</u> α. Also (2.3) requires that $f(\omega_\infty)$ be well defined.

Let us mention three cases in which (2.3) is satisfied. In each case we assume $X_t^1 = t$, i.e. $b^1(x,u) = 1$, so $\omega_t^1 = t$ almost surely for <u>all</u> α.

i) If $D = (-1,T)\times \mathbf{R}^{d-1}$, then $\rho(\omega) = T$ and (2.3) holds provided f is lower semicontinous.

ii) If $D = \mathbf{R} \times G$ with G open, and if $f(x) = x^1$, then $f(\omega_\rho) = \rho_G$, the first exit time from G, and so is lower semicontinuous. Note that there is no objection to $\rho_G(\omega) = +\infty$.

iii) If $\omega \to \rho(\omega)$ is continuous, e.g. if

$$n^*(x)a(x,u)n(x) + |b(x,u)\cdot n(x)| \neq 0 \qquad \forall u \in U, \ \forall x \in \partial D$$

where ∂D is the boundary of D and n(x) is a normal to D at x, and if $f(x) = g(x)\exp(\lambda x_1)$ with $\lambda < 0$, g bounded lower semicontinuous, then (2.3) is satisfied. Observe that $f(\omega_\infty) = 0$.

The following theorem, whose proof follows in the next section, states that there exists an optimal control for (2.1) and that it can be taken to be Markov.

Theorem 2.1 Assume (2.2), (2.3). Then there exists $\hat{\alpha}$ in U_{0x} and a Borel measurable, U-valued function $\hat{u}$ such that $\hat{\alpha}$ solves (2.1) and

$$u^{\hat{\alpha}}(t,\omega) = \hat{u}(X_t^{\hat{\alpha}}(\omega)) \qquad \text{a.s.}$$

Let us transform the problem somewhat in preparation for the proof. We work with the canonical space

$$\Omega = C(0,\infty;\mathbf{R}^d), \qquad F_t = M_t = \sigma\{\omega_\theta : \theta \le t\}.$$

Define also $M_t^s = \sigma\{\omega_\theta : s \le \theta \le t\}$. For ω in Ω define $K(t,\omega) = \tilde{K}(\omega_t)$. For $s \ge 0$ let c be a $\{M_t^s\}$ progressively measurable selector of the multifunction K on $[s,\infty)$ i.e. of $(t,\omega) \to K(t,\omega)$, $t \ge s$, so we can write $c(t,\omega) = (a(t,\omega), b(t,\omega))$ with $a(t,\omega)$ in $\mathbf{S}^d$, the symmetric d×d matrices, and $b(t,\omega)$ in $\mathbf{R}^d$. Let ∂_i denote differentiation with respect to x_i, let a^{ij}, b^i denote the components of a, b, use the convention of summing over repeated indices, and let

$$L_{t\omega}^c = \frac{1}{2} a^{ij}(t,\omega)\partial_i\partial_j + b^i(t,\omega)\partial_i .$$

We denote by $P_{sx}(c)$ the set of all solutions of the martingale problem corresponding to c starting at (s,x), i.e. if $\phi \in C_0^\infty(\mathbf{R}^d)$, the set of infinitely differentiable functions on $\mathbf{R}^d$ of compact support, and if

$$M_t^{cs}\phi(\omega) = \phi(\omega_t) - \int_s^t L_{\theta\omega}^c \phi(\omega_\theta)d\theta, \ t \ge s,$$

then $P \in P_{sx}(c)$ if and only if

$$P\{\omega_t = x,\ t \le s\} = 1$$

and $\{M_t^{cs}\phi\}$ is a (P, M_t) martingale for $t \ge s$. Now define $A(s,x) = \bigcup_{c \in S_s} P_{sx}(c)$, where S_s is the set of all $\{M_t^s\}$ progressively measurable selectors of K on $[s,\infty)$.

The method of proof of Bismut (1976), Proposition IV-3, and Theorem 4.5.2 of Stroock and Varadhan (1979) (which reference we abbreviate henceforth to S.V.(1979)) imply that $P \in A(s,x)$ if and only if $P \in P_{sx}(c)$ for some c of the form.

$$c(t,\omega) = \begin{cases} (a(\omega_t, u^\alpha(t,\omega)),\ b(\omega_t, u^\alpha(t,\omega))) & \text{if } t < \rho(\omega) \\ 0 & \text{otherwise} \end{cases}$$

a.s. for some α in U_{sx}. If we write E^P for expectation with respect to P, it now follows that the problem (2.1) can be expressed concisely as

$$\min\{E^P(f(X_\rho)):\ P \in A(0,x)\} \tag{2.4}$$

where $X_t(\omega) = \omega_t$. It is for this problem that we shall establish the existence of an optimal Markovian control.

3. The Proof

The proof consists of three steps. First we show that for any (s,x), the problem

$$\min\{E^P\ f(X_\rho):\quad P \in A(s,x)\}$$

has an optimal solution. But if $P_n \to P$ weakly and if g is bounded and lower semicontinuous, then $\liminf \int g dP_n \ge \int g dP$, so that (2.3) and the monotone convergence theorem imply that $\liminf E^{P_n} f(X_\rho) \ge E^P f(X_\rho)$. This means that

$$P \to E^P\ f(X_\rho)$$

is lower semicontinuous, so all we need to show is that $A(s,x)$ is non void and compact. Next, if $A'(s,x)$ denotes the set of all solutions of the above problem, we use Krylov's Markovian selection theorem as presented in S.V.(1979), §12.2, to show that we can choose $\hat{P}_{sx} \in A'(s,x)$ such that $\{\hat{P}_{sx}\}$ is a Markov process. In the final step we show that this process corresponds to a Markov control $\hat{u}(x)$. We consider $A(s,x)$ as a subset of the metric space of probability measures on Ω; the metric is given by weak convergence.

Proposition 3.1 $A(s,x) \neq \emptyset$ and is compact.

Proof: For u in U, let $\bar{c}(t,\omega) = (a(\omega_t,u), b(\omega_t,u))$. Then $P_{sx}(\bar{c}) \neq \emptyset$ by the continuity of b,σ, and if we set $c(t,\omega) = \bar{c}(t,\omega)\, 1_{\{t<\rho(\omega)\}}$ then c is in S and $P_{sx}(c) \neq \emptyset$ since $P \in P_{sx}(\bar{c})$ implies that $P \otimes_{\rho(\omega)} \delta_{\omega_{\rho(\omega)}} \in P_{sx}(c)$. Hence $A(s,x) \neq \emptyset$. Note that $\otimes$ is defined in S.V.(1979), §6.1.

For ϕ in $C_0^\infty(\mathbf{R}^d)$ let

$$K_\phi = \sup\{|L^c_{t\omega}\phi(x)| : t \geq s,\ \omega \in \Omega,\ c \in S_s,\ x \in \mathbf{R}^d\} < \infty.$$

Now if $\phi \geq 0$, if y is in $\mathbf{R}^d$ and if P is in $A(s,x)$, then $\phi(\omega_t+y) + K_\phi t$ is a (P,M_t) non-negative submartingale, and it follows from S.V.(1979), Theorem 1.4.6, that $A(s,x)$ is precompact.

It remains to show that it is closed. Suppose $P^n \to P$ and $P^n \in P_{sx}(c^n)$. Since the multifunction $\tilde{K}$ is bounded then there exists a closed, bounded convex set Γ in $L^\infty([s,\infty); S^d \times \mathbf{R}^d)$ such that for each n $c^n(\cdot,\omega)$ is in Γ a.s. But Γ under the induced weak * topology is compact and metrizable. Let $\tilde{P}^n$ be the law of $(\omega, c^n(\cdot,\omega))$ on $\tilde{\Omega} = \Omega \times \Gamma$ induced by P^n. Since $P^n \to P$ and Γ is compact then $\{\tilde{P}^n\}$ is tight, i.e. there exists a convergent subsequence, again denoted $\{\tilde{P}^n\}$, with a limit $\tilde{P}$. Then $\tilde{P} = P$ on M, i.e. $\tilde{P}(A\times\Gamma) = P(A)$ for A in M.

We wish to show that there exists c in S_s such that $\{M^{cs}_t\phi\}$ is a (P, M_t) martingale for $t \geq s$. We begin by defining a process on $\Omega \times \Gamma$ by

$$M_t\phi(\omega,\gamma) = \phi(\omega_t) - \int_s^t L^\gamma_\theta\, \phi(\omega_\theta)\, d\theta$$

where $L^\gamma_\theta = \frac{1}{2} a^{ij}(\theta)\, \partial_i\partial_j + b^i(\theta)\partial_i$ if $\gamma(\cdot) = (a(\cdot),b(\cdot))$. Now we complete the proof as a sequence of lemmata.

Lemma 3.1 $\{M_t\phi\}$ is a $(\tilde{P}, \tilde{M}_t)$ martingale for $t \geq s$.

Here $\tilde{M}_t = M_t \otimes G_t$ where $\{G_t\}$ is the canonical filtration on Γ, i.e. is generated by

$$\{\{\gamma: \int_s^t \langle \ell(\tau),\gamma(\tau)\rangle\, d\tau \le k\}: k\in \mathbf{R},\ \ell \in L^1([0,t]; \mathbf{S}^d\times\mathbf{R}^d)\}$$

with $\langle(a,b),(\tilde{a},\tilde{b})\rangle = \text{trace}(a\tilde{a}^*) + b\cdot\tilde{b}$. Thus $M_t\phi$ is $\{\tilde{M}_t\}$ adapted.

Proof: Since $\gamma(\cdot) = c^n(\cdot,\omega)$ a.s. $(\tilde{P}^n)$, then $\{M_t\phi\}$ is a $(\tilde{P}^n, \tilde{M}_t)$ martingale. If $\Phi: \tilde{\Omega} \to R$ is bounded, continuous and $\tilde{M}_t$ measurable, then for $t \ge s$

$$0 = E^{\tilde{P}^n}\{\Phi\cdot(M_{t+h}\phi - M_t\phi)\} \to E^{\tilde{P}}\{\Phi\cdot(M_{t+h}\phi - M_t\phi)\}$$

by weak convergence since $(\omega,\gamma) \to M_t\phi(\omega,\gamma)$ is continuous. Hence the lemma follows.

Lemma 3.2. There is a $\{M_t^s\}$ progressively measurable function $c: [s,\infty)\times\Omega \to \mathbf{S}^d\times \mathbf{R}^d$ such that $P \in P_{sx}(c)$.

Proof: Since $(t,\omega,\gamma) \to \int_s^t \gamma_\theta\, d\theta$ is $\{\tilde{M}_t\}$ adapted and continuous hence progressive, there exists $\tilde{c}(t,\omega,\gamma) = (\tilde{a},\tilde{b})$, progressively measurable, such that $\tilde{c}(t,\omega,\gamma) = \gamma_t$ a.e.(t). Now we can replace L_θ^γ by $L_{\theta\omega\gamma}^{\tilde{c}}$ in the definition of $M_t\phi$. From S.V.(1979), Theorem 4.5.2, it follows that ω_t is a $(\tilde{P}, \tilde{M}_t)$ semi-martingale (the probability space may have to be enlarged), and by a result of Wong (1971), Theorem 4.3, it can be written as

$$\omega_t = x + \int_s^t \bar{b}_\theta\, d\theta + \int_s^t \bar{\sigma}\, d\bar{w}_\theta, \quad t \ge s \qquad (3.1)$$

where $\bar{b}$, $\bar{\sigma}$, $\bar{w}$ are $\{M_t^s\}$-adapted processes, the last being a standard Brownian motion. Note that we write M_t^s for $M_t^s \times$\{trivial σ-algebra\}, i.e. for the σ-algebra induced by M_t^s on the larger space. Moreover $\tilde{a} = \bar{\sigma}\,\bar{\sigma}^*$ is $\{M_t^s\}$-adapted and $\bar{b}_t = E^{\tilde{P}}\{\tilde{b}_t | M_t^s\}$. Since $\tilde{P} = P$ on M, we can take $\tilde{a}_t$ and $\bar{b}_t$ as defined on (Ω, M, P). If $\bar{c} = (\tilde{a},\bar{b})$ then it follows that $M_t^{\bar{c}s}\phi$ is a (P, M_t^s) martingale by (3.1). Now $\bar{c}$ is measurable and $\{M_t^s\}$ adapted, hence has a progressively measurable version. This is the desired c.

Lemma 3.3. $P \in A(s,x)$.

Proof: We have just established that $P \in P_{sx}(c)$ with c $\{M_t^s\}$ progressively measurable. It remains to show that $c(t,\omega) \in K(t,\omega)$.

Let us return to $\Omega \times \Gamma$. Then $c^n(\cdot,\omega) = \gamma$ a.s.$(\tilde{P}^n)$ and $\tilde{c}(\cdot,\omega,\gamma) = \gamma$ a.s.$(\tilde{P})$. Let Λ be a countable dense subset of

$\mathbf{S}^d \times \mathbf{R}^d$ and fix λ in Λ, $t \geq s$, and $h > 0$. Then

$$h^{-1} \int_t^{t+h} \langle \lambda, \gamma_\theta \rangle \, d\theta = h^{-1} \int_t^{t+h} \langle \lambda, c_\theta^n \rangle \, d\theta \quad \text{a.s.}(\tilde{P}^n)$$
$$\leq \sup_{t \leq \theta \leq t+h} \; \sup_{\gamma \in \tilde{K}(\omega_\theta)} \langle \lambda, \gamma \rangle. \tag{3.2}$$

But $\gamma \to h^{-1} \int_t^{t+h} \langle \lambda, \gamma_\theta \rangle d\theta$ is bounded, continuous, as is

$$\omega \to \sup_{t \leq \theta \leq t+h} \; \sup_{\gamma \in \tilde{K}(\omega_\theta)} \langle \lambda, \gamma \rangle$$

since $\omega \to \bigcup_{t \leq \theta \leq t+h} \tilde{K}(\omega_\theta)$ is continuous. By weak convergence (3.2) holds a.s.$(\tilde{P})$. Now let $h \to 0$ to obtain

$$\langle \lambda, \gamma_t \rangle \leq \sup_{\gamma \in \tilde{K}(\omega_t)} \langle \lambda, \gamma \rangle \quad \text{a.s. } dt \times d\tilde{P}$$

and the null set can be made independent of λ since Λ is countable. It follows that γ_t is in $\tilde{K}(\omega_t)$ a.s. $dt \times d\tilde{P}$ since $\tilde{K}(\omega_t)$ is convex, compact. But $\gamma_t = \tilde{c}(t,\omega)$ a.e.(t), and $\bar{c}_t = E\{\tilde{c}_t | M_t^s\}$ so again by convexity of $\tilde{K}(\omega_t)$, $\bar{c}(t,\omega)$ is in $K(t,\omega)$ a.s. $dt \times dP$ as is c. Since $K(t,\omega)$ is $\{M_t^s\}$ progressive as is c then the exceptional null set is also and hence c can be modified to be $\{M_t^s\}$ progressively measurable and to be a selector of K. There is one caveat in the above proof: $\omega \to \cup \tilde{K}(\omega_\theta)$ is only continuous if $\rho(\omega) \notin [t,t+h]$. We circumvent this difficulty by extending b, σ to $\mathbf{R}^d \times U$ (bounded and continuous) and defining

$$\tilde{K}_0(x) = \{(a(x,u), b(x,u)): \; u \in U\}.$$

Now for any c in S there exists c_0, a progressively measurable selector of $K_0(t,\omega) = \tilde{K}_0(\omega_t)$, such that $c(t,\omega) = c_0(t,\omega) \, 1_{\{t<\rho(\omega)\}}$. The above proof is valid for c^n replaced by c_0^n. The resulting c_0 gives rise to a c in S by stopping at ρ. This completes the proof of the Lemma and the Proposition.

We begin the second step of the proof with a lemma.

Lemma 3.4.

a) $(s,x) \to A(s,x)$ is measurable,

b) $P \in A(0,x)$ if and only if $P\circ\Phi_s^{-1} \in A(s,x)$ for all $s \geq 0$ where $(\Phi_s\omega)(t) = \omega((t-s)\vee 0)$,

c) if $P \in A(0,x)$, if τ is a finite stopping time such that $\tau \leq \rho$, if $\{P_\omega\}$ is a regular conditional probability distribution (i.e. r.c.p.d.) of $P|M_\tau$, then there exists a P-null set $N \in M_\tau$ such that

$$\bar{P}_\omega \equiv \delta_{\omega_{\tau(\omega)}} \otimes_{\tau(\omega)} P_\omega \in A(\tau(\omega), \omega_{\tau(\omega)}) \text{ for } \omega \notin N,$$

d) if $P \in A(0,x)$, if τ is a finite stopping time, $\tau \leq \rho$, and if $\{Q_\omega\}$ is a family of probability measures such that $\omega \to Q_\omega$ is M_τ measurable and $\bar{Q}_\omega \equiv \delta_{\omega_{\tau(\omega)}} \otimes_{\tau(\omega)} Q_\omega \in A(\tau(\omega),\omega_{\tau(\omega)})$ then $P \otimes_\tau Q_\cdot \in A(0,x)$.

Remark. We have used here the notation $P\otimes_t Q$ of S.V.(1979), Theorem 6.1.2. Condition (c) says that $P \to P|M_\tau$ is an injection of $A(0,x)$ into $A(\tau,\omega_\tau)$, whereas (d) says that $A(0,x)$ is closed under strong concatenation, i.e. concatenation at a stopping time. In control terms this means that if u is a control on $[0,\infty)$ then it is also a control on $[\tau,\infty)$ (given the past), and if v is a control on $[\tau,\infty)$, then u concatenated with v (at τ) is a control on $[0,\infty)$.

Proof: To establish (a) we need to show, according to S.V.(1979), Lemma 12.1.8, that if $P^n \in A(s^n,x^n)$, $(s^n,x^n) \to (s,x)$, then there exists $P \in A(s,x)$ such that for a subsequence $P^{n_k} \to P$. As in Proposition 3.1 $\{P^n\}$ is precompact hence has a convergent subsequence with limit P. It can be shown as in Proposition 3.1 that P is in $A(s,x)$.

To prove (b) we define Ψ_s by $(\Psi_s\omega)_t = \omega_{t+s}$ and $\tilde{c}$ by $\tilde{c}(t,\omega) = c(t-s, \Psi_s\omega)$ for $t \geq s$ so that $c(t,\omega) = \tilde{c}(t+s, \Phi_s\omega)$. It can now be shown that $M_t^{\tilde{c}s}\phi(\omega) = M_{t-s}^{c0}\phi(\Psi_s\omega)$ so that $M_t^{\tilde{c}s}\phi$ under $P\circ\Phi_s^{-1}$ has the same distribution as $M_t^{c0}\phi$ under P, i.e. $P \in P_{0x}(c)$ if and only if $P\Phi_s^{-1} \in P_{sx}(\tilde{c})$. The conclusion of (b) follows readily.

Turning to part (c), we fix $\bar{\omega} \in \Omega$, $s > 0$ and define

$$(\bar{\omega}\otimes_s\omega)_t = \begin{cases} \bar{\omega}_t & \text{if } t < s \\ \omega_t & \text{if } t \geq s. \end{cases}$$

Let $P \in P_{0x}(c)$ and let $P_{\bar{\omega}}$ be a r.c.p.d of $P|M_s$. Then $\bar{\omega}\otimes_s\omega \in \Omega$ for $\omega \notin N_0$, a $P_{\bar{\omega}}$-null set. Define

$$\tilde{c}(t,\omega) = c(t,\bar{\omega}\otimes_s\omega).$$

Now S.V.(1979), Theorem 6.1.3 implies that for $\omega \notin N \in M_s$, a P-null set, $M_t^{cs}\phi$ is a $(P_{\bar{\omega}}, M_t)$ martingale, $t \geq s$, and hence so is $\tilde{M}_t^{cs}\phi$ since $c = \tilde{c}$ $P_{\bar{\omega}}$ a.s. for $t \geq s$. It follows as in S.V.(1979), Theorem 6.2.1, that $\delta_{\bar{\omega}_s} \otimes_s P_{\bar{\omega}} \in P_{s,\bar{\omega}_s}(\tilde{c})$, and now by Theorem 6.1.3, that the same inclusion holds with s replaced by $\tau(\bar{\omega})$. Clearly $\tilde{c}$ is a selector of K, and $\tilde{c}$ is $\{M_t^{\tau(\bar{\omega})}\}$-progressive, hence (c) follows.

Finally for part (d), given $P \in P_{0x}(c) \subset A(0,x)$, $\bar{Q}_\omega \in P_{\tau(\omega)\omega_{\tau(\omega)}}(c_\omega)$ for $\omega \notin N$, define

$$\tilde{c}(t,\omega) = \begin{cases} c(t,\omega) & \text{if } t < \tau(\omega) \\ c_\omega(t,\omega) & \text{if } t \geq \tau(\omega). \end{cases}$$

Since $M_{t\wedge\tau}^{\tilde{c}0}\phi = M_{t\wedge\tau}^{c0}\phi$ is a (P, M_t) martingale, and since $M_t^{\tilde{c}0}\phi - M_{t\wedge\tau(\omega)}^{\tilde{c}0}\phi = M_t^{c_\omega\tau(\omega)}\phi - \phi(\omega_{\tau(\omega)})$ is a $(\bar{Q}_\omega, M_t)$ martingale after time $\tau(\omega)$ (trivially it is one up to time τ), then by S.V.(1979), Theorem 1.2.10, $P \otimes_{\tau(\cdot)} Q_\cdot \in P_{0x}(\tilde{c})$. Clearly $\tilde{c}(t,\omega) \in K(t,\omega)$ so we need only show that $\tilde{c}$ is progressive, i.e. $1_{\{t\geq\tau(\omega)\}}(t,\omega)\, c_\omega(t,\omega)$ is, or that $\tilde{c}$ has a progressive version. Writing $c_\omega = (a_\omega, b_\omega)$ we have for $h > 0$ and for N_ω a Lebesgue null set

$$h^{-1}E^{P\otimes_\tau Q}\{1_{t\geq\tau}(\omega_{t+h} - \omega_t) | M_t\}$$

$$= 1_{t\geq\tau}\, E^{Q_\omega}\{h^{-1}\int_t^{t+h} b_\omega(\theta,\omega)d\theta | M_t\}$$

$$\to 1_{t\geq\tau}\, E^{Q_\omega}\{b_\omega(t,\cdot) | M_t\}, \quad t \notin N_\omega$$

$$= 1_{t\geq\tau}\, b_\omega(t,\omega).$$

Since the left side is measurable and adapted and since the convergence set of a sequence of measurable functions is measurable, then the right side has a measurable, adapted version, i.e. a progressive version. Similarly

$$h^{-1} E^{P\otimes_\tau Q}\{1_{t\geq\tau}[\omega_{t+h}\omega^*_{t+h} - \omega_t\omega^*_t - \int_t^{t+h}(b_\omega(\theta,\omega)\omega^*_\theta + \omega_\theta b_\omega(\theta,\omega)^*)d\theta]|M_t\}$$

$$= 1_{t\geq\tau} E^{Q_\omega}\{h^{-1} \int_t^{t+h} a_\omega(\theta,\omega)d\theta|M_t\},$$

and hence the result follows. This completes the proof of the lemma.

We now follow more or less the procedure of S.V.(1979), §12.2. For $s \geq 0$, $x \in D$ let us introduce the notation $X_t(\omega) = \omega_t$ and define the value function

$$V(s,x) = \inf_{P\in A(s,x)} E^P\{f(X_\rho)\}$$

and the set of optimal measures

$$A'(s,x) = \{P \in A(s,x): E^P\{f(X_\rho)\} = V(s,x)\}$$

Note that if $V(0,x) = +\infty$ then every control is optimal and the theorem is trivial, so we assume that $V(0,x) < \infty$.

<u>Lemma 3.5</u>. For all $s \geq 0$, x in D

i) $V(s,x) = V(0,x)$,

ii) $A'(s,x)$ is non-empty, compact

iii) $\{A'(s,x)\}$ satisfies (a)-(d) of Lemma 3.4.

Proof:

To establish (i) observe that for $P \in A(0,x)$,

$$E^P f(X_\rho) = E^{P\circ\Phi_s^{-1}} f(X_{\rho(\Psi_s\omega)}(\Psi_s\omega))$$
$$= E^{P\circ\Phi_s^{-1}} f(X_\rho)$$

since $X_t(\Psi_s\omega) = X_{t+s}(\omega)$ and

$$\rho(\Psi_s\omega) + s = \rho(\omega) \qquad \text{a.s. } P\circ\Phi_s^{-1}. \tag{3.3}$$

Now (b) of the previous lemma implies that $V(0,x) \geq V(s,x)$.

Conversely if $P \in A(s,x)$ then (b) implies that $P\circ\Psi_s^{-1} \in A(0,x)$ and since for $t < s$, $X_t = x$ a.s. P, then

$$E^P f(X_\rho) = E^{P\circ\Psi_s^{-1}} f(X_{\rho(\Phi_s\omega)}(\Phi_s\omega))$$

$$= E^{P\circ\Psi_s^{-1}} f(X_\rho)$$

$$\geq V(0,x)$$

because $X_t(\Phi_s\omega) = X_{(t-s)\vee 0}(\omega)$ and

$$[\rho(\Phi_s\omega)-s]\vee 0 = \rho(\omega). \tag{3.4}$$

This implies that $V(s,x) \geq V(0,x)$ and hence (i) follows.

Since $A(s,x) \neq \phi$, compact, then $A'(s,x) \neq \phi$ as remarked at the beginning of the section and is compact.

It remains to establish (iii). For A' (a) follows from (a) for A and from S.V.(1979), Lemma 12.1.7. Similarly (b) follows from (b) for A and from (i), (3.3), (3.4).

Turning to (c), we are given $P \in A'(0,x)$, a stopping time $\tau \leq \rho$, and P_ω a r.c.p.d. of $P|M_\tau$. Let

$$N = \{\omega: \bar{P}_\omega \notin A(\tau(\omega), \omega_{\tau(\omega)})\}$$

$$A = \{\omega \notin N: \bar{P}_\omega \notin A'(\tau(\omega), \omega_{\tau(\omega)})\}.$$

Since A satisfies (c), then $N \in M_\tau$, $P(N) = 0$. Moreover $N \cup A = \{\omega: \bar{P}_\omega \notin A'(\tau(\omega), \omega_{\tau(\omega)})\} \in M_\tau$ by S.V.(1979), Lemma 12.1.9. Hence $A \in M_\tau$ and it remains to show that $P(A) = 0$. Since A' satisfies (a), then according to S.V.(1979), Theorem 12.1.10 there exists a measurable selector $R(s,x)$ of $A'(s,x)$. Now $\omega \to R_\omega = \delta_\omega \otimes_{\tau(\omega)} R(\tau(\omega), \omega_{\tau(\omega)})$ is M_τ measurable and if

$$Q_\omega = \begin{cases} R_\omega & \text{if } \omega \in N \cup A \\ P_\omega & \text{otherwise.} \end{cases}$$

then $Q \equiv P \otimes_\tau Q_\cdot \in A(0,x)$ since A satisfies (d). Since $\tau \leq \rho$, and since $P_\omega = \bar{P}_\omega$ on M_∞^τ then

$$V(0,x) \leq E^{Q}\{f(X_\rho)\}$$

$$= E^P E^{Q_\omega}\{f(X_\rho)\}$$

$$= E^P (1-1_{N\cup A})E^{P_\omega}\{f(X_\rho)\}$$

$$+ E^P 1_{N\cup A} E^{R_\omega}\{f(X_\rho)\}$$

$$= V(0,x) + E^P 1_A[E^{R_\omega}\{f(X_\rho)\} - E^{P_\omega}\{f(X_\rho)\}]$$

$$= V(0,x) + E^P 1_A\left[V(\tau(\omega), \omega_{\tau(\omega)}) - E^{\bar{P}_\omega}\{f(X_\rho)\}\right].$$

By the definition of A it follows that $1_A = 0$ a.s. P.

Finally for (d), let $P \in A'(0,x)$, and let $\omega \to Q_\omega$ be an M_τ measurable map such that for all ω

$$\bar{Q}_\omega \in A'\left(\tau(\omega), \omega_{\tau(\omega)}\right).$$

Set $Q = P \otimes_\tau Q_\cdot$. Then $Q \in A(0,x)$ since A satisfies (d). Moreover $\tau \leq \rho$ so that $(X_\rho, Q_\omega) \sim (X_\rho, \bar{Q}_\omega)$ and hence

$$V(0,x) \leq E^{Q}\{f(X_\rho)\}$$

$$= E^P E^{Q_\omega}\{f(X_\rho)\}$$

$$= E^P V\left(\tau(\omega), \omega_{\tau(\omega)}\right)$$

$$\leq E^P E^{P_\omega}\{f(X_\rho)\}$$

$$= E^P\{f(X_\rho)\}$$

$$= V(0,x).$$

Hence $Q \in A'(0,x)$. Note that the above is equivalent to the principle of optimality in dynamic programming.

<u>Proposition 3.2</u> There is a measurable map $(s,x) \to P_{sx} \in A(s,x)$ such

that $\{P_{sx}\}$ is a strong Markov process on D and P_{sx} solves

$$\inf\{E^P f(X_\rho): P \in A(s,x)\}.$$

Proof: This is identical to S.V.(1979), Theorem 12.2.3 except that we take $\mathcal{C}_0(s,x) = A(s,x)$ and $u_0(s,x) = V(s,x)$, and we restrict x to lie in D. In fact one defines inductively $\{\mathcal{C}_n(s,x)\}_{n=1}^{\infty}$ and shows that for each n the multifunction $\mathcal{C}_n$ satisfies (a) - (d). Note that $\mathcal{C}_1(s,x) = A'(s,x)$. Then one shows that $\mathcal{C}_\infty(s,x) = \bigcap_n \mathcal{C}_n(s,x)$ contains a unique element, P_{sx}, and of course satisfies (a) - (d). The result follows.

The final step in the proof is to show that the Markov process $\{P_{sx}\}$ is generated by a Markov control $\hat{u}$. Since $P_{0x} \in \mathcal{P}_{0x}(c)$ for some $c(t,\omega) = (a(t,\omega), b(t,\omega))$ then there exists a Brownian motion (possibly on some enlarged probability space) such that $X_t(\omega) \equiv \omega_t$ (possibly $X_t(\omega,\omega') \equiv \omega_t$) satisfies

$$dX_t = b(t,X)dt + a^{1/2}(t,X)\, dw_t, \tag{3.5}$$

and since the process is Markov, then

$$c(t,X) = (a(t,X),\ b(t,X)) = \tilde{c}(t,X_t) \quad \text{a.s., a.e.}$$

for some Borel measurable function $\tilde{c}$. Indeed note that a.s. dtdP

$$b(t,\omega) = \lim_{h\downarrow 0} h^{-1}\ E^P\{\omega_{t+h} - \omega_t \mid M_t\}$$

$$= \lim_{h\downarrow 0} h^{-1}\ E^{P_{t\omega_t}}\{\omega_{t+h} - \omega_t\}$$

by the Markov property. Moreover if we set $Q_{ty} = P_{ty} \circ \Psi_t^{-1}$ with $(\Psi_t\omega)_\cdot = \omega_{t+\cdot}$, then $P_{ty} = Q_{ty} \circ \Phi_t^{-1}$ so by property (b) satisfied by $\mathcal{C}_\infty(s,y) = \{P_{sy}\}$ we conclude $Q_{ty} = P_{0y}$. Hence

$$E^{P_{ty}}\{\omega_{t+h} - \omega_t\} = E^{P_{0y}}\{\omega_h - \omega_0\},$$

and so $b(t,\omega)$ is independent of the initial condition $X_0 = x$. The function a is treated similarly (using $\omega\ \omega^*$) to conclude that

$$\tilde{c}(t,X_t) = \tilde{c}(X_t)$$

is independent of the initial condition, and that $\tilde{c}(\omega_t) \in \tilde{K}(\omega_t)$ a.s. dtdP. Let R be the set of full measure on which $c(t,\omega) = \tilde{c}(\omega_t)$, and let $R_0 = \{\omega_t: (t,\omega) \in R\}$. Then

$$R_0 \subset \{x: \tilde{c}(x) \in \tilde{K}(x)\} \equiv R_1$$

and the latter set is a Borel set. By redefining $\tilde{c}$ off R_1 we can obtain a Borel measurable selection $\hat{c}$ of $\tilde{K}$ such that $\hat{c} = \tilde{c}$ on R_1, i.e.

$$c(t,\omega) = \hat{c}(\omega_t) \qquad \text{a.s. dtdP.}$$

By Lemma 5 of Beneš (1971) we conclude the existence of $\hat{u}$, Borel measurable, U-valued such that $\hat{c}(x) = (a(x,\hat{u}(x)), b(x,\hat{u}(x)))$. This completes the proof.

Remarks. Since $\tilde{c}$ is independent of $X_0 = x$, then $\hat{u}$ is also optimal for the problem where X_0 is given an arbitrary initial distribution.

We add that the most bothersome hypothesis is that of convexity of $\tilde{K}(x)$. Without it, we can convexify in the usual manner to obtain an optimal "control" which is randomized, i.e. u(x) is a probability distribution on U (and can be taken to consist of at most m point masses with $m = 1 + (d^2+3d)/2$). One can now apply the theory of strong extremals, cf. Haussmann (1985), Corollary 6.5, under the requisite added hypotheses, to dispense with the convexity of $\tilde{K}(x)$, as was done by Bismut, (1976) chapter IV, §5.

REFERENCES

Becker, H. and Mandrekar, V. (1969) On the existence of optimal random controls, J. Math. Mech. 18, 1151-1166.

Beneš, V.E. (1971), Existence of optimal stochastic controls, SIAM J. Control 9, 446-472.

Bismut, J.M. (1976), Théorie probabiliste du controle des diffusions, Memoir, Amer. Math. Soc., No. 176.

Davis, M.H.A. (1973), On the existence of optimal policies in stochastic control, SIAM J. Control, 11, 587-594.

El Karoui, N. (1981), Les aspects probabiliste du contrôle stochastique, Lecture Notes in Mathematics 876, 74-239.

Fleming, W.H. and Nisio, M. (1966), On the existence of optimal stochastic controls, J. Math. Mech. 15, 777-794.

Fleming, W.H. and Rishel, R.W. (1975), Deterministic and Stochastic Optimal Control, Springer Verlag, New York.

Haussmann, U.G. (1985), A stochastic maximum principle for optimal control of diffusions, preprint.

Krylov, N.V. (1980), Controlled Diffusion Processes, Springer Verlag, New York.

Kushner, H.J. (1975), Existence results for optimal stochastic controls, J.O.T.A. 15, 347-359.

Lions, P.L. (1981), On the Hamilton-Jacobi-Bellman equations, Acta Applicandae Math. 1, 17-41.

Stroock, D. and Varadhan, S.R.S. (1979), Multidimensional Diffusion Processes, Springer Verlag, New York.

Wong, E. (1971), Representation of martingales, quadratic variations and applications, SIAM J. Control 9, 621-633.

Yamada, K. (1973), Continuity of cost functionals in diffusion processes and its application to an existence theorem of optimal controls, Proceedings of I.E.E.E. Conference on Decision and Control, San Diego, 1973.

ON LEVY'S AREA PROCESS *

K. Helmes

Institute of Applied Mathematics
University of Bonn
D-5300 Bonn, FRG

We prove a law of the iterated logarithm for the Euclidean norm of a particular vector process in $\mathbb{R}^3$ and give formulae for its characteristic and conditional characteristic functions. The conditional characteristic function yields an explicit expression for the propagator of the Schrödinger operator with constant magnetic field.

1. INTRODUCTION

In this paper I would like to present a result concerning the asymptotic behaviour of the sample paths of a particular stochastic process in $\mathbb{R}^3$ together with formulae for its characteristic and conditional characteristic functions. The term asymptotic refers here to the time dependence of the trajectories at zero and at infinity.

The process (α_t) in which we are interested, cf. also [10], is derived from 3-dimensional Brownian motion $\beta_t = (b_1(t), b_2(t), b_3(t))$ by summing the vector products of β_s with its increments over some interval $[0,t]$, i.e.

$$\alpha_t := \int_0^t \beta_s \times d\beta_s \;; \qquad (1.1)$$

(α_t) could be described more abstractly as being the stochastic line integral of a differential form along (β_t), cf. [6]. This vector process

* This work was supported by the Deutsche Forschungsgemeinschaft (DFG), Sonderforschungsbereich 72 (SFB 72), at the University of Bonn, Bonn, West Germany.

is closely related to Lévy's area process - cf. e.g. [5] and the literature cited therein - in so far as we can write α_t as

$$\alpha_t = (L_{2,3}(t), L_{3,1}(t), L_{1,2}(t)), \tag{1.2}$$

where $L_{i,j}(t)$, $1 \leq i,j \leq 3$, denotes the area process associated with the 2-dimensional Wiener process $(b_i(t), b_j(t))$ and is defined by

$$L_{i,j}(t) := \int_0^t b_i(s)db_j(s) - b_j(s)db_i(s). \tag{1.3}$$

In the following we will encounter expressions of the form $\langle \lambda, \alpha_t \rangle$, $\lambda \in \mathbb{R}^3$, where $\langle \cdot, \cdot \rangle$ denotes the scalar product in $\mathbb{R}^3$. Such linear forms of α_t can be represented as a "generalized" area process as introduced in [5]; viz. for $\lambda \in \mathbb{R}^3$, put

$$A: = A(\lambda) = \begin{pmatrix} 0 & -\lambda_3 & \lambda_2 \\ \lambda_3 & 0 & -\lambda_1 \\ -\lambda_2 & \lambda_1 & 0 \end{pmatrix}; \tag{1.4}$$

then,

$$\langle \lambda, \alpha_t \rangle = \int_0^t \langle A\beta_s, d\beta_s \rangle = : L_t^A .$$

To motivate our investigation of the process (α_t) we shall show in Section 2 how the formula for the conditional characteristic function of α_t, given β_t, yields an explicit expression for Green's function $k_t(x,y)$ of the Schrödinger operator with constant magnetic field and zero potential; for different ways to derive this kernel see [1,p. 860]; cf. also [2], [3] and [11]. The explicit formula for the kernel $k_t(x,y)$ allows us, in particular, to check the validity of the property

$$\lim_{t\to\infty} \frac{1}{t} \log \left(\sup_{x,y\in\mathbb{R}^3} |k_t(x,y)| \right) < 0,$$

which was shown by Malliavin to hold for a large class of processes including (1.1); for general results of this kind see [7] and [9]. In Section 3 we shall prove the law of the iterated logarithm for $|\alpha_t|$, (where $|\cdot|$ is the Euclidean norm in $\mathbb{R}^3$), at zero and at infinity.

2. SCHRÖDINGER OPERATOR WITH CONSTANT MAGNETIC FIELD

We shall see in this section that a motivation for studying the process (α_t) comes from physics. To be specific, consider the quantum mechanical energy operator for a particle in a magnetic field $\vec{B}(x)$ with vector potential $\vec{a}(x)$ $(\vec{B} = \vec{\nabla} \times \vec{a})$,

$$H(\vec{a}, V): = \frac{1}{2}(-i\vec{\nabla} - \vec{a})^2 + V,$$

where V describes, for instance, the potential due to an electric field. By the Feynman - Kac formula the action of the semi-group $exp\{-tH(\vec{a},V)\}$ is given by

$$(f,e^{-tH(\vec{a},V)}g) = \int_\Omega d\mu(w)e^{F(w,t)}\overline{f(w(o))}g(w(t)) ,$$

$f,g\in L^2(\mathbb{R}^3)\cap L^\infty(\mathbb{R}^3)$, $w(t): = x+\beta(t)$, $x\in\mathbb{R}^3$, where

$$F(w,t): = -i\int_o^t \langle\vec{a}(w(s)),dw(s)\rangle -\frac{i}{2}\int_o^t div\vec{a}(w(s))ds - \int_o^t V(w(s))ds \tag{2.1}$$

and "$d\mu$" denotes the product measure of 3-dimensional Lebesgue and 3-dimensional Wiener measure on $\Omega = \mathbb{R}^3\times C([o,\infty),\mathbb{R}^3)$, see e.g. [11, pp.159]. Thus, if

$$\vec{a}(x) = \frac{1}{2}\begin{pmatrix}-Bx_2\\ Bx_1\\ o\end{pmatrix},$$

i.e. if there is a constant field in "z-direction" ($\vec{B} = \vec{\nabla}\times\vec{a} = (0,0,B)$), then

$$div\vec{a} \equiv 0 .$$

If, moreover, we also assume $V \equiv 0$ then the expression (2.1) reduces to just one integral and the propagator $k_t(x,y)$ of $exp\{-tH(\vec{a},0)\}$ is thus given by

$$k_t(x,y) = \left(\frac{1}{\sqrt{2\pi t}}\right)^3 exp\{-\frac{1}{2t}|x-y|^2\} exp\{-i\langle\vec{a}(x),y\rangle\} \cdot \cdot E[exp\{\frac{iB}{2}L_{1,2}(t)\}|\beta_t=y-x], \tag{2.2}$$

where E denotes expectation with respect to Wiener measure. Therefore, by the formula for the conditional characteristic function of $L_{1,2}(t)$ given $(b_1(t),b_2(t))$ - note that $L_{1,2}$ is independent of b_3 -, already derived by Lévy in [8], i.e.

$$E[exp\{\frac{i\lambda}{2}L_{1,2}(t)\}|(b_1(t),b_2(t))=(\xi_1,\xi_2)] = \frac{t\lambda}{2sinh(t\lambda/2)} exp\{\frac{|\xi|^2}{2t}(1-\frac{t\lambda}{2}coth(\lambda t/2))\},$$

we obtain

$$k_t(x,y) = \frac{B}{4\pi sinh(Bt/2)}\left(\frac{1}{2\pi t}\right)^{1/2} exp\{-\frac{1}{2t}(x_3-y_3)^2-\frac{B}{4}coth(Bt/2) \cdot \cdot [(x_1-y_1)^2+(x_2-y_2)^2] -\frac{iB}{2}(x_1y_2-x_2y_1)\} . \tag{2.3}$$

Now let us consider the case of an arbitrary constant magnetic field

$$\vec{B} = (\lambda_1, \lambda_2, \lambda_3) .$$

The constant field $\vec{B}$ can be described by the vector potential

$$\vec{a}(x) = \frac{1}{2}(\vec{B} \times x) . \qquad (2.4)$$

In the same manner as for the special case considered above we get the following expression for the kernel in the general case, i.e. $\vec{a}$ as in (2.4)

$$k_t(x,y) = \left(\frac{1}{\sqrt{2\pi t}}\right)^3 exp\{-\frac{1}{2t}|y-x|^2\} exp\{-\frac{i}{2}<\vec{B}\times x, y>\} \cdot$$

$$\cdot E[exp\{\frac{i}{2}(\lambda_1 L_{2,3}(t)+\lambda_2 L_{3,1}(t)+\lambda_3 L_{1,2}(t))\} | \beta_t = y-x]$$

$$= \left(\frac{1}{\sqrt{2\pi t}}\right)^3 exp\{-\frac{1}{2t}|y-x|^2\} exp\{-i<\vec{a}(x), y>\} \cdot$$

$$\cdot E[exp\{\frac{i}{2}<\lambda, \alpha_t>\} | \beta_t = y-x] . \qquad (2.5)$$

This expression motivates our first result.

THEOREM 2.1. For $\lambda, \gamma \in \mathbb{R}^3$, $t>0$,

$$E[exp\{i<\lambda,\alpha_t>+i<\gamma,\beta_t>\}]$$

$$= \frac{1}{cosh(|\lambda|t)} exp\{-\frac{1}{2}[(O^*\gamma)_1^2+(O^*\gamma)_2^2]\frac{th(|\lambda|t)}{|\lambda|} - \frac{1}{2}(O^*\gamma)_3^2 \, t\}, \qquad (2.6)$$

where

$$O^* = \begin{pmatrix} -\frac{\lambda_1\lambda_3}{|\lambda|\sqrt{\lambda_1^2+\lambda_2^2}} & -\frac{\lambda_2\lambda_3}{|\lambda|\sqrt{\lambda_1^2+\lambda_2^2}} & \frac{\lambda_1^2+\lambda_2^2}{|\lambda|\sqrt{\lambda_1^2+\lambda_2^2}} \\ \frac{\lambda_2}{\sqrt{\lambda_1^2+\lambda_2^2}} & -\frac{\lambda_1}{\sqrt{\lambda_1^2+\lambda_2^2}} & 0 \\ \frac{\lambda_1}{|\lambda|} & \frac{\lambda_2}{|\lambda|} & \frac{\lambda_3}{|\lambda|} \end{pmatrix} . \qquad (2.7)$$

Proof. Formula (2.6) is a special case of a more general formula for the joint characteristic function of a d-dimensional, $d \geq 2$, Brownian motion and an associated "generalized" area process which was derived in [5, Corollary 2]; note that for O we should read O^* and that for an odd dimension d the term $exp\{-t(O^*\gamma)_d^2/2\}$ is missing there. Since $<\lambda,\alpha_t> = L_t^A$, A given by (1.4), and since O as in (2.7) reduces A to its normal form, i.e.

$$O^*AO = \begin{pmatrix} 0 & -|\lambda| & 0 \\ |\lambda| & 0 & 0 \\ 0 & 0 & 0 \end{pmatrix},$$

(2.6) follows from the formula given in [5].

□

If we let $\gamma = 0$ in (2.6) we get the characteristic function of α_t.

Corollary 2.2.

$$E[\exp\{i<\lambda,\alpha_t>\}] = \frac{1}{\cosh(|\lambda|t)} .$$

Corollary 2.3. The distribution of the 3-dimensional random variable α_t has the density

$$f_t(x) = \frac{1}{8t^2}\frac{1}{|x|} \tanh(\tfrac{\pi}{2t}|x|)/\cosh(\tfrac{\pi}{2t}|x|) . \qquad (2.8)$$

Proof. By Corollary 2.2, taking the inverse Fourier transform and using spherical coordinates we have

$$f_t(x) = \frac{1}{2\pi^2}\int_0^\infty \frac{\sin(r|x|)}{r|x|}\frac{r^2}{\cosh(rt)}dr .$$

Formula (2.8) now follows by formula 4.111(3) in Gradshteyn, I.S. and Ryzhik, I.M.: Table of Integrals, Series and Products (1980).

□

Corollary 2.4.

$$E[\exp\{i<\lambda,\alpha_t>\}|\beta_t=\xi] = (2\pi t)^{3/2}\exp\{|\xi|^2/2t\}\frac{1}{\sqrt{2\pi t}}\exp\{-\frac{1}{2t}<\frac{\lambda}{|\lambda|},\xi>^2\} \cdot$$

$$\cdot \frac{|\lambda|}{2\pi\sinh(|\lambda|t)} \exp\{-\frac{1}{2}|\lambda\times\xi|^2\frac{\coth(|\lambda|t)}{|\lambda|}\} \qquad (2.9)$$

Proof. Let $\varphi(\xi)$ denote the left hand side of (2.9). Since

$$E[\exp\{i<\lambda,\alpha_t>+i<\gamma,\beta_t>\}] = \left(\frac{1}{\sqrt{2\pi t}}\right)^3 \int_{\mathbb{R}^3} d\xi\ \exp\{-i<\gamma,\xi>\}\varphi(\xi) \cdot$$

$$\cdot \exp\{-|\xi|^2/2t\}$$

we can take the inverse Fourier transform on both sides of the preceding equation and get by Theorem 2.1

$$\varphi(\xi) = (2\pi t)^{3/2}\exp\{|\xi|^2/2t\}\frac{1}{\sqrt{2\pi t}}\exp\{-\frac{1}{2t}(O^*\xi)_3^2\}\frac{|\lambda|}{2\pi\sinh(|\lambda|t)} .$$

$$\cdot \ exp\{-\frac{1}{2}[(O^*\xi)_1^2 + (O^*\xi)_2^2]|\lambda| coth(|\lambda| t)\} \quad .$$

Since

$$|\lambda\times\xi|^2 = |A\xi|^2 = |OO^*AOO^*\xi|^2 = |O^*AOO^*\xi|^2$$

the assertion now follows by the definition of O^* (see (2.7)).

□

If we insert the right hand side of (2.9) into the expression (2.5) for $k_t(x,y)$, we get the explicit formula for the Green function of the Schrödinger operator with constant magnetic field and zero potential:

$$k_t(x,y) = \frac{1}{\sqrt{2\pi t}} \frac{|\vec{B}|}{4\pi sinh(|\vec{B}|t/2)} exp\{-G_t(x,y)\} , \tag{2.10}$$

where

$$G_t(x,y) = \frac{1}{2t} \langle\frac{\vec{B}}{|\vec{B}|}, x-y\rangle^2 - \frac{i}{2}\langle\vec{B}, x\times y\rangle + \frac{|\vec{B}|}{4} coth(|\vec{B}|t/2) \left|\frac{\vec{B}}{|\vec{B}|}\times(x-y)\right|^2 .$$

3. LAW OF THE ITERATED LOGARITHM

The precise asymptotic behaviour of the Euclidean norm of (α_t) is given by the following result.

THEOREM 3.1.

$$\limsup_{t\to 0} \frac{|\alpha_t|}{\frac{2}{\pi} t \log\log(1/t)} = 1 \quad \text{a.e.} , \tag{3.1}$$

$$\limsup_{t\to\infty} \frac{|\alpha_t|}{\frac{2}{\pi} t \log\log(t)} = 1 \quad \text{a.e.} . \tag{3.2}$$

Proof. We shall prove (3.1) only, since the proof for (3.2) runs along the same lines. If (3.1) is false, then there is a $\delta > 0$ such that

$$\overline{\lim_{t\to 0}} \frac{|\alpha_t|}{\frac{2}{\pi} t \log\log(1/t)} \geq 1 + \delta$$

with positive probability. Given any $\varepsilon > 0$, cover the unit sphere in $\mathbb{R}^3$ by a finite number of spherical regions D_i with opening ε with respect to the origin. Then, for some i,

$$P[\overline{\lim_{t\to 0}} \frac{|\alpha_t|}{\frac{2}{\pi} t \log\log(1/t)} \geq 1 + \delta , \frac{\alpha_t}{|\alpha_t|} \in D_i] > 0 .$$

But if λ is the centre of D_i and $(1 + \delta)cos(\varepsilon) > 1$ then

$$P[\overline{\lim_{t\to o}} \frac{\langle\lambda,\alpha_t\rangle}{\frac{2}{\pi}t\log\log(1/t)} \geq (1+\delta)\cos(\varepsilon) > 1] > 0. \tag{3.3}$$

This is, however, impossible for the following reason:
Firstly (see Introduction)

$$\langle\lambda,\alpha_t\rangle = L_t^A .$$

Now, by the law of the iterated logarithm for "generalized" area processes, see [4, Theorem 4.1],

$$\overline{\lim_{t\to o}} \frac{L_t^A}{\frac{2}{\pi}\sigma t\log\log(1/t)} = 1 \quad \text{a.e.} , \tag{3.4}$$

where $\sigma = max\{|a_k| \mid a_k \text{ eigenvalue of } A\}$.
But for $A(\lambda)$ - note $\lambda \in D_i$ -

$$\sigma(A) = |\lambda| = 1 ,$$

so (3.4) contradicts (3.3).

□

REFERENCES

[1] AVRON, J., HERBST, I. and SIMON, B., Schrödinger operators with magnetic fields. I. General interactions, *Duke Mathematical Journal* 45 (1978), 847-883.

[2] FEYNMAN, R. P. and HIBBS, Q., *Quantum mechanics and path integrals*, McGraw-Hill, New York, 1965.

[3] HABA, Z., Behaviour in strong fields of Euclidean gauge theories, II, *Physical Review D*, 29(1984), 1718-1743.

[4] HELMES, K., The local law of the iterated logarithm for processes related to Lévy's stochastic area process, *Studia Mathematica* 84,1 (1985 or 1986).

[5] HELMES, K. and SCHWANE, A., Lévy's stochastic area formula in higher dimensions, *J. Fct. Analysis* 54 (1983), 177-192.

[6] IKEDA, N. and MANABE, S., Integral of differential forms along the path of diffusion processes, *Publ. RIMS, Kyoto Univ.* 15(1979), 827 - 852.

[7] IKEDA, N., SHIGEKAWA, I. and TANIGUCHI, S., The Malliavin calculus and long time asymptotics of certain Wiener integrals, to appear in Proc. of the Conf. on *"Linear Analysis and Function Spaces"* (Canberra, 1984), Centre for Mathem. Analysis (ANU).

[8] LEVY, P. Wiener's random functions, and other Laplacian random functions, *Proc. 2nd Berkeley Symposium on Mathem. Statistics and Prob.*, University of California Press, Berkeley, Ca., vol. 2 (1951).

[9] MALLIAVIN, P., Sur certains intégrals stochastiques oscillantes, *C. R. Acad. Sci. Paris*, *295(1982), 295 - 300.*

[10] PRICE, G.C., ROGERS, L.C.G. and WILLIAMS, D., 'BM($\mathbb{R}^3$) and its area integral $\int \beta \wedge d\beta$ ', *Lecture Notes in Mathematics, vol. 1095(1984), 155 - 165.*

[11] SIMON, B., *Functional Integration and Quantum Physics*, Academic Press, New York, 1979.

CENTRAL LIMIT THEOREMS AND RANDOM CURRENTS

by Nobuyuki Ikeda[1)] and Yoko Ochi

Department of Mathematics, Osaka University
Toyonaka 560, Osaka, Japan

1. Introduction

The study of asymptotic behavior of stochastic processes is a central problem in probability theory and its applications. Central limit theorems for additive functionals of diffusion processes and dynamical systems under various circumstances are among the most well-known examples, (cf. [2], [4], [8], [20], [21], [22], [23] and [24]). In several cases, by using the framework given by Itô [12], we can reformulate these as limit theorems for current valued stochastic processes. To do this, we need to show that a trajectory of continuous semi-martingale on a manifold defines a random current. This is a slight generalization of the main idea of Ochi [18]. In the proof of this fact, the same idea as in de Rham [6], (Chapter III, §8, Example 1) and Stratonovich's integral play an important role. In this article, we will show that in case of diffusion processes on a compact manifold with boundary, various central limit theorems can be discussed in a unified way by using the above framework. In this process, we will also give a brief survey on related results to our formulation which have been obtained during last years.

The organization of the paper is as follows. In Section 2, we prepare several notions and fundamental facts which will be needed latter. Next, in Section 3, we state our main results and give a sketch of the proof. Some results of this section have been announced in Ochi [19]. Section 4 will be devoted to some typical examples of the theorem stated in Section 3. Finally in Section 5, we will give the proof of Lemma 2.1.

2. Preliminaris

Let M be a d-dimensional, connected, compact manifold with smooth boundary ∂M and D_1 be the space of all smooth differential 1-forms on M endowed with the Schwartz topology, ([6], [9] and [12]). Here the boundary ∂M may be empty. We now consider a family of standard Sobolev seminorms $\{\|\ \|_p\}_{p\in Z_+}$ on D_1 which are similar to those in [18]. For details of $\|\ \|_p$, see § 3. This family satisfies the following properties: $\|\ \|_p \prec \|\ \|_q$ if $p < q$ and $\|\ \|_p \prec_{HS} \|\ \|_q$ if $q > p+\frac{d}{4}$. Here, for the definition of "$\prec$" and "$\prec_{HS}$", see Definition 1.1.2 of [12]. We denote

1) Part of the research of the first author was carried out while he was visiting the Centre for Mathematical Analysis at the Australian National University.

by D_{1p} the completion of $D_1/\mathrm{Ker}\,\|\ \|_p$ with respect to $\|\ \|_p$. For simplicity we also denote by the same notation $\|\ \|_p$ the induced norm on D_{1p}. It is well known that the topology determined by $\{\|\ \|_p\}_{p\in Z^+}$ consists with the Schwartz topology τ and the topological space $\{D_1,\tau\}$ is a nuclear multi-Hilbertian space. We denote by D_1' and D_{1p}' the dual spaces of D_1 and D_{1p} respectively. For details, see [9] and [12].

We now consider a continuous M-valued semi-martingale $\{x(t);\ 0 \leq t < \infty\}$ defined on a probability space $\{\Omega,F,P\}$ with a reference family $\{F_t\}$, i. e., for every $f \in C^\infty(M)$, $\{f(x(t));\ 0 \leq t < \infty\}$ is a real valued quasi-martingale in the sense of [11] where $C^\infty(M)$ is the space of all infinitely differentiable functions on M. Now, following Ikeda-Manabe [10], for every $\alpha \in D_1$ we define the stochastic line integral:

$$x_t(\alpha) = \int_{x[0,t]} \alpha$$

of α along the curve $x[0,t]$ $(= \{x(s);\ 0 \leq s \leq t\})$. Then the stochastic process $\{x_t(\alpha);\ 0 \leq t < \infty\}$ is again a quasi-martingale, ([10] and [11]). Hence, for every $\alpha \in D_1$, $\{x_t(\alpha);\ 0 \leq t < \infty\}$ is decomposed into the martingale part $m(\alpha) = \{m_t(\alpha)\}$ and the bounded variation part $a(\alpha) = \{a_t(\alpha)\}$:

$$x_t(\alpha) = m_t(\alpha) + a_t(\alpha), \qquad \text{for } t \geq 0 \qquad \text{a. s.}$$

We are ready to state our fundamental lemma.

Lemma 2.1. There are continuous D_1'-valued stochastic processes $X = \{X_t\}$, $M = \{M_t\}$ and $A = \{A_t\}$ such that

(i) for every $\alpha \in D_1$

(2.1) $\quad X_t(\alpha) = x_t(\alpha),\ M_t(\alpha) = m_t(\alpha),\ A_t(\alpha) = a_t(\alpha) \qquad$ for $t \geq 0$, a. s.,

(ii) (the decomposition of Doob and Meyer),

(2.2) $\quad X = M + A, \qquad$ a. s.,

(i. e., $X_t(\alpha) = M_t(\alpha) + A_t(\alpha)$ for $t \geq 0$, $\alpha \in D_1$, a. s.).

Furthermore, for sufficiently large p, X, M and A are continuous D_{1p}'-valued stochastic processes.

Remark 2.1. For the definition of continuous D_1'-valued stochastic process, see Itô [12].

This lemma means that a continuous M-valued semi-martingale can be regarded as a continuous D_1'-valued stochastic process in the sense of [12]. Furthermore for sufficiently large p, it is a continuous D_{1p}'-valued stochastic process. Various problems related to diffusion processes and dynamical systems can be discussed in this framework, (e.g., homogenization problems for periodic diffusion processes, asymptotic properties of the homological position of geodesic flows on compact manifolds of constant negative curvature and statistical properties of the position of particle of Lorentz gas with periodic configuralation of scatterers). (See [1], [2], [4], [7] and [8]). For example, the law of large numbers and the central limit theorem for the homological position of the geodesic $\gamma(t)$ on a compact Riemannian manifold of constant negative curvature can be reduced to the those for the line integral of harmonic forms along the geodesic γ. Hence these can be regarded as the problems related to asymptotic properties of the continuous D_1'-valued stochastic process associated with the geodesic γ. These problems has been studied by many authors, (see Gel'fand and Pyateckiĭ-Šapiro [8], Sinai [24] and Arnold and Avez [1] etc.).

Finally we should emphasize that X, M and A take values in a Hilbert space D_{1p}' for sufficiently large p. This makes our framework a more convenient tool for analytical treatment and for applications.

3. Central limit theorem

In this and next sections we will restrict ourselves in case where $\{x(t);\ 0 \leq t < \infty\}$ is a diffusion process on M. Before proceeding, following [11], we will summarize some of basic facts on the theory of diffusion processes. Let g be a smooth Riemannian metric on M and $O(M)$ be the bundle of orthonormal frames over M. We denote by π the natural projection: $O(M) \longrightarrow M$. We also consider the second order differential operator L given by

$$L = \frac{1}{2}\Delta + b$$

where Δ is the Laplace-Beltrami operator associated with the Riemannian metric g and b is a smooth vector field on M. Then, letting $\{L_1, L_2, \cdots, L_d\}$ be the system of standard horizontal vector fields on $O(M)$ with respect to the Riemannian connection and L_0 be the horizontal lift of b with respect to the Riemannian connection, we have

$$Lf = \frac{1}{2}\sum_{i=1}^{d}(L_i)^2(f\circ\pi) + L_0(f\circ\pi) \qquad \text{for every } f \in C^\infty(M),$$

(see [11]). We now consider the following stochastic differential equation in the form of the Stratonovich differentials defined on the d-dimensional Wiener space $\{W_0^d, B(W_0^d), P^W\}$:

$$(3.1)\qquad \begin{cases} dr(t) = \sum_{k=1}^{d} L_k(r(t))\circ dw^k(t) + L_0(r(t))dt + \tilde{n}(r(t))d\psi(t) \\ r(0) = r_0 \end{cases}$$

where $\psi(t)$ is a continuous non-decreasing process such that $\psi(0) = 0$ and

$$(3.2)\qquad \int_0^t I_{\partial M}(\pi(r(s)))d\psi(s) = \psi(t), \qquad t \geq 0,\ a.\ s.,$$

(for details, see [11]). Here $\tilde{n}$ is the horizontal lift with respect to the Riemannian connection of the inward unit normal vector field n on ∂M. For simplicity, throughout this section, we fix a point $r_0 \in O(M)$ and set $x_0 = \pi(r_0)$. Let $\{r(t);\ 0 \leq t < \infty\}$ be the solution of the stochastic differential equation (3.1) with (3.2) and we set

$$x(t) = \pi(r(t)), \qquad t \geq 0.$$

Then $\{x(t);\ 0 \leq t < \infty\}$ is a diffusion process on M generated by the operator L and subject to the boundary condition: $nu = 0$ on ∂M, (i. e., the reflecting boundary condition), ([11]). Furthermore it is a continuous M-valued semi-martingale and we have

$$\text{(3.3)} \qquad \begin{aligned} m_t(\alpha) &= \sum_{k=1}^{d}\int_0^t \bar{\alpha}_k(r(s))dw^k(s) \\ a_t(\alpha) &= \int_0^t (\alpha(b) - \tfrac{1}{2}\delta\alpha)(x(s))ds + \int_0^t \alpha(n)(x(s))d\psi(s) \end{aligned}, \qquad \alpha \in D_1,$$

where $\bar{\alpha} = (\bar{\alpha}_1, \bar{\alpha}_2, \cdots, \bar{\alpha}_d)$ is the scalarization of α and δ is the formal adjoint operator of the exterior differential operator d with respect to g, i. e., $\delta = (-1)^{2d+1} * d *$, where $*$ denotes the usual duality operator carrying a differential p form into one of complementary degree $d - p$, (see [5], [10] and [11]). Here $m_t(\alpha)$ and $a_t(\alpha)$ are the martingale part and the bounded variation part of the line integral $x_t(\alpha)$ of α along the curve $x[0,t]$ respectively.

Letting μ be the invariant probability measure of the diffusion process $\{x(t);\ 0 \leq t < \infty\}$, we define an element e of D_1' by

$$\text{(3.4)} \qquad e(\alpha) = \int_M (\alpha(b) - \frac{1}{2}\delta\alpha)(x)\mu(dx) + \frac{1}{2}\int_{\partial M} \alpha(n)(\xi)\tilde{\mu}(d\xi), \qquad \alpha \in D_1,$$

where $\tilde{\mu}$ is the measure on ∂M induced by μ, i. e., the marginal measure of μ on ∂M. We note that μ has a smooth density with respect to the Riemannian volume element. For fixed $\alpha \in D_1$, we consider the following equation:

$$\text{(3.5)} \qquad \begin{aligned} Lu(x) &= (\alpha(b) - \frac{1}{2}\delta\alpha)(x) - e(\alpha) && \text{on } M \\ nu(x) &= \alpha(n)(x) && \text{on } \partial M. \end{aligned}$$

Then, the equation (3.5) has a unique solution u_α up to an additive constant. For details, see [2], [13], [14] and [26].

Next, by Lemma 2.1, there is a continuous D_1'-valued stochastic process $X = \{X_t\}$ associated with $\{x(t);\ 0 \leq t < \infty\}$. We denote by $M = \{M_t\}$ and $A = \{A_t\}$ the martingale part and bounded variation part of X in the sense of (2.2) respectively. For every $\lambda > 0$, we now define continuous D_1'-valued stochastic processes $M^{(\lambda)} = \{M_t^{(\lambda)}\}$ and $X^{(\lambda)} = \{X_t^{(\lambda)}\}$ by

$$\text{(3.6)} \qquad \begin{aligned} M_t^{(\lambda)}(\alpha) &= \frac{1}{\sqrt{\lambda}} M_{\lambda t}(\alpha) \\ X_t^{(\lambda)}(\alpha) &= \frac{1}{\sqrt{\lambda}}(X_{\lambda t}(\alpha) - \lambda t e(\alpha)) \end{aligned} \qquad \text{for } t \in [0,\infty),\ \alpha \in D_1$$

respectively. We also set

$$\langle\!\langle \alpha,\beta \rangle\!\rangle = \int_M \langle\alpha,\beta\rangle(x)\mu(dx), \qquad \alpha,\ \beta \in D_1.$$

We can now state our main result.

Theorem 3.1. (a) As $\lambda \longrightarrow \infty$, the stochastic process $M^{(\lambda)}$ converges in the law sense to the D_1'-valued Wiener process η with zero mean and the covariance functional $(t \wedge s)\langle\!\langle \alpha,\beta \rangle\!\rangle$ in $C([0,\infty) \longrightarrow D_1')$.[1)]

(b) As $\lambda \longrightarrow \infty$, the stochastic process $X^{(\lambda)}$ converges in the law sense to the D_1'-valued Wiener process ξ with zero mean and the covariance functional $(t \wedge s)\langle\!\langle \alpha - du_\alpha, \beta - du_\beta \rangle\!\rangle$ in $C([0,\infty) \longrightarrow D_1')$.

Remark 3.1. (a) Following Itô [12], a continuous D_1'-valued stochastic process $\{B_t\}$ with stationary independent increments is called a D_1'-valued Wiener process, if $B_0 = 0$. It is characterized by its mean functional and covariance functional. For details, see Itô [12].

(b) If $b = 0$, the invariant probability measure μ is equal to the normalized Riemannian volume element m and so $e = 0$, (see [16]). In this case, the decomposition $\alpha = du_\alpha + (\alpha - du_\alpha)$ of $\alpha \in D_1$ gives an orthogonal decomposition discussed by Conner [5].

(c) In case of $\partial M = \phi$ and $b = 0$, Theorem 3.1, (ii) implies that a. s., for every $t \geqq 0$, $\xi_t(\alpha) = 0$ if α is exact, (also see Ochi [18]). This means that with probability 1, for every $t \geqq 0$, the current ξ is *coclosed*, i. e., $\delta\xi_t = 0$, $t \geqq 0$ a. s. For the definition of the operator δ acting on currents, see de Rham [6].

Proof of Theorem 3.1. For fixed n and $\alpha^1, \alpha^2, \cdots, \alpha^n \in D_1$, we consider the n-dimensional continuous stochastic process $M_n^{(\lambda)}$ given by

$$M_n^{(\lambda)} = \{(M_t^{(\lambda)}(\alpha^1), M_t^{(\lambda)}(\alpha^2), \cdots, M_t^{(\lambda)}(\alpha^n)); 0 \leqq t < \infty\}.$$

Now we note that by (3.3) and (3.6), we have

$$\langle M_\cdot^{(\lambda)}(\alpha), M_\cdot^{(\lambda)}(\beta)\rangle_t = \frac{1}{\lambda}\int_0^{\lambda t} \langle\alpha,\beta\rangle(x(s))ds, \text{ for } \alpha,\beta \in D_1.$$

Hence we can use the ergodic property of $\{x(t);\ 0 \leqq t < \infty\}$ to show that with probability 1, as $\lambda \longrightarrow \infty$, the quadratic variation process of $M_n^{(\lambda)}$ converges to the constant process $\{(\langle\!\langle \alpha^i, \alpha^j \rangle\!\rangle t);\ i,j = 1,2,\cdots,n,\ 0 \leqq t < \infty\}$ which is the quadratic variation process of the n-dimensional diffusion process

$$\eta_n = \{(\eta_t(\alpha^1), \eta_t(\alpha^2), \cdots, \eta_t(\alpha^n));\ 0 \leqq t < \infty\}.$$

Before proceeding, we need to give details of the system of Sobolev seminorms $\{\|\ \|_p\}_{p\in Z_+}$ on D_1 stated in Section 2. For example, we can define a system of Sobolev seminorms on D_1 as follows: First we choose a finite open covering $\{U_n\}_{n=1}^m$ of the smooth manifold M which satisfies the following properties:

1) $a \wedge b = \min\{a,b\}$

(i) For every $n = 1,2,\cdots,m$, U_n is a coordinate neighbourhood. (ii) For every $n = 1,2,\cdots,m$, U_n is homeomorphic to an open subset K_n with smooth boundary of the closed half space $\overline{R^d_+} = \{x ; x = (x^1,x^2,\cdots,x^d) \in R^d, x^d \geqq 0\}$. Let $\{\psi_n\}_{n=1}^m$ be a partition of unity subordinate to $\{U_n\}_{n=1}^m$. If $\alpha \in D_1$ is expressed in the form

$$\alpha = \sum_{i=1}^d \alpha_i^{(n)}(x)dx^i \qquad \text{on } U_n ,$$

we can regarded $\psi_n \alpha_i^{(n)}$ as a smooth function with compact support of the coordinate $(x^1,x^2,\cdots,x^d) \in K_n$. Now combining this fact and the notion of a standard Sobolev space on R^d, we define a norm $\|\alpha\|_p$ on D_1 by

$$\|\alpha\|_p^2 = \sum_{n=1}^m \sum_{i=1}^d \|\psi_n \alpha_i^{(n)}\|_{2p,K_n}^2 .$$

Here $\| \ \|_{2p,K_n}$ denotes the Sobolev norm of the Sobolev space $H_{2p}(K_n)$. For details of $H_{2p}(K_n)$, see, for example, Kumano-go [14], Chapter 6, §2. By using several familiar properties of the Sobolev spaces, we can show the properties of the norm $\| \ \|_p$ stated in Section 2.

Then, we can use the same idea as the one in Ochi [18] to show that for $p > d/4$ there is a positive constant $K_1 = K_1(p)$ satisfying

$$\text{(3.7)} \quad E^W[|M_t^{(\lambda)}(\alpha) - M_s^{(\lambda)}(\alpha)|^4] \leqq K_1\|\alpha\|_p^4|t-s|^2 \quad \text{for } t,s \in [0,\infty) \text{ and } \lambda > 0.$$

Here E^W denotes the expectation with respect to P^W. Since, as stated in Section 2, $\| \ \|_{q_1} \prec_{HS} \| \ \|_{q_2}$ if $q_2 > q_1 + d/4$, we can conclude, by combining (3.7) and the same idea as the one in Ochi [18], that for $p > d/2$ there exists a positive constant $K_2 = K_2(p)$ satisfying

$$\text{(3.8)} \quad E^W[\|M_t^{(\lambda)} - M_s^{(\lambda)}\|_p^4] \leqq K_2|t-s|^2, \text{ for } t,s \in [0,\infty) \text{ and } \lambda > 0$$

where $\| \ \|_p$ in the left hand denotes the norm of the dual space D'_{1p}. Here the constant $K_2 = K_2(p)$ can depend on p. Therefore, for every fixed positive integer n, the stochastic process $M_n^{(\lambda)}$ converges in the law sense to the diffusion process η_n in $C([0,\infty) \longrightarrow R^n)$ as $\lambda \longrightarrow \infty$. We now choose $p,q \in Z_+$ such that $q > p > d/2$. Then, since the natural inclusion mapping $\iota : D'_{1p} \longrightarrow D'_{1q}$ is a compact operator and (3.8) holds, the family of the probability laws of $M^{(\lambda)}$, $\lambda > 0$, on $C([0,\infty) \longrightarrow D'_{1q})$ is tight in the space of all probability measures on $C([0,\infty) \longrightarrow D'_{1q})$. For details, see Lemma 4.1 of [18]. Now, by using the same method as in [18], we can complete the proof of (a).

By (3.3), (3.5) and (3.6), we have

$$\text{(3.9)} \quad \begin{cases} X_t^{(\lambda)}(\alpha) - \frac{1}{\sqrt{\lambda}} X_{\lambda t}(du_\alpha) = M_t^{(\lambda)}(\alpha - du_\alpha) \\ X_t(du_\alpha) = u_\alpha(x(t)) - u_\alpha(x(0)) \end{cases}, \quad \alpha \in D_1 .$$

Hence, by using the same method as in Ochi [18] and combining the fact in Theorem 3.1, (a) and some properties of u_α with (3.9), we obtain Theorem 3.1, (b).

4. Examples

In this section, we will provide two typical examples which are closely related to Theorem 3.1.

Example 4.1 (homogenization problems). Let D_i, $i = 1,2,\cdots,m$, be open balls distributed disjointly in the d-dimensional torus T^d. Set $M = T^d \setminus \bigcup_{i=1}^{m} D_i$ and $M^* = p^{-1}(M)$, where p is the natural projection from the universal covering space R^d of T^d to T^d. We consider the diffusion process $\{x(t) ; 0 \leqq t < \infty\}$ on M given in Section 3 and denote by $X = \{X_t\}$ the continuous D_1'-valued stochastic process associated with $\{x(t) ; 0 \leqq t < \infty\}$. We set

$$Y_t = (X_t(\alpha^1), X_t(\alpha^2), \cdots, X_t(\alpha^d)), \quad t \geqq 0$$

where α^i, $i = 1,2,\cdots,d$, are the differential 1-forms given by $\alpha^i = dx^i$, $i = 1,2,\cdots,d$ and $(x^1,x^2,\cdots,x^d)$ is the standard coordinate in R^d. Then it is easy to see that $Y = \{Y_t\}$ is the d-dimensional reflecting diffusion process on M^* generated by the periodic extension of L. Now, for every $\lambda > 0$, we define $Y^{(\lambda)} = \{Y_t^{(\lambda)}\}$ by

$$Y_t^{(\lambda)} = \frac{1}{\sqrt{\lambda}}(X_{\lambda t}(\alpha^1) - \lambda t e(\alpha^1), X_{\lambda t}(\alpha^2) - \lambda t e(\alpha^2), \cdots, X_{\lambda t}(\alpha^d) - \lambda t e(\alpha^d))$$

for $t \geqq 0$, where e is the element of D_1' given by (3.4). As stated in Section 3, $Y^{(\lambda)}$ converges in the law sense to the diffusion process on R^d generated by

$$A = \frac{1}{2}\sum_{i,j=1}^{d} q^{ij}\frac{\partial^2}{\partial x^i \partial x^j}, \quad q^{ij} = \langle\!\langle \alpha^i - du_{\alpha^i}, \alpha^j - du_{\alpha^j} \rangle\!\rangle,$$

in $C([0,\infty) \longrightarrow R^d)$, as $\lambda \longrightarrow \infty$. This is a generalization of Theorem 3 of Bhattacharya [3].

Furthermore, in case when the vector fields b and n satisfy the *centering condition*, i. e., $e(\alpha^i) = 0$, $i = 1,2,\cdots,d$, this for instance was discussed by Bensoussan, Lions and Papanicolaou [2], Papanicolaou and Varadhan [21] and [22] etc., (see also Nakao [17]). However, the framework of Section 3 does not fit for the homogenization problem in the case discussed by Tanaka [25].

Example 4.2 As shown in Manabe [15], limit theorems for the homological position of $x(t)$ in case of a compact Riemannian surface M with genus κ ($\geqq 1$) is closely related to Theorem 3.1. We consider the Brownian motion $\{x(t); 0 \leqq t < \infty\}$ on M and denote by $X = \{X_t\}$ the continuous D_1'-valued stochastic process associated with $\{x(t); 0 \leqq t < \infty\}$. Let $\{(c_i, c_{i+\kappa}); i = 1,2,\cdots,\kappa\}$ be a canonical homological basis of M and $z(t) = (z^1(t), z^2(t), \cdots, z^{2\kappa}(t))$ be the homological position of $x(t)$ with respect to $\{(c_i, c_{i+\kappa}); i = 1,2,\cdots,\kappa\}$ in the sense of Manabe [15]. It is easy to see that the study of asymptotic properties of $z(t)$ can be reduced to the one of $\{(X_t(\alpha^1), X_t(\alpha^2), \cdots, X_t(\alpha^{2\kappa})); 0 \leqq t < \infty\}$ where $(\alpha^i, \alpha^{i+\kappa})$, $i = 1,2,\cdots,\kappa$, are the harmonic differential 1-forms on M corresponding to $(c_i, c_{i+\kappa})$, $i = 1,2,\cdots,\kappa$:

$$\int_{c_j} \alpha^i = \delta_{ij}, \qquad i,j = 1,2,\cdots,2\kappa.$$

In this case, since $\partial M = \phi$, $b = 0$ and α^i, $i = 1,2,\cdots,2\kappa$, are harmonic, we have

$$A_t(\alpha^i) = 0, \qquad i = 1,2,\cdots,2\kappa,$$

where $A = \{A_t\}$ is the bounded variation part of X.

For a relation between occupation time laws for recurrent diffusions on M and Theorem 3.1, see Remark 2.3 of Ochi [18].

5. Proof of Lemma 2.1

We now return to the proof of Lemma 2.1. Let $\{W_n\}_{n=1}^m$ and $\{U_n\}_{n=1}^m$ be the systems of coordinate neighbourhoods such that

$$M = \bigcup_{n=1}^m W_n = \bigcup_{n=1}^m U_n, \quad U_n \subset \bar{U}_n \subset W_n, \quad n = 1,2,\cdots,m$$

and $\tilde{\phi}_n = (\tilde{\phi}_n^1, \tilde{\phi}_n^2, \cdots, \tilde{\phi}_n^d)$ be a coordinate function from W_n to R^d for $n = 1,2,\cdots,m$. We may assume that the system of coordinate neighbourhoods $\{U_n\}$ has the properties mentioned in Section 3. We modify $\tilde{\phi}_n$ to the R^d-valued C^∞-function ϕ_n defined on M such that

$$\phi_n(x) = \begin{cases} \tilde{\phi}_n(x), & \text{for} \quad x \in \bar{U}_n, \\ 0, & \text{for} \quad x \in W_n^c. \end{cases}$$

Then for each i and n, $\{\phi_n^i(x(t));\ 0 \leq t < \infty\}$ is a quasi-martingale decomposed into

$$\phi_n^i(x(t)) = \phi_n^i(x(0)) + m_n^i(t) + a_n^i(t)$$

where $\{m_n^i(t)\}$ and $\{a_n^i(t)\}$ are the martingale part and the bounded variation part of $\{\phi_n^i(x(t))\}$ respectively. For every i and n, we set

$$y_n^i(t) = \max\{\, |m_n^i(t)|\, ,\ \langle m_n^i\rangle_t,\ |a_n^i|_t \,\}$$

where $|a_n^i|_t$ is the total variation of the function a_n^i on the interval $[0,t]$. For every $N \in Z_+$, we define a stopping time σ_N by

$$\sigma_N = \inf\{t;\ \max\{y_n^i(t)\ ;\ i = 1,2,\cdots,d\ ,\ n = 1,2,\cdots,m\} \geq N\}.$$

Then, letting $\{\psi_n\}_{n=1}^m$ be the partition of unity subordinate to $\{U_n\}_{n=1}^m$, we have, for every $N \in Z_+$,

$$x_{t\wedge\sigma_N}(\alpha) = m_{t\wedge\sigma_N}(\alpha) + a_{t\wedge\sigma_N}(\alpha)$$

$$(5.1) \qquad m_{t\wedge\sigma_N}(\alpha) = \sum_{n=1}^m \sum_{i=1}^d \int_0^{t\wedge\sigma_N} (\psi_n\alpha_i^n)(x(s))\,dm_n^i(s)$$

$$a_{t\wedge\sigma_N}(\alpha) = \sum_{n=1}^m \sum_{i=1}^d \int_0^{t\wedge\sigma_N} (\psi_n\alpha_i^n)(x(s))\,da_n^i(s) + \sum_{n=1}^m \sum_{i,j=1}^d \int_0^{t\wedge\sigma_N} (\partial_j(\psi_n\alpha_i^n))(x(s))\,d\langle m_n^i, m_n^j\rangle(s)/2$$

where $(x_n^1, x_n^2, \cdots, x_n^d) = (\tilde{\phi}_n^1(x), \tilde{\phi}_n^2(x), \cdots, \tilde{\phi}_n^d(x))$, $\partial_j = \partial/\partial x_n^j$ and $\alpha_i^n(x)$ is the i-th component of α on W_n, i. e.,

$$\alpha = \sum_{i=1}^{d} \alpha_i^n(x) dx_n^i \quad \text{on} \quad W_n.$$

Then, there exists a positive constant C_1 such that

$$\begin{aligned} &E[|m_{t\wedge\sigma_N}(\alpha)|^2] \\ &\leqq C_1 \sum_{n=1}^{m} \sum_{i=1}^{d} E[(\int_0^{t\wedge\sigma_N} (\psi_n \alpha_i^n)(x(s)) dm_n^i(s))^2] \\ &\leqq C_1 \sum_{n=1}^{m} \sum_{i=1}^{d} (\sup_{x\in M} |(\psi_n \alpha_i^n)(x)|^2) E[\langle m_n^i \rangle_{t\wedge\sigma_N}] \end{aligned}$$

where E denotes the expectation with respect to P. Hence, in the same way as in [18], we can use the Sobolev lemma to show that for $p > d/4$, there exists a positive constant $C_2 = C_2(p)$ satisfying

$$(5.2) \qquad E[|m_{t\wedge\sigma_N}(\alpha)|^2] \leqq C_2 \|\alpha\|_p^2 \max_{\substack{i=1,2,\cdots,d \\ n=1,2,\cdots,m}} E[\langle m_n^i \rangle_{t\wedge\sigma_N}],$$

(cf. [14]). By (5.1), there also is a positive constant C_3 such that

$$\begin{aligned} &E[|a_{t\wedge\sigma_N}(\alpha)|^2] \\ &\leqq C_3 (\sum_{n=1}^{m} \sum_{i=1}^{d} \sup_{x\in M} |\psi_n \alpha_i^n(x)|^2 + \sum_{n=1}^{m} \sum_{i,j=1}^{d} \sup_{x\in M} |(\partial_j(\psi_n \alpha_i^n))(x)|^2) \\ &\times \max_{\substack{i=1,2,\cdots,d \\ n=1,2,\cdots,m}} (E[|a_n^i|^2_{t\wedge\sigma_N}] \vee E[|\langle m_n^i \rangle_{t\wedge\sigma_N}|^2]), \end{aligned}$$

where $a \vee b = \max\{a,b\}$. Hence, by the Sobolev lemma, for $p > \frac{d}{4} + \frac{1}{2}$ there is a constant $C_4 = C_4(p)$ such that

$$(5.3) \qquad E[|a_{t\wedge\sigma_N}(\alpha)|^2] \leqq C_4 \|\alpha\|_p^2 \max_{\substack{i=1,2,\cdots,d \\ n=1,2,\cdots,m}} \{E[|a_n^i|^2_{t\wedge\sigma_N}] \vee E[|\langle m_n^i \rangle_{t\wedge\sigma_N}|^2]\}.$$

Now, for p and q such that $p > \frac{d}{4} + \frac{1}{2}$ and $q > p + \frac{d}{4}$, we choose an ONB e_k on D_{1q} such that for every k, $e_k \in D_1$. Next for every $N \in Z_+$, we define $M^N = \{M_t^N\}$ and $A^N = \{A_t^N\}$ by

$$(5.4) \qquad \begin{aligned} M_t^N(\alpha) &= \sum_{k=1}^{\infty} (\alpha, e_k)_q \, m_{t\wedge\sigma_N}(e_k) \\ A_t^N(\alpha) &= \sum_{k=1}^{\infty} (\alpha, e_k)_q \, a_{t\wedge\sigma_N}(e_k) \end{aligned} \qquad \text{for} \quad \alpha \in D_1$$

where $(\cdot,\cdot)_q$ is the inner product on D_{1q} defined by the Hilbertian norm $\| \; \|_q$. Then, by (5.2) and (5.3), we have

$$E[|M_t^N(\alpha)|^2] \leqq C_2\|\alpha\|_q^2 (\sum_{k=1}^{\infty} \|e_k\|_p^2) N$$

$$E[|A_t^N(\alpha)|^2] \leqq C_4\|\alpha\|_q^2 (\sum_{k=1}^{\infty} \|e_k\|_p^2) N^2.$$

Since,

$$\sum_{k=1}^{\infty} \|e_k\|_p^2 < \infty,$$

M_t^N and A_t^N is well defined. Furthermore, for every $N \in Z_+$, there exists a subset $\Omega_N^{(1)}$ of Ω such that $P(\Omega_N^{(1)}) = 1$ and on $\Omega_N^{(1)}$

$$M_t^N\ ,\quad A_t^N \in D'_{1q}\ .$$

Also, since $\{m_{t\wedge\sigma_N}(e_k);\ 0 \leqq t < \infty\}$ and $\{a_{t\wedge\sigma_N}(e_k);\ 0 \leqq t < \infty\}$, $(k = 1,2,\cdots)$, are continuous and the convergence of (5.4) is uniform with respect to t, for every $N \in Z_+$, there exists a subset $\Omega_N^{(2)}$ of Ω such that $P(\Omega_N^{(2)}) = 1$ and on $\Omega_N^{(2)}$

$$M^N, A^N \in C([0,\infty) \longrightarrow D'_{1q}).$$

It is easy to see that for $N_1 > N_2$

$$M_t^{N_1} = M_t^{N_2} \quad \text{and} \quad A_t^{N_1} = A_t^{N_2} \quad \text{for} \quad t < \sigma_{N_2} \quad \text{on} \quad \bigcap_{N=1}^{\infty} \Omega_N^{(2)}.$$

This implies that there exist continuous D'_{1q}-valued stochastic processes $M = \{M_t\}$ and $A = \{A_t\}$ satisfying the following: for every $\alpha \in D_{1q}$,

$$M_t(\alpha) = m_t(\alpha)\ ,\quad A_t(\alpha) = a_t(\alpha)\ ,\quad t \geqq 0\ ,\quad \text{a. s.},$$

which completes the proof.

Acknowledgements The authors are deeply indebted to Professor K. Itô for his valuable suggestions. We also wish to acknowledge helpful comments for the proof of Lemma 2.1 from Professor S. Nakao.

References

[1] V. A. Arnold and A. Avez, Problémes ergodiques de la méchanique classique, Paris, 1967.

[2] A. Bensoussan, J. L. Lions and G. C. Papanicolaou, Asymptotic analysis for periodic structure, North-Holland, 1978.

[3] R. Bhattacharya, A central limit theorem for diffusions with periodic coefficients, Ann. Prob., 13 (1985), 385-396.

[4] L. A. Bunimovich and Ya. G. Sinai, Statistical properties of Lorentz gas with periodic configuration of scatterers, Comm. Math. Phys., 78 (1981), 479-497.

[5] P. E. Conner, The Neumann's problem for differential forms on Riemannian manifolds, Memoirs of the Amer. Math. Soc., 20 (1956).

[6] G. de Rham, Differentiable manifolds, Springer, 1984.

[7] I. M. Gel'fand and S. V. Formin, Geodesic flow on manifold of constant negative curvature, Uspehi Mat. Nauk, 47 (1952), 118-137, (Amer. Math. Soc. Transl. Vol.2 (1955), 49-67).

[8] I. M. Gel'fand and I. I. Pyateckiĭ-Šapiro, A theorem of Poincaré, Dokl. Akad. Nauk, 127 (1959), 490-493.

[9] I. M. Gel'fand and N. Ya. Vilenkin, Generalized functions, Vol.4, Academic Press, 1964.

[10] N. Ikeda and S. Manabe, Stochastic integral of differential forms and its applications, Stochastic Analysis, ed. by A. Friedman and M. Pinsky, 175-185, Academic Press, 1978.

[11] N. Ikeda and S. Watanabe, Stochastic differential equations and diffusion processes, Kodansha/ North-Holland, 1981.

[12] K. Itô, Foundation of stochastic differential equations in infinite dimensional spaces, CBMS-NSF, Regional Conference Series in Applied Mathematics, 1984.

[13] S. Itô, Foundamental solutions of parabolic differential equations and boundary value problems, Jap. J. Math., 20 (1957), 55-102.

[14] H. Kumano-go, Pseudo-differential operators, MIT Press, 1981.

[15] S. Manabe, Stochastic intersection number and homological behaviors of diffusion processes on Riemannian manifolds, Osaka Jour. Math., 19 (1982), 429-457.

[16] M. Nagasawa, The adjoint process of a diffusion with reflecting barrier, Kodai Math. Seminar Reports, 13 (1961), 235-248.

[17] S. Nakao, Stochastic calculus for continuous additive functionals of zero energy, Z. Wahr. verw Geb., 68 (1985), 557-578.

[18] Y. Ochi, Limit theorems for a class of diffusion processes, to appear in "Stochastics".

[19] Y. Ochi, Limit theorems for diffusion processes on compact manifolds, to appear in "Stochastic Processes and their Applications", (Abstract of the talk at 15-th Conference on Stochastic Processes and their Applications of Bernoulli Society for Math. Statist. and Prob.).

[20] G. C. Papanicolaou, D. Stroock and S. R. S. Varadhan, Martingale approach to some limit theorems, 1976 Duke Turbulence Conference, Duke Univ. Math. Series III, 1977.

[21] G. C. Papanicolaou and S. R. S. Varadhan, Diffusions with random coefficients, Statist. and Prob.: Essays in Honor of C. R. Rao, ed. by G. Kallianpur, P. R. Krishnaiah and J. K. Glosh, 547-552, North-Holland, 1982.

[22] G. C. Papanicolaou and S. R. S. Varadhan, Boundary value problems with rapidly oscillating random coefficients, Colloquia Mathematica Societaties, János Bolyai: ed. by J. Fritz, Lebowitz and D. Szász, 1981.

[23] M. Ratner, The central limit theorem for geodesic flows on n-dimensional manifolds of negative curvature, Israel J. Math., 16 (1973), 181-197.

[24] Ya. G. Sinai, The central limit theorem for geodesic flows on manifolds of constant negative curvature, Soviet Math. Dokl., 1 (1960), 983-987.

[25] H. Tanaka, Homogenization of diffusion processes with boundary conditions, Stochastic Analysis and Applications, ed. by M. Pinsky, 411-437, Marcel Dekker, 1984.

[26] H. Watanabe, Potential operator of a recurrent strong Feller process in the strict sense and boundary value problem, J. Math. Soc. Japan, 16 (1964), 83-95.

ON GIRSANOV SOLUTIONS OF INFINITE DIMENSIONAL SDEs

M. Jerschow

Department of Mathematics
University of Essen
D-4300 Essen
Federal Republic of Germany

1. The paper is devoted to an auxiliary result on the existence of solutions of countable dimensional SDEs with unit diffusion matrix which may be of some interest in its own right. This result is intended to be applied for constructing a model for infinite number of Wiener particles interacting according to a potential of finite range under weaker conditions than those in [3] an [1] . The author is thankful to A. Wakolbinger (University of Linz, Austria) for drawing his attention to that model, the essential literature and for many stimulating discussions on that subject.

2. Let $(x_o^i)_{i\in\mathbb{N}}$ be a fixed initial configuration in $\mathbb{R}^d$, $d > 1$, that is x_o^i is the position of the i-th particle at time 0. The movement of the particles should be "governed" by the dynamics of the form

$$dx_t^i = b^i((x_t^j)_{j\in\mathbb{N}})dt + dW_t^i \qquad (i\in\mathbb{N}) \tag{1}$$

where $(W^i)_{i\in\mathbb{N}}$ are independent Wiener processes. Natural physical arguments imply that $b^i((x^j)_{j\in\mathbb{N}})$ is approximately proportional to

$\mu^i((x^j))$:=the number of particles $x^j \in$ the unit sphere around x^i .

This means that, if there exists a solution of (1), there must be no explosions ($\mu^i((x_t^j))$ must be finite for all t).

The classical method of successive approximations does not work for obvious reasons: it makes use of Lipschitz conditions on the coefficients which can not be even formulated here since there is no norm in $(\mathbb{R}^d)^{\mathbb{N}}$. The usual way of avoiding this obstacle even for more general countable dimensional SDEs consists in introducing weights $\alpha_i, i\in\mathbb{N}, \alpha_i\to 0$, for the components thus embedding a large part of $(\mathbb{R}^d)^{\mathbb{N}}$ in a Hilbert space. However, this method seems to be unnatural in the problem of

interacting particles since those must be treated equally and every weighting neglects particles with large numbers.
By "freezing" the particles outside large cubes, Lang [3] and Fritz [1] reduced equation (1) to finite dimensional ones, solved them and proved the convergence of the partial solutions to a (strong) solution of (1). The conditions on b^i are in both the cases are rather strong (but reasonable physically). Still, it is interesting to find out what precisely causes explosions of (1). For that purpose, solving of SDEs by the Maruyama-Girsanov absolute continuous transformation of the measure (cf. [2], [4]) seems to be the most adequate methods since it requires the weakest conditions on the drift. This method gives, of course, not always strong solutions.
3. A direct application of this method is unfortunately not possible for even the measures in the function space corresponding to $(W^i)_{i\in\mathbb{N}}$ and $(W^i$ + unit drift$)_{i\in\mathbb{N}}$ are orthogonal. On the other hand, the Maruyama-Girsanov transformation for Hilbert-spaced drifts is inadequate because it assumes particles to be "unequal".
4. Now we formulate the auxiliary result. It concerns a slightly more general equation than (1):

$$dx^i_t = b^i_t((x^j_\cdot)_{j\in\mathbb{N}})dt + dW^i_t \qquad (i\in\mathbb{N}) \tag{2}$$

where $b^i_\cdot(\cdot)$ are measurable functions which depends causally on the functions $[0,1]\ni s \longmapsto x^j_s$, $j\in\mathbb{N}$, that is $b^i_t((x^j_\cdot)_{j\in\mathbb{N}}$ depends only on the "past" up to t of these functions.
Theorem. Let, for each $i\in\mathbb{N}$, there exists a $c_i\in\mathbb{R}_+$ such that (a) $b^i_\cdot((x^j_\cdot)_{n\in\mathbb{N}}$ depends only on $x^j_\cdot$ with $j<c_i$ and (b) $|b^i_t((x^j_\cdot)_{j\in\mathbb{N}}| < c_i$ for all $t\in[0,1]$.
Then there exists a ("weak") solution of (2).
Remark. Condition (a) for general SDEs is, of course, very restictive. However, it is resonable in application to interacting Wiener particles: intuitively, it must be fulfilled at least up to a strictly positive stopping time if the system does not explode.
Condition (b) can be easily weakened, for instance, by

$$\int_0^1 [b^i_t((x^j_\cdot)_{j\in\mathbb{N}})]^2 dt < c_i.$$

5. Here is the idea of the proof.
One considers truncated equations (with $b^i=0$ for $i>n$). These equations

are then solved (in the distribution sense) by using the Maruyama-Girsanov transformation. Condition (b) implies that, for any finite number of $x^i_\cdot$'s the corresponding sequences of distributions are relatively compact in the Vitali-Hahn-Saks sense ($\Leftrightarrow$ the densities w.r. to the Wiener measure are relatively $\sigma(L_1,L_\infty)$-compact. By the diagonal procedure, one chooses a subsequence of infinite-dimensional distributions such that all its finite-dimensional projections converge. The finite-dimensional limiting distributions are consistent and thus define an infinite-dimensional distribution. This distribution solves (2) because of the "finite-dimensional" Vitali-Hahn-Saks convergence and condition (a). (Note that because of this very strong convergence no regularity conditions on b^{i}'s are needed. Note also that, in contrast to the finite-dimensional case, we do not automatically have here the uniqueness in the distribution sense.)

References

[1] J. Fritz: Gradient dynamics of finite point systems, Preprint No. 15 (1984), Math. Inst. Hungar. Acad.Sci. Budapest

[2] M.Jerschow (Ershow): On absolute continuity of measures corresponding to diffusion type processes, Theory of Probability and Its Applic., 17,1 (1972)

[3] R. Lang: Unendlich-dimensionale Wiederprozesse mit Wechselwirkung, Z. für Wahrscheinlichkeitstheorie und verw. Gebiete, 38 (1977), 55-72 (for a correction see T. Shiga: A Remark on Infinite-Dimensional Wiener Process with Interactions, ibid, 47 (1979), 299-304)

[4] R.S. Liptser & A.N. Shiryayev: Statistics of Random Processes, Vol I, Springer-Verlag (1977)

EXPLICIT SOLUTION OF A GENERAL CONSUMPTION/INVESTMENT PROBLEM

Ioannis Karatzas
Columbia University

Suresh P. Sethi
University of Toronto

John P. Lehoczky
Carnegie-Mellon University

Steven E. Shreve
Carnegie-Mellon University

ABSTRACT: This talk shows how to solve a general consumption and investment decision problem in closed form. An investor seeks to maximize total expected discounted utility of consumption. There are N distinct risky investments, modelled by dependent geometric Brownian processes, and one riskless (deterministic) investment. The analysis allows for a general utility function and general rates of return. The model and analysis take into consideration the inherent nonnegativity of consumption and consider bankruptcy. The value function is determined explicitly, as are the optimal consumption and investment policies.

§1. INTRODUCTION

Consider an agent who, at time t, has wealth $x(t)\geq 0$ and must determine a rate of consumption $c(t)\geq 0$ and also choose an investment portfolio for his unconsumed wealth. There are N + 1 distinct investments available, one of which is riskless with rate of return $r>0$. The other N investments are risky and, following Merton [7]-[9] and Black and Scholes [1], are modelled in such a way that relative price changes are drifted Brownian motions:

$$\frac{dP_i(t)}{P_i(t)} = \alpha_i\, dt + \underset{\sim}{e}_i\, \underset{\sim}{D}\, d\underset{\sim}{w}^T(t),\ i=1,\ldots\ldots,N, \qquad [1.1]$$

where P_i is the price of one share of the i-th risky asset, $\underset{\sim}{e}_i$ is the unit row vector with a one in the i-th position, $\underset{\sim}{D}$ is an NxN matrix with $\underset{\sim}{\Sigma} = \underset{\sim}{D}\,\underset{\sim}{D}^T$ positive definite, and $\{\underset{\sim}{w}(t), \mathscr{F}_t;\ 0\leq t<\infty\}$ is a standard, N-dimensional Wiener process.

If $\underset{\sim}{\pi}(t) = (\pi_1(t), \ldots, \pi_N(t))$ is the vector of wealth proportions invested in the risky investments, so $\pi_0(t) = 1 - \sum_{i=1}^{N} \pi_i(t)$ is the fraction of wealth invested in the risk-free investment, then the agent's wealth process obeys the stochastic differential equation

$$dx(t) = (\underset{\sim}{\alpha} - r\,\underset{\sim}{1})\,\underset{\sim}{\pi}^T(t)\, x(t)dt + (rx(t) - c(t))dt + x(t)\,\underset{\sim}{\pi}(t)\,\underset{\sim}{D}\, d\underset{\sim}{w}^T(t), \qquad [1.2]$$

where $\underset{\sim}{\alpha} = (\alpha_1,\ldots,\alpha_n)$ and $\underset{\sim}{1} = (1,\ldots,1)$. See [3], [4] and [5] for more detail on the derivation of [1.2].

If the agent consumes too much or has unfortunate investment experience, he can see his wealth fall to zero. We call this state <u>bankruptcy</u> and define

$$T_o = \inf\{ t \geq 0: \ x(t) = 0 \} \qquad [1.3]$$

to be the first time bankruptcy occurs, if it does.

Let $U: (0,\infty) \to \mathbb{R}$ be a strictly increasing, strictly concave, <u>utility function</u> which is three times continuously differentiable. We set $U(0) = \lim_{c \downarrow 0} U(c)$, $U(\infty) = \lim_{c \uparrow \infty} U(c)$ and adopt the same conventions for U'. The agent wishes to choose consumption rate $\{c(t): 0 \leq t < \infty\}$ and investment portfolio $\{\underset{\sim}{\pi}(t): 0 \leq t < \infty\}$, both adapted to $\{\mathscr{F}_t\}$, so as to maximize

$$V_{c(\cdot), \underset{\sim}{\pi}(\cdot)}(x) \triangleq E_x[\int_0^{T_o} e^{-\beta t} U(c(t))dt + P e^{-\beta T_o}], \qquad [1.4]$$

where $x(o) = x, \beta > 0$ is a <u>discount factor</u>, and P is a <u>payment</u> (which may be negative) received upon bankruptcy if it occurs. In this model, P is a free parameter, but the value of P equal to $\frac{1}{\beta} U(0)$ has special standing because it corresponds to forcing the agent to consume at rate zero from the time of bankruptcy onward. We denote by

$$V(x) = \max_{c(\cdot), \underset{\sim}{\pi}(\cdot)} V_{c(\cdot), \underset{\sim}{\pi}(\cdot)}(x), \quad x \geq 0, \qquad [1.5]$$

the <u>value function</u> for this problem.

Several variants of this consumption/investment problem have been studied extensively ([5]-[11]). Most of this literature is devoted to drawing conclusions from the Bellman equation about the nature of optimal policies, and most authors have not attended to the difficulties created by the consumption constraint $c(t) \geq 0$ or the possibility of bankruptcy. Instead, it has been tacitly assumed that under an optimal policy, the consumption constraint will not be active and bankruptcy will not occur. We report here on Karatzas, Lehoczky, Sethi and Shreve [5], in which these assumptions are examined closely. In contrast to many previous works, it is not necessary in [5] to assume a priori that V is concave and V" is continuous; we provide an explicit representation for V which allows verification of these properties.

More precisely, we have the following results. Let

$$\gamma = \frac{1}{2} (\underset{\sim}{\alpha} - r\underset{\sim}{1}) \underset{\sim}{\Sigma}^{-1} (\underset{\sim}{\alpha} - r\underset{\sim}{1})^T, \qquad [1.6]$$

and assume that $\underset{\sim}{\alpha} \neq r\underset{\sim}{1}$, so $\gamma > 0$. (The case $\gamma = 0$ is treated in [6].) The quadratic equation

$$\gamma\lambda^2 - (r - \beta - \gamma)\lambda - r = 0 \qquad [1.7]$$

has roots $\lambda_- < 1$ and $\lambda_+ > 0$. We assume that

$$\int_1^{\infty} \frac{d\theta}{(U'(\theta))^{\lambda_-}} < \infty \quad . \tag{1.8}$$

(It can be shown that if U(c) is of the form $(c+\eta)^{\delta}$ for $0<\delta<1$ and $\eta \geq 0$, then [1.8] is equivalent to assuming that the value function [1.5] is finite.) We consider only values of P which satisfy

$$\frac{1}{\beta} U(0) \leq P < \frac{1}{\beta} U(\infty), \tag{1.9}$$

because for P outside this range the problem can be shown to be degenerate. Then there are optimal consumption and investment policies $c^*(x)$ and $\underset{\sim}{\pi}^*(x)$ in feedback form. For arbitrary positive initial wealth, the optimal wealth trajectory depends on the model parameters as we now detail. The probability of bankruptcy is zero if and only if $P = \frac{1}{\beta} U(0)$. If $P > \frac{1}{\beta} U(0)$, then the probability of bankruptcy is one if and only if $\beta \geq r + \gamma$. When $U'(0) = \infty$, then $c^*(x) > 0$ for all $x > 0$, and $\inf_{x>0} c^*(x) = 0$ if and only if $P = \frac{1}{\beta} U(0)$. When $U'(0) < \infty$, we define $P^* > \frac{1}{\beta} U(0)$ by

$$P^* \triangleq \frac{1}{\beta} U(0) - \frac{(U'(0))^{1+\lambda_-}}{\beta \lambda_-} \int_0^{\infty} (U'(c))^{-\lambda_-} dc$$

and distinguish three cases:

1) If $\frac{1}{\beta} U(0) \leq P < P^*$, then there exists $\bar{x} > 0$ such that $c^*(x) = 0$ for $0 < x \leq \bar{x}$ and $c^*(x) > 0$ for $x > \bar{x}$;

2) If $P = P^*$, then $c^*(x) > 0$ for all $x > 0$, but $\inf_{x>0} c^*(x) = 0$;

3) If $P > P^*$, then $\inf_{x>0} c^*(x) > 0$.

§2. THE BELLMAN EQUATION AND THE MUTUAL FUND REDUCTION

The following theorem can be proved by standard methods.

Theorem 2.1: With P a finite number, let us assume that $Z: (0,\infty) \to (P,\infty)$ is a C^2 function satisfying the Bellman equation

$$\beta Z(x) = \max_{c \geq 0, \underset{\sim}{\pi}} [(\underset{\sim}{\alpha} - r\underset{\sim}{1})\underset{\sim}{\pi}^T x Z'(x) + (rx-c) Z'(x) + \frac{1}{2} \underset{\sim}{\pi} \underset{\sim}{\Sigma} \underset{\sim}{\pi}^T x^2 Z''(x) + U(c)], \quad x>0. \tag{2.1}$$

If U(0) is finite or if

$$E_x \int_0^{T_0} e^{-\beta t} \max\{ U(c(t)), 0\} dt < \infty \tag{2.2}$$

holds for every admissible $c(\cdot)$ and $\underset{\sim}{\pi}(\cdot)$, then $Z(x) \geq V(x)$, $x>0$. □

When there is only one risky asset, the wealth equation [1.2] takes the form

$$dx(t) = (\alpha - r)\,\pi(t)x(t)\,dt + (rx(t)-c(t))\,dt + x(t)\pi(t)\sigma\,dw(t), \qquad [2.3]$$

where $\pi(t)$ denotes the proportion of wealth invested in this risky asset, and the Bellman equation [2.1] becomes

$$\beta Z(x) = \max_{c\geq 0,\pi} \left[(\alpha-r)\pi x Z'(x) + (rx-c)Z'(x) + \frac{1}{2}\pi^2\sigma^2 x^2 Z''(t) + U(c)\right],\ x>0. \qquad [2.4]$$

It is a straightforward matter to verify that if

$$\gamma \triangleq \frac{1}{2}\,(\underset{\sim}{\alpha}-r\underset{\sim}{1})\,\underset{\sim}{\Sigma}^{-1}(\underset{\sim}{\alpha}-r\underset{\sim}{1})^T = \frac{(\alpha-r)^2}{2\sigma^2}\ , \qquad [2.5]$$

then any strictly concave solution to [2.4] also solves [2.1]. By this device, the problem with N risky investments can be reduced to a problem with one risky investment whose α and σ^2 satisfy [2.5]. One way to create this single investment is to form a continuously trading mutual fund which maintains proportions of the risk-free and N risky assets given by the (N+1)-dimensional vector $(1-(\underset{\sim}{\alpha}-r\underset{\sim}{1})\,\underset{\sim}{\Sigma}^{-1}\underset{\sim}{1}^T,\ (\underset{\sim}{\alpha}-r\underset{\sim}{1})\,\underset{\sim}{\Sigma}^{-1})$. This fund has average rate of return $\alpha=r+2\gamma$ and variance $\sigma^2=2\gamma$, so [2.5] holds.

§3. SOLVING THE BELLMAN EQUATION (CONSUMPTION CONSTRAINT INACTIVE)

The maximizations indicated in the reduced Bellman equation [2.4] are accomplished by setting

$$c^*(x) = I(Z'(x)),\quad \pi^*(x) = -\frac{(\alpha-r)\,Z'(x)}{\sigma^2 x Z''(x)} \qquad [3.1]$$

where $U'(I(y)) = y$ for $0 < y < U'(0)$, $I(y) = 0$ for $y \geq U'(0)$, and Z is assumed to be strictly concave. Substituting [3.1] back into [2.4], the Bellman equation becomes

$$\beta Z(x) = -\gamma\frac{(Z'(x))^2}{Z''(x)} + [rx - I(Z'(x))]\,Z'(x) + U(I(Z'(x))). \qquad [3.2]$$

We expect Z to be strictly increasing, so we postulate the existence of an inverse function X, i.e.,

$$Z'(X(y)) = y,\quad y>0. \qquad [3.3]$$

Differentiating [3.2] with respect to x and rewriting the result in terms of X, we obtain the second-order <u>linear</u> equation

$$\gamma y^2 X''(y) - (r-\beta-2\gamma)yX'(y) - rX(y) = -I(y), \qquad [3.4]$$

which has the general solution

$$X(y) = By^{\lambda+} + \hat{B}y^{\lambda_-} + \frac{1}{r} I(y) - \frac{1}{\gamma(\lambda_+ - \lambda_-)} \left\{ \frac{y^{\lambda+}}{\lambda+} \int_a^{I(y)} \frac{d\theta}{(U'(\theta))^{\lambda+}} + \frac{y^{\lambda-}}{\lambda-} \int_{I(y)}^{\infty} \frac{d\theta}{(U'(\theta))^{\lambda-}} \right\}. \quad [3.5]$$

For reasons which become apparent later, we set $\hat{B} = 0$. We shall later choose $a \geq 0$ and $B \leq 0$, and then X will be strictly increasing. We denote its inverse by Z', so [3.3] holds.

The goal now is to choose $a \geq 0$ and $B \leq 0$ so that Z' is the derivative of the value function V. The parameter a plays the role of $\inf_{x>0} c^*(x)$, where $c^*(x)$ is the optimal consumption law in feedback form. We shall find $c^*(x)$ to be nondecreasing in x, so $a = \lim_{x \downarrow 0} c^*(x)$. When the consumption constraint is inactive, $c^*(x)>0$ for all $x>0$, and so [3.1] leads to the equation $U'(c^*(x)) = Z'(x)$, $x>0$, and letting $x \downarrow 0$, we obtain

$$U'(a) = Z'(0). \quad [3.6]$$

For each initial wealth $x^*(0)>0$ and each choice of $a \geq 0$ and $B \leq 0$ for which [3.6] holds, we may substitute the feedback law [3.1] into the wealth equation [2.3] to obtain a wealth trajectory $x^*(t)$. We define

$$y(t) = Z'(x^*(t)), \quad [3.7]$$

and after some manipulations involving Ito's rule, we obtain the surprisingly simple stochastic differential equation

$$dy(t) = -(r-\beta)\, y(t)dt - \sqrt{2\gamma}\; y(t)dw(t). \quad [3.8]$$

With initial condition $y(0) = y$, the solution to this equation is

$$y(t) = y \exp\left[(\beta - r - \gamma)t - \sqrt{2\gamma}\, w_t \right]. \quad [3.9]$$

Let us define $\tau = \inf\{t \geq 0:\ y(t) = U'(a)\}$, and set

$$G(y) = E^y \left[\int_0^{\tau} e^{-\beta t} U(I(y_t))\, dt + P e^{-\beta\tau} \right],\ y>0. \quad [3.10]$$

When $y = Z'(x^*(0))$, then because of [3.6] and [3.7], we have $\tau = T_o \triangleq \inf\{t \geq 0: x^*(t) = 0\}$, and comparison of [3.10] with [1.4] shows that

$$V_{c^*, \pi^*}(x) = G(Z'(x)),\ x>0. \qquad [3.11]$$

It follows immediately that

$$V(x) \geq G(Z'(x)),\ x>0. \qquad [3.12]$$

According to the Feynman-Kac formula, G defined by [3.9] and [3.10] satisfies the second-order, linear equation

$$\beta G(y) = \gamma y^2 G''(y) - (r-\beta)\, yG'(y) + U(I(y)),\ 0<y<U'(a), \qquad [3.13]$$

which has general solution

$$G(y) = Ay^{\lambda_+ +1} + \hat{A}y^{\lambda_- +1} + \frac{1}{\beta} U(I(y)) \qquad [3.14]$$

$$- \frac{1}{\gamma(\lambda_+ - \lambda_-)} \left\{ \frac{y^{\lambda_+}}{\lambda_+ +1} \int_a^{I(y)} \frac{d\theta}{(U'(\theta))^{\lambda_+}} + \frac{y^{\lambda_-}}{\lambda_- +1} \int_{I(y)}^{\infty} \frac{d\theta}{(U'(\theta))^{\lambda_-}} \right\}.$$

By considering the rate of growth in [3.10] as $y \downarrow 0$, one can show that $\hat{A}$ in [3.14] must be zero. Furthermore, [3.10] implies that

$$G(U'(a)) = P. \qquad [3.15]$$

Finally, a bit of calculus shows that if

$$(\lambda_+ +1)\, A = \lambda_+ B, \qquad [3.16]$$

then $G(Z'(x))$ solves the Bellman equation [2.4].

One can show that, except when $U'(0)$ is finite and $\frac{1}{\beta}U(0) \leq P < P^*$ (see [1.10]), equations [3.6], [3.15] and [3.16] uniquely determine $a \geq 0$, $B \leq 0$ and $A \leq 0$. Proceeding under these assumptions with these choices of a, B and A, we see that not only does the composite function $G'(Z'(x))$ solve the Bellman equation, but $G(Z'(0))=P$ and $G(Z'(x))$ is strictly increasing and of class C^2. By Theorem 2.1 and [3.12], we conclude that $V(x) = G(Z(x))$, $x \geq 0$.

With some additional computations, one can show that $V'=Z'$, and so the process

$$y(t) = Z'(x^*(t)) = V'(x^*(0)) \exp\,[(\beta-r-\gamma)t - \sqrt{2\gamma}\ w_t] \qquad [3.17]$$

provides a convenient tool for studying the optimal wealth trajectory. We have $y(0)=$

$V'(x^*(0)) < U'(a)$, and bankruptcy occurs if and only if $y(\cdot)$ reaches $U'(a)$. Now $y(\cdot)$ rises to every finite level with positive probability, and this probability is one if and only if $\beta-r-\gamma\geq 0$. Bankruptcy cannot occur if $U'(a) = \infty$, which is the case only when $U'(0) = \infty$ and $a=0$. Under the condition $U'(0) = \infty$, we have $a=0$ if and only if $P = \frac{1}{\beta}U(0)$. Note, however, that when $U'(0)$ is finite and $\frac{1}{\beta} U(0) \leq P < P^*$, the analysis of this section is not applicable.

§4. SOLVING THE BELLMAN EQUATION (CONSUMPTION CONSTRAINT ACTIVE)

In the previous section, we gave a complete solution to the consumption/investment problem posed in §1 except when $U'(0)$ is finite and $\frac{1}{\beta} U(0) \leq P<P^*$. Here we indicate the modifications required to treat this case.

We define X by [3.5] as before, but now we set a=0. Equation [3.6] is no longer valid, but there is a unique $\bar{x} > 0$ for which $U'(0) = Z'(\bar{x})$. (See case 1) at the end of §1.) We set

$$\bar{y} = Z'(0), \qquad [4.1]$$

$\tau = \inf \{t \geq 0: y(t) = \bar{y}\}$, and we define G by [3.10]. We again have $T_0 = \tau$, so [3.11] is valid. In place of [3.15], we require

$$G(\bar{y}) = P. \qquad [4.2]$$

Equations [4.2] and [3.16] uniquely determine $B \leq 0$ and $A \leq 0$ so that $\bar{y} > 0$. We may now proceed as before to show that $V(x) = G(Z(x))$, $x \geq 0$. Bankruptcy can occur only if $\bar{y}$ is finite, which is the case when $\frac{1}{\beta} U(0) < P < P^*$; when $P = \frac{1}{\beta} U(0)$, we have $\bar{y} = \infty$.

REFERENCES

[1] F. Black and M. J. Scholes, The pricing of options and corporate liabilities, J. Pol. Econ. 81 (1973), 637-654.

[2] E. B. Dynkin, Markov Processes, Vol. II, Academic Press, New York, 1965.

[3] J. M. Harrison and D. M. Kreps, Martingales and arbitrage in multiperiod security markets, J. Econom. Theory 20 (1979), 381-408.

[4] J. M. Harrison and S. R. Pliska, Martingales, stochastic integrals, and continuous trading, Stoch. Proc. Appl. 11 (1981), 215-260.

[5] I. Karatzas, J. Lehoczky, S. Sethi, and S. Shreve, Explicit solution of a general consumption/investment problem, Math. Operations Research, to appear.

[6] J. Lehoczky, S. Sethi, and S. Shreve, Optimal consumption and investment policies allowing consumption constraints and bankruptcy, Math. Operations Research 8, (1983), 613-636.

[7] R. C. Merton, Lifetime portfolio selection under uncertainty: The continuous-time case, Rev. Econ. Statist. 51 (1969), 247-257.

[8] R. C. Merton, Optimum consumption and portfolio rules in a continuous time model. J. Economic Theory 3 (1971), 373-413. Erratum, J. Economic Theory 6 (1973), 213-214.

[9] R. C. Merton, An intertemporal capital asset pricing model. Econometrica 41 (1973), 867-887.

[10] S. F. Richard, A generalized captial asset pricing model. In Portfolio Theory, 25 Years After, E. J. Elton and M. J. Gruber, eds., TIMS Studies in the Management Sciences 11 (1979), 215-232.

[11] P. A. Samuelson, Lifetime portfolio selection by dynamic stochastic programming, Rev. Econ. Statist. 51 (1969), 239-246.

Viscosity solutions in partially observed control.

Michael Kohlmann, Universität Konstanz.

We are concerned with a "simple" partially observed control problem first considered by Beneš and Karatzas in [1], where the relation between Zakai's equation and Mortensen's equation was revealed. In [1] it was pointed out that the dynamic programming techniques do not give us an existence theorem (- as it would be in the completely observed case -), but it was shown that the value function is a lower bound for the solution of Mortensen's equation, if such a solution exists.

In this contribution, we do not attack the existence problem, and our results do not contribute any idea towards such a result, as for as we see.

We consider Mortensen's equation and use the concept of viscosity solution for an attempt towards a better understanding of the important role this equation plays in stochastic control. A viscosity solution is defined to be a subsolution and a supersolution of the problem. We are thus trying to describe the sets of subsolutions and supersolutions and give some control-theoretical interpretation of these sets: Subsolutions may be seen as solutions of a suboptimal problem and supersolutions are related with superoptimal solutions of the control problem. Our results are related to those of Bensoussan in [2], and a lot of techniques are taken from [2] and [3]. The reference on viscosity solutions related to our problem is [6].

This article treats a slimmed problem which was described in more detail in [4]. Some of the technical details which are left out here, can be found there.

1. The control problem

The dynamics under consideration are of the form

$$\begin{aligned} dx_t &= (g(x_t) + v_t)dt + dw_t \quad , \quad t \in [0,T] \\ x(0) &= x_0 \end{aligned} \tag{1}$$

and the observation is given by

$$dy_t = h(x_t)dt + d\eta_t \quad , \quad t \in [0,T]$$
$$y(0) = 0 \tag{2}$$

where $g: \mathbb{R}^n \to \mathbb{R}^n$ is a bounded continuous Lipshitz-function with first derivative in L^∞, and $h: \mathbb{R}^n \to \mathbb{R}^d$ is a bounded function, $h \in W^{2,\infty}(\mathbb{R}^n)$. (w_t) and (η_t) are independent Brownian motions.

The cost is given by (T=1)

$$J(v(\ )) = E^{v(\ )}\left[\int_0^1 f(x(t),v(t))dt + \ell(x(1))\right] . \tag{3}$$

Now let (Ω,F,P) be a probability space on which the solution of (2) is a d-dimensional Wiener process. The history of the observations is given by

$$Y_t = \sigma(y_s: s \leq t) .$$

With this we define an admissible control

$$u: [0,1] \times \Omega \to \mathbb{R}^k \tag{4}$$

with compact and convex range U_{ad} to be an element $z(t)$ of $L^2_Y((0,1),V)$, which is defined as the subspace of $L^2((0,1) \times \Omega;\ dt \times dP;\ V)$, such that $z(t)$ a.e. in t lies in $L^2(\Omega,Y_t,P;V)$, where $V = \mathbb{R}^k$. Besides several technical assumptions we mainly assume:
$f: \mathbb{R}^n \times \mathbb{R}^k \to \mathbb{R}$ satisfies $f_v(\ ) = f(\ ,v) \in H = L^2(\mathbb{R}^n)$ for all $v \in U_{ad}$, with $|f_v|_H \leq C$ independent of $v \in U_{ad}$. $v \mapsto f_v$ is assumed to be continuous as a mapping from U_{ad} to H. For ℓ, finally, we assume $\ell \in H$.

The filtering equation associated with (5) and (6) is given as a stochastic p.d.e. which will be referred to as Zakai's from now on:

$$dp = -A^{v(\ )^*} p\,dt + p\,h\,dy$$
$$p(0) = \Pi \in H \ , \quad \Pi \geq 0 \ . \tag{5}$$

Here A^* is the dual of the operator A associated with the diffusion (1):

$$A = -\frac{1}{2}\Delta - \{g + v\}'\nabla$$

that is

$$A^* = -\tfrac{1}{2}\Delta + \{g + v\}'\nabla + \mathrm{tr}\,(\nabla g)$$

The cost criterion is now transformed into

$$J(\Pi, v(\ )) = E\left[\int_0^1 (f_{v(t)}, p_\Pi^{v(\ )}(t))dt + (\ell, p_\Pi^{v(\ )}(1))\right] \qquad (6)$$

We should note here, that from [1,2,4] this functional may be viewed as linear in Π.

We now consider the control problem (5), (6) with the above defined class of controls. In [1] the analogue of the Hamilton-Jacobi equation was elaborated

$$V_t(\tau,T) = \tfrac{1}{2} V_{\Pi\Pi}(\tau,p)\,[h\,p,\ h\,p]$$

$$+ \min_{a\in U_{ad}} \left\{ \left(- A^a\, V_\Pi(\tau,p)(x) + f(x,a), p\right) \right. \qquad (7)$$

$$V(0,p) = (\ell, p)$$

and Beneš and Karatzas could prove that - if a solution exists - this solution is a lower bound on the value function

$$V(\tau,\Pi) \le E\left[\int_{T-\tau}^{1} (f_{v(t)}, p_\Pi^{v(\ )}(t))dt + (\ell, p_\Pi^{v}(1))\right] \qquad (8)$$

for all admissible v.

2. Viscosity solutions

Lions, Crandall, and Evans did the pioneering work on a generalized concept of solutions for Hamiltonians [5,6,7]. We follow their outline to generalize their concept for infinite dimensional problems. In this paper we shortly go into existence and uniqueness problems [see: 4,7], but use the techniques 'formally' to describe their use in partially observed control.

Let ψ be a continuous function on $L^2(\mathbb{R}^n) =: E$, and consider for $p \in E$ the sub- and superdifferentials of ψ :

$$\partial_+ \psi(p) = E\left\{(A,\xi)/\limsup\,\{\psi(q) - \psi(p) - (\xi, q-p) - \tfrac{1}{2}A[q-p, q-p]\}\ ||q-p||^{-2} \le 0\right\} \qquad (9).$$

and $\partial_- \psi(p)$ similarly.

Now write Mortensen's equation (7) as

$$-\frac{\partial}{\partial t} V + \min_{a \in U_{ad}} H(V_{\Pi\Pi}, V_{\Pi}, V, \Pi, a) = 0 \tag{10}$$

Definition 2.1: A function u on $[0,1] \times E$ which is differentiable in t and continuous on E, $u \in C^{1,0}$, is called

(i) a subsolution of (10), if

$$-\frac{\partial u}{\partial t} + \min_{a \in U_{ad}} H(A, \xi, u, a) \leq 0 \tag{11}$$

(ii) a supersolution of (10), if

$$-\frac{\partial u}{\partial t} + \min_{a \in U_{ad}} H(A, \xi, u, a) \geq 0$$

for all $p \in E$ and all $(A,\xi) \in \partial_+ u(p)$ and $\partial_- u(p)$, respectively.

(iii) u is viscosity solution if (i) and (ii) hold.

A main problem now is whether $\partial_+\psi$ and $\partial_-\psi$ are nonempty at least on a dense set of $p \in E$. For general B-spaces there is no positive answer to this question. But as long as we are working on $E = L^2(\mathbf{R}^n)$ a result of Ekeland [8] gives us the desired result. This is deduced from the existence of a Frêchet differentiable function $\phi: E \to \mathbf{R}$ with $\phi(0) > 0$ and $\phi(\ell) \leq 0$ outside the unit ball. In our case just take $\phi(\ell) = 1 - (\ell,\ell)$. For more general conditions, see [7].

For the above defined control problem, we know that an ε-optimal control, u^ε, always exists [see 9,10]. We now consider the value function for the ε-problem

$$V^\varepsilon(\tau,\Pi) = \min_{v \in \mathcal{U}^\varepsilon_{ad}} E\left[\int_{T-\tau}^{T} (f_{v(t)} p_\Pi^{v(\ )}(t))dt + (\ell, p_\Pi^{v(\ )}(1))\right], \tag{12}$$

where $\mathcal{U}^\varepsilon_{ad}$ is the set of admissible controls, which are not better than a given ε-optimal control.

It is not clear here, whether we could derive Hamilton-Jacobi equations for V^ε. Consider V^ε to be the solution of the Hamilton-Jacobi form of the Mortensen equation with the minimization left out:

$$V_t^\varepsilon(\tau,\Pi) = \frac{1}{2} V_{\Pi\Pi}^\varepsilon(\tau,\Pi)\ [h\ \Pi, h\Pi]$$

$$(- A^{u_\varepsilon} V_\Pi(\tau,\Pi)(x) + f(x,u_\varepsilon),\Pi) \quad . \tag{13}$$

Then Itô's formula gives us

$$d\ V^\varepsilon(T - t,p_t) = - (f(\ ,u_\varepsilon),p_t)dt \ + \ V_\Pi^\varepsilon(T - t,p_t)[h\ p]dy_t$$

and integrating over $[T - \tau,T]$ and taking the expectation gives us

$$V^\varepsilon(T,\Pi) = J(u^\varepsilon,\Pi) = \min_{u\in\mathcal{U}_{ad}^\varepsilon} E\left[\left(\int_0^T (f_u,p_\Pi^u)dt + (\ell,p_\Pi^u(T)\right)\right] \ .$$

Theorem 2.2: Let $V^\varepsilon(\tau,p)$ satisfy (13) and be a twice differentible function of $\Pi \in E$, and differentiable in τ. Then V^ε is a subsolution of (10).

Proof: First note that the differentiability of V^ε implies

$$\partial_+ V^\varepsilon(p) = \partial_- V^\varepsilon(p) = \{(V_{\Pi\Pi}^\varepsilon, V_\Pi^\varepsilon)\} \ . \qquad \text{Then}$$

$$- \frac{\partial V^\varepsilon}{\partial t} + \min_{a\in U_{ad}} \mathcal{H}(V_{\Pi\Pi}^\varepsilon, V_\Pi^\varepsilon, V^\varepsilon,\Pi,a)$$

$$= - \frac{\partial V^\varepsilon}{\partial t} + \min_{a\in U_{ad}} \mathcal{H}(V_{\Pi\Pi}^\varepsilon, V_\Pi^\varepsilon, V^\varepsilon,\Pi,a) + \min_{v\in\mathcal{U}_{ad}^\varepsilon} \mathcal{H}(V_{\Pi\Pi}^\varepsilon, V_\Pi^\varepsilon, V^\varepsilon,\Pi,v)$$

$$- \min_{v\in\mathcal{U}_{ad}} \mathcal{H}(V_{\Pi\Pi}^\varepsilon, V_\Pi^\varepsilon, V^\varepsilon,\Pi,v)).$$

$$= \min_{a\in U_{ad}} \mathcal{H}(V_{\Pi\Pi}^\varepsilon, V_\Pi^\varepsilon, V^\varepsilon,\Pi,a) - \min_{v\in\mathcal{U}_{ad}^\varepsilon} \mathcal{H}(V_{\Pi\Pi}^\varepsilon, V_\Pi^\varepsilon, V^\varepsilon,\Pi,v) \leq 0$$

as V^ε solves the Mortensen equation (13) for the ε-problem:

$$- V_t^\varepsilon + \min_{v\in\mathcal{U}_{ad}^\varepsilon} \mathcal{H}(V_{\Pi\Pi}^\varepsilon, V_\Pi^\varepsilon, V^\varepsilon,\Pi,v) = 0 \quad . \tag{14}$$

But we really do not need the differentiability of $V^\varepsilon(\tau,\Pi)$

Theorem 2.3: Each subsolution $V^\varepsilon(\tau,\Pi) \in C^{1,0}$ of (14) is a subsolution of (10).

Proof: For $\Pi \in E$, $(A,\xi) \in \partial_+ V^\varepsilon(\tau,\Pi)$ we have

$$- V_t^\varepsilon + \min_{a \in U_{ad}} H(A,\xi,V^\varepsilon,\Pi,a) + \min_{v \in U_{ad}^\varepsilon} H(A,\xi,V^\varepsilon,\Pi,v) - \min_{v \in U_{ad}^\varepsilon} H(A,\xi,V^\varepsilon,\Pi,v)$$

$$\leq \min_{a \in U_{ad}} H - \min_{v \in U_{ad}^\varepsilon} H \leq 0 .$$

This implies

$$- V_t^\varepsilon + \min_{a \in U_{ad}} H(A,\xi,V^\varepsilon, ,a) \leq 0 ,$$

so that V^ε is a viscosity subsolution of (10).

The next question would be, whether any subsolution can be expressed by some V^ε. There will be a positive answer to this, but here we do not go into this problem [see: 4].

3. Minimal subsolutions and maximal supersolutions.

We have seen that any V^ε is a subsolution of (10). We now turn to the question of existence of minimal subsolutions and their properties. To this end we use the concept of [2], slightly generalize this to show that in a set of supersolutions there is a maximal element which in a sense coincides with a minimal subsolution. In this way we derive the existence of a viscosity solution for (10). Most of the techniques used here are taken from [2,3], so that we shortly summarize the results. For details, see [4].

Let C be the space of functionals on $[0,1] \times L^2(\mathbb{R}^n)$ which are differentiable in t and uniformly continuous in Π, satisfying some linear growth condition. Furthermore, consider the subset in C :

$$\tilde{S} = \{T \in C \,/\, T(1-s,\Pi) \leq \int_s^{1-\tau} (f_v, p_\Pi^v)dt + T(\tau, p_\Pi^v)$$

$$\text{for all } 0 \leq s \leq 1-\tau \leq 1,\ \Pi \in L^2(\mathbb{R}^n), v \in U_{ad}\} \quad . \tag{15}$$

Then it is easily seen [4] that any supersolution of (10) satisfies the conditions in the definition of $\tilde{S}$. On the other hand any $T \in \tilde{S}$ is a supersolution of (10). To see this, first assume that $T \in \tilde{S}$ is differentiable, $T \in C^{1,2}$.
Itô's formula for T gives us

$$d\,T(1-t,p_t) = \{-T_t + T_\Pi\,[A^{v*}\,p_t] + \tfrac{1}{2}\,T_{\Pi\Pi}\,[hp_t,hp_t]\}\,dt$$
$$+ T_\Pi\,[hp_t]dy \geq -(f_v,p_t) \tag{16}$$

and hence

$$-T_t + \min_{a \in U_{ad}} H(T_{\Pi\Pi}, T_\Pi, T, \Pi, a) \geq 0 \quad . \tag{17}$$

If T is only continuous in $\Pi \in E$, fix some $p \in E$, $t \in [0,1]$ and $(A,\xi) \in \partial_-\,T(p)$. Then there is a twice differentiable function $\psi(t,\Pi)$, such that

$$\begin{aligned} T\,(1-s,\Pi) &\geq \psi(s,\ ) \qquad \forall_{s,\Pi} \\ T\,(1-t,p) &= \psi(t,p) \\ \psi_t(t,\Pi) &= T_t\,(1-t,\Pi) \\ \psi_\Pi(t,\Pi) &= \xi \\ \psi_{\Pi\Pi}(t,\Pi) &= A \quad . \end{aligned} \tag{18}$$

This is taken from an equivalent definition of supersolutions [6].
Then

$$0 \leq -T(1-s,\Pi) + T(t,p_\Pi^v) + \int_s^{1-t} (f_v,p_\Pi^v)dt'$$

$$\leq -\psi(s,\Pi) + \psi(t,p) + \int_s^{1-t} (f_v,p_\Pi^v)dt'$$

Now we apply Itô's formula as above and conclude

$$- \psi_t + \min_{a\in U_{ad}} H(\psi_{\Pi\Pi},\psi_\Pi,\psi,p,a)$$

$$= - \psi_t + \min_{a\in U_{ad}} H(A,\xi,T,p,a) \geq 0 \quad .$$

As this can be done for any p,A,ξ, we see that T is a supersolution:

<u>Theorem</u> 3.1: Any $T \in \tilde{S}$ is a supersolution of (10), and every supersolution of (10) is in $\tilde{S}$.

Bensoussan considers a set $S(\Pi)$ which is analogously defined to our $\tilde{S}$. As he treats a discounted cost criterion, he does not treat the dependence on t. The results however 'generalize' to our case, and especially the main result in [2] carries over:

<u>Theorem</u> 3.2 [2]: There is a maximal element in $\tilde{S}$, say F_{max}, and

$$F_{max}(T,\Pi) = \inf_v J_\Pi(v) \ . \qquad (19)$$

In our terms this means that a maximal supersolution F_m exists, namely $F_m = F_{max}$.

Following Bensoussan's derivation of this result, this maximal element is constructed by an approximation of F_m by functionals F_h which are value functions

$$F_h(T,\Pi) = \inf_{w\in W_h} J_\Pi(w) \quad , \qquad (20)$$

where W_h is a set of piecewise constant controls.

Without loss of generality, let us assume that no piecewise constant control is optimal for the original problem. Then $F_h(t,\Pi)$ may be identified with some $V^\varepsilon(t,\Pi)$, the value function of some ε-optimal control. Again in our terms this means that F_m may be viewed as a limit of subsolutions of the form V^ε. Making use of a 'weak' stability result, we arrive at the desired result, that a maximal supersolution is a minimal subsolution, so that the following holds:

Theorem 3.3: The function F_m is a viscosity solution of Mortensen's equation in the sense that the following inequalities hold:

$$\forall_\Pi \quad \forall_{A,\xi\in\partial_+ F_m(\Pi)} \quad -\frac{\partial F_m}{\partial t} + \min_{a\in U_{ad}} H(A,\xi,F_m,\Pi,a) \leq 0$$

$$\forall_\Pi \quad \forall_{A,\xi\in\partial_- F_m(\Pi)} \quad -\frac{\partial F_m}{\partial t} + \min_{a\in U_{ad}} H(A,\xi,F_m,\Pi,a) \geq 0 \qquad (20)$$

Finally, $F_m(T,\Pi)$ is the value of the original control problem.

Remarks 3.4: (i) At this point I was going to point out the relevance of the above derivation for computing upper and lower values of games. But, just recently I received a preprint from W. Fleming where this is made rigorous.

(ii) All the above consideration are not going to give us some existence results for the problem. An optimal control should be of the form [see 1]

$$a(\tau,p) = \arg\min_{a\in K} \{a' \int \nabla V_\Pi(\tau,p)(z)p(z)dz + \int \phi(a,z)p(z)dz\}$$

$$u(t) = a(T - t, p_t) \qquad (21)$$

This control would be measurable w.r.t. the history of (p_t) and not of (y_t). These histories may be different, as may be seen from Cirelson's counterexample. This also is the key problem, if one wants to show that the value V^ε, (12), satisfies the Hamilton-Jacobi equations (13).

(iii) Although we think, that the above derivations might (again) illustrate the relations between Zakai's and Mortensen's equation, and especially the use of the viscosity solution concept, we feel that at the end we should mention some positive result: In [4] we derived from the above considerations some maximum principles, similar to those in [2] but for ε-optimal controls. A typical such result is:

Theorem 3.5: There exists an ε-optimal control v, ε > 0, such that for all $u \in U_{ad}$

$$-\varepsilon \leq \frac{1}{|u|} \sum_j \left[\int_{\mathbf{R}^n} \frac{\partial f}{\partial u_j}(x,v(t)) + \sum_i \frac{\partial \lambda}{\partial x_i} \frac{\partial g_i}{\partial x_j}(x,v(t)) \right] p(t,x)dx \leq \varepsilon \quad (22)$$

a.e. in t, a.s., where the adjoint λ is given by an explicit differential equation.
Again, this uses a deep result from [8].

Acknowledgement. We would like to thank P.L.Lions for sending us recent preprints on the subject of this article.

(1) V.E. BENEŠ, I. KARATZAS, On the relation of ZAKAI's and MORTENSEN's equations, SIAM.J. Control and Opt. 21 (3) 1983, 472-489

(2) A. BENSOUSSAN, Maximum principle and dynamic programming approaches of the optimal control of partially observed diffusions, Stochastics, 9 (3) 1983, 169-222

(3) A. BENSOUSSAN, M. ROBIN, On the convergence of the discrete time dynamic programming equation for general semi groups, SIAM.J. Control and opt. 20 (5) 1982, 722-746

(4) M. KOHLMANN, ε - optimal controls in partially observed control: a maximum principle and viscosity solutions for Mortensen's equation, preprint Universität Konstanz

(5) P.L. LIONS, Generalized solutions of Hamilton-Jacobi equations, Pitman APP, Boston-London-Melbourne, 1982

(6) P.L. LIONS, Optimal control of diffusion processes and Hamilton-Jacobi-Bellman equations, part 2: viscosity solutions and uniqueness, Comm. P.D.E 8 (11) 1983, 1229-1276

(7) M.G. CRANDALL, P.L. LIONS, Hamilton-Jacobi equations in infinite dimensions, part I: uniqueness of viscosity solutions, preprint Université de Paris IX - Dauphine (1984)

(8) I. EKELAND, Nonconvex minimization problems, Bull.AMS 1 (3) 1979, 443-474

(9) M. KOHLMANN, Concepts for the derivation of optimal partially observed controls, preprint Universität Konstanz 1984

(10) R.J. ELLIOTT, M. KOHLMANN, The variational principle and stochastic optimal control, Stochastics 3, 1980, 229-241

+) This work was supported by the SFB 72 of the DFG at the Universität Bonn, by the British Council during a visit to the Mathematical Institute at the University of Warwick, and by the AFF at the Universität Konstanz

Michael Kohlmann
Universität Konstanz
Postfach
D-7750 Konstanz, FRGermany

ON NECESSARY AND SUFFICIENT CONDITIONS FOR THE CONVERGENCE TO QUASICONTINUOUS SEMIMARTINGALES

K.Kubilius (Vilnius)

1. Introduction. In this paper the extended weak convergence introduced by D.Aldous [1] is considered. This convergence is stronger than the one considered in [4], [5]. We'll compare these both convergences and the convergence in probability of the predictable characteristics of semimartingales. The results of this article generalise the results of D.Aldous [1], R.Liptser and A.Shiryaev [7], [8], A.Butov [2], L.Slominski [9].

2. Extended stable weak convergence

Definition 1 ([3]). We say that a sequence of random elements (X^n), defined on a probability space $(\Omega, \mathcal{F}, P)$ and taking values in a Polish space $\mathcal{X}$, converges $\mathcal{G}$-stable weakly to X $(X^n \xrightarrow{\mathcal{D}(\mathcal{G})} X)$ if for any $A \in \mathcal{G} \subset \mathcal{F}$ and any bounded continuous function g on $\mathcal{X}$ $(g \in C(\mathcal{X}))$

$$EI_A g(X^n) \longrightarrow EI_A g(X).$$

Let $(\Omega, \mathcal{F}, P)$ be a complete probability space with a filtering family $\mathbb{F}=\{\mathcal{F}_t, t \geq 0\}$ of σ-algebras satisfying usual conditions. By $(D(S), \mathcal{D}(S))$ we denote the measurable space of all càdlàg functions $x : [0, \infty) \longrightarrow S$ endowed with Skorohod's J_1 topology, where S is a Polish space. Let $\mathcal{M}(S)$ denote the set of probability measures on $(D(S), \mathcal{D}(S))$. Introduce in $\mathcal{M}(S)$ the topology of weak convergence which makes $\mathcal{M}(S)$ a Polish space.

For each $t \geq 0$ and $\mathbb{F}$-adapted process X with trajectories in

D(S) there exists a regular conditional distribution $Z_t: \Omega \to \mathcal{M}(S)$ such that for each $A \in \mathcal{D}(S)$

$$Z_t(A)=P(X\in A \mid \mathcal{F}_t) \qquad P - a.e. \qquad (1)$$

and $Z \in D(\mathcal{M}(S))$ ([1]). Moreover for the process Z the equality (1) extends from constants t to all finite $\mathbb{F}$-stopping times

$$Z_\tau(A)=P(X\in A \mid \mathcal{F}_\tau) \qquad P - a.e.$$

Following D.Aldous, we call the process $(Z, \mathbb{F})$ the prediction process of $(X, \mathbb{F})$.

Later on we shall write $X^n \xrightarrow{\mathcal{D}_f(\mathcal{G})} X$, if for any k ($k \geq 1$), any $t_1, \ldots, t_k \in J(X)$ and any $A \in \mathcal{G}$, $g \in C(S^k)$

$$EI_A g(X^n_{t_1}, \ldots, X^n_{t_k}) \to EI_A g(X_{t_1}, \ldots, X_{t_k}),$$

where J(X) is the set of stochastic continuity points of the process X.

Let $\mathbb{F}^n=\{\mathcal{F}^n_t, t\geq 0\}$, $n\geq 1$, be a sequence of filtering families of σ-algebras satisfying usual conditions, let $((X^n, \mathbb{F}^n))$ a sequence of processes with trajectories in D(S) and prediction processes Z^n.

Definition 2 ([5]). We say that $(X^n, \mathbb{F}^n) \xrightarrow{\mathcal{D}_c(\mathcal{G})} (X, \mathbb{F})$, if $Z^n \xrightarrow{\mathcal{D}_f(\mathcal{G})} Z$.

Definition 3 ([1]). We say that $((X^n, \mathbb{F}^n))$ extended $\mathcal{G}$-stable weakly convergence to $(X, \mathbb{F})$ $((X^n, \mathbb{F}^n) \xrightarrow{(\mathcal{G})} (X, \mathbb{F}))$, if $(X^n, Z^n) \xrightarrow{\mathcal{D}(\mathcal{G})} (X, Z)$ on $D(S) \times \mathcal{M}(S)$.

3. Semimartingales

Let $(X, \mathbb{F})$ be d-dimensional semimartingale, defined on a probability space $(\Omega, \mathcal{F}, P)$ with filtration $\mathbb{F}$. Let us fix a continuous function $h: R^d \to R^d$, such that $h(x)=x$ for $|x| \leq \frac{1}{2}$, $=0$ for $|x| > 1$ and $|h(x)| \leq 1$. It is well known that the semimartingale $(X, \mathbb{F})$ can be uniquely represented in the form

$$X_t=X_0+\alpha_t+X_t^c+\int_0^t\int_{R_0^d} h(x)(p-\Pi)(ds,\ dx)+\int_0^t\int_{R_0^d}(x-h(x))p(ds,\ dx),\quad t\geq 0,$$

where α is a predictable process with locally integrable variation, X^c is a continuous local martingale, $p(dt,\ dx)$ is the measure of jumps of the process X, $\Pi(dt,\ dx)$ is the dual predictable projection of the measure $p(dt,\ dx)$, $R_0^d=R^d\setminus\{0\}$. We shall call (α, B, Π) the triplet of modified characteristics (t.m.ch.), where $B=(B^{(i,\ j)})$, $B^{(i,\ j)}=\langle X^{(i)c},\ X^{(j)c}\rangle$. Let us denote

$$\Gamma_t^{(i,j)}=B_t^{(i,j)}+\int_0^t\int_{R_0^d} h^{(i)}(y)h^{(j)}(y)\ \Pi(ds,\ dy)-$$

$$-\sum_{s\leq t}\int_{R_0^d} h^{(i)}(y)\ \Pi(\{s\},\ dy)\int_{R_0^d} h^{(j)}(y)\ \Pi(\{s\},\ dy),$$

$$1\leq i,\ j\leq d$$

$$\Pi_t(\varphi)=\int_0^t\int_{R_0^d}\varphi(y)\ \Pi(ds,\ dy)\qquad \varphi\in\tilde{C}_+(R_0^d),$$

where $\tilde{C}_+(R_0^d)$ is the set of nonnegative continuous bounded functions on R^d which vanish in the neighborhood of zero.

The measures p and Π can be considered as cádlág increasing processes with values in the space $R^+(R_0^d)$ of nonnegative Radon measures on R_0^d. The process B can be considered as a continuous increasing process with values in the space S_d^+ of nonnegative symmetric matrices.

4. The main result

Let us denote $\bar{\Omega}=\Omega\times D(R^d)$, $\bar{\mathcal{G}}=\mathcal{G}\otimes\mathcal{D}(R^d)$, $\mathbb{G}_+=\{\bar{\mathcal{G}}_{t+}=$ $=\mathcal{G}\otimes\mathcal{D}_{t+},\ t\geq 0\}$, $\mathbb{D}_+=\{\mathcal{D}_{t+},\ t\geq 0\}$, where $\mathcal{G}\subseteq\bigcap_n\mathcal{F}_0^n$, $\mathcal{D}_t=$

$= \sigma(x_s, s \leq t)$, $\mathcal{D}_{t+} = \bigcap_{\varepsilon > 0} \mathcal{D}_{t+\varepsilon}$, $t \geq 0$.

Let $(\hat{\alpha}, \hat{B}, \hat{\Pi})$ be a triplet of continuous $\mathbb{G}_+$-measurable processes defined on $\bar{\Omega}$ and taking values in $R^d \times S_d^+ \times R^+(R_0^d)$, where $\hat{\alpha}$ is a process with locally integrable variation, $\hat{B}$ and $\hat{\Pi}$ are increasing processes and for each $t > 0$

$$\int_{R_0^d} |x|^2 \wedge 1 \; \hat{\Pi}_t(dx) < \infty.$$

Let $(X^n, \mathbb{F}^n)$, $n \geq 1$, and $(X, \mathbb{F})$ be semimartingales with t.m.ch. (α^n, B^n, Π^n), $n \geq 1$, and $(\hat{\alpha}(\bar{X}), \hat{B}(\bar{X}), \hat{\Pi}(\bar{X}))$ correspondingly, where $\bar{X} : w \to (w, X(w))$, i.e. $\bar{X}$ is the mapping from Ω into $\bar{\Omega}$.

Further we shall make use of the following conditions:

(O) $X_0^n \xrightarrow{d(\mathcal{S})} X_0$,

(A) $\forall t \notin I$, $\varepsilon > 0$ $\quad \Pi^n(t, |x| > \varepsilon) \xrightarrow{P} 0$,

(sup B) $\forall t > 0$ $\quad \sup_{s \leq t} |\alpha_s^n - \hat{\alpha}_s(\bar{X}^n)| \xrightarrow{P} 0$,

(C) $\forall t \notin I$, $z \in R^d$, $\quad |(\Gamma_t^n z, z) - (\hat{\Gamma}_t(\bar{X}^n) z, z)| \xrightarrow{P} 0$,

(D) $\forall t \notin I$, $\varphi \in \tilde{C}_+(R_0^d)$ $\quad |\Pi_t^n(\varphi) - \hat{\Pi}_t(\bar{X}^n, \varphi)| \xrightarrow{P} 0$,

(E) for all $t > 0$ and $\varepsilon > 0$

$$\lim_{l \to \infty} \; \overline{\lim_n} \; P(\Pi^n(t, \{|x| > l\}) > \varepsilon) = 0,$$

(F) for all $z \in R^d$, $\varphi \in \tilde{C}_+(R_0^d)$ the sequences of measures corresponding to processes $(\hat{\alpha}(\bar{X}^n))_n$, $((\hat{\Gamma}(\bar{X}^n) z, z))_n$, $(\hat{\Pi}(\bar{X}^n, \varphi))_n$ are relatively compact,

(H) for P-a.e. w, each $z \in R^d$, $\varphi \in \tilde{C}_+(R_0^d)$, $t > 0$ the processes $\hat{\alpha}_t(w, x)$, $(\hat{\Gamma}_t(w, x) z, z)$, $\hat{\Pi}_t(w, x, \varphi)$ are continuous with respect to x. Every where in these conditions I denotes a countable subset of R_+.

Let $Q=Q(s,w, f)$, $f\in D(R^d)$, be a probability measure on $D(R^d)$ such that $Q(x_u=f_u,\ u\le s)=1$ and $(x_u)_{u\ge s}$ is a semimartingale with t.m. ch. $(\hat{\alpha}_u(w, x)-\hat{\alpha}_s(w, x),\ \hat{B}_u(w, x)-\hat{B}_s(w, x),\ \hat{\Pi}(w, x, (s, u]\times dy))_{u\ge s}$

We shall say that condition (MP_s) is satisfied, if for each w and each $f\in D(R^d)$ the measure $Q=Q(s,w, f)$ is unique and the mapping $(s,w, f)\to Q(s,w, f)$ is $\mathcal{B}([0,\infty))\otimes \mathcal{G}\otimes \mathcal{D}(R^d)$-measurable.

Theorem. Suppose that the conditions $(MP_s)_{s\ge 0}$, (H) are satisfied.

Then the following statements are equivalent:

1) $(X^n, \mathbb{F}^n)\xrightarrow{(\mathcal{G})}(X, \mathbb{F})$ and $(\alpha^n)\in(RC)$, i.e. the sequence of distributions corresponding to processes $(\alpha^{n,\cdot})$ is relatively compact in $D(R^d)$ and limit points are concentrated on the subset of continuous function with locally bounded variation,

2) $(X^n, \mathbb{F}^n)\xrightarrow{\mathcal{D}c(\mathcal{G})}(X, \mathbb{F})$ and (α^n), $((\Gamma^n z, z))$, $(\Pi^n(\varphi))\in(RC)$ for all $z\in R^d$, $\varphi\in\tilde{C}_+(R_0^d)$,

3) (O), (sup B), (C), (D), (E), (F).

Remark 1. If the processes α^n, $n\ge 1$, are increasing then the condition $(\alpha^n)\in(RC)$ can be omitted in 1).

Corollary 1. Let $(X, \mathbb{F})$ be a continuous semimartingale with t.m.ch. $(\hat{\alpha}(\bar{X}), \hat{B}(\bar{X}), 0)$. Suppose that the conditions $(MP_s)_{s\ge 0}$, (H) are satisfied. Then the following statements are equivalent:

1) $(X^n, \mathbb{F}^n)\xrightarrow{(\mathcal{G})}(X, \mathbb{F})$, $(\alpha^n)\in(RC)$,

2) $(X^n, \mathbb{F}^n)\xrightarrow{\mathcal{D}c(\mathcal{G})}(X, \mathbb{F})$, $(\alpha^n)\in(RC)$,

3) (sup B), $X^n\xrightarrow{\mathcal{D}(\mathcal{G})}X$,

4) (O), (A), (sup B), (C), (F).

Remark 2. The statements 3) and 4) of the corollary 1 are equivalent even though the conditions $(MP_s)_{s\geq 0}$ are exchange by condition (MP_0) (cf. [2], [8]).

Example (see [2]). Let conditions (0), (A), (sup B), (C) are satisfied. Let (X, F) be a continuous one-dimensional semimartingale with t.m.ch. $(\hat{\mathcal{L}}(X), \hat{B}(X), 0)$ and $\mathcal{G} = \{\emptyset, \Omega\}$. Suppose that for any $T>0$ there exists a continuous increasing function $K_t=K_t(T)<\infty$, $K_0=0$, such that for any $p, q\leq t\leq T$ with $|p-q|\leq\Delta$

$$|\hat{\mathcal{L}}_p(x)-\hat{\mathcal{L}}_q(x)|+|B_p(x)-B_q(x)|^{\frac{1}{2}}\leq K_\Delta\,(1+\sup_{s\leq t}|x_s|).$$

Then 4) in corollary 1 is satisfied.

Corollary 2 (cf.[4]). Suppose the t.m.ch. $(\hat{\mathcal{L}}, \hat{B}, \hat{\Pi})$ doesn't depend on x. Then the following statments are equivalent:

1) $(X^n, \mathbb{F}^n) \xrightarrow{(\mathcal{G})} (X, \mathbb{F})$, $(\alpha^n)\in(RC)$,

2) $(X^n, \mathbb{F}^n) \xrightarrow{(\mathcal{G})} (X, \mathbb{F})$, (sup B),

3) (0), (sup B), (C), (D). ∎

Let

$$dX_t^n=b_n(X_t^n)dt+\sigma_n(X_t^n)dW_t^n+\int F_n(X_t^n, y)q^n(dt, dy),\ X_0^n=x,\ n\geq 0, \qquad (2)$$

where W^n, $n\geq 0$, are standart Wiener processes, $p^n(dt, dy)$, $n\geq 0$, are Poisson measures on $[0,\infty)\times R$, $q^n(dt, dy)=p^n(dt, dy)-\pi(dy)dt$, $n\geq 0$, are martingale measures, b_0, σ_0 are continuous functions and for all x

$$\int|F_0(x, y)-F_0(x', y)|^2\pi(dy)\rightarrow 0,\ x'\rightarrow x$$

and

$$|b_n(x)|^2+|\sigma_n(x)|^2+\int|F_n(x, y)|^2\pi(dy)\leq C(1+|x|^2),\ n\geq 0.$$

Let X^0 be a unique weak solution of the equation (2).

Corollary 3. The convergence $(X^n, \mathbb{F}^n)\rightarrow(X, \mathbb{F})$ holds iff for each $t>0$, $\varphi\in\tilde{C}_+(R_0)$

$$\sup_{s\leq t}\left|\int_0^s [b_n(X_u^n)-b_0(X_u^n)]du\right| \xrightarrow{P} 0,$$

$$\left|\int_0^t [\sigma_n^2(X_s^n)-\sigma_0^2(X_s^n)]ds+\right.$$

$$\left.\int_0^t\int [h^2(F_n(X_s^n, y))-h^2 F_0(X_s^n, y))]\pi(dy)ds\right| \xrightarrow{P} 0,$$

$$\left|\int_0^t\int [\varphi(F_n(X_s^n, y))-\varphi(F_0(X_s^n, y))]\pi(dy)ds\right| \xrightarrow{P} 0.$$

5. We say that a sequence of processes $((Y^n, \mathbb{F}^n))$ belongs to the class (AUQ) $(((Y^n, \mathbb{F}^n))\in$ (AUQ)) if for each $T>0$, any sequences (δ_n) and (T_n)

$$Y^n_{T_n+\delta_n}-Y^n_{T_n} \xrightarrow{P} 0,$$

where $\delta_n\downarrow 0$, (T_n) is a sequence of $\mathbb{F}^n$-stopping times, $T_n\leq T$. $((Y^n, \mathbb{F}^n))\in$ (T) if for each $T>0$ the sequence $(\sup_{t\leq T}|Y_t^n|)$ is tight in R.

Let $M_b^+(\Omega\times\mathfrak{X})$ be a space of finite non-negative measures on $(\Omega\times\mathfrak{X}, \mathcal{G}\otimes\mathcal{B}(\mathfrak{X}))$ with the weakest topology such that the mappings

$$\mu\rightarrow\mu(g)\equiv\int g(w, x)\mu(dw, dx)$$

are continuous for each bounded measurable function g which is continuous in x for all fixed $w\in\Omega$ $(g\in B_{mc}(\Omega\times\mathfrak{X}))$, where $\mathfrak{X}$ is a Polish space (see [3]). Denote $\mu_X(dx, dx)=P(dw)\,\varepsilon_{X(w)}(dx)$, where ε_a is a Dirac measure. According to [3] (X^n) is $\mathcal{G}$-stably weakly convergent iff (μ_{X^n}) is convergent in $M_b^+(\Omega\times\mathfrak{X})$. Note that measures μ_X and their limits belongs to such closed subset $M(P)$

of probability measures that $\mu(A\times\mathcal{X})=P(A)$, $A\in\mathcal{G}$.

Proof of theorem. The theorem we shall prove in some steps. At first we shall prove that 1) $\Rightarrow$ 2). Note that it is enough to prove that $(((\Gamma^n z, z), \mathbb{F}^n)), ((\Pi^n(\varphi), \mathbb{F}^n)) \in (AUQ)\cap(T)$ for all $z\in R^d$, $\varphi\in\tilde{C}_+(R_0^d)$ (see proposition 2 in [4]).

For fix $\varphi\in\tilde{C}_+(R_0^d)$ put

$$f(x)=x-x_0-\sum_{t\leq\cdot}(\Delta x_t-h(\Delta x_t)),$$

$$g(x)=\sum_{t\leq\cdot}\varphi(\Delta x_t).$$

It is not difficult to see that f and g are continuous mappings from $D(R^d)$ to $D(R^d)$. So from the convergence $(X^n, \mathbb{F}^n) \xrightarrow{(\mathcal{G})} (X, \mathbb{F})$ and continuous mapping theorem 43.1 in [1] it follows that $(f(X^n), \mathbb{F}^n) \xrightarrow{(\mathcal{G})} (f(X), \mathbb{F})$ and $(g(X^n), \mathbb{F}^n) \xrightarrow{(\mathcal{G})} (g(X), \mathbb{F})$. Since f(X) and g(X) are quasicontinuous processes, then from theorem 38.5 in [1] it follows that $(f(X^n), \mathbb{F}^n), (g(X^n), \mathbb{F}^n)\in(AUQ)$. Now from Lenglart's inequality and the fact that (see [6])

$$(X^n, \mathbb{F}^n) \xrightarrow{\mathcal{D}c(\mathcal{G})} (X, \mathbb{F}),\ (\alpha^n)\in(RC) \Rightarrow (\sup B)$$

it is easy to prove that $(((\Gamma^n z, z), \mathbb{F}^n))\in(AUQ)$, $((\Pi^n(\varphi), \mathbb{F}^n))\in(AUQ)$ for any $z\in R^d$, $\varphi\in\tilde{C}_+(R_0^d)$.

So it's remain to prove that $(((\Gamma^n z, z), \mathbb{F}^n)), ((\Pi^n(\varphi), \mathbb{F}^n))\in(T)$. From lemma 2 in [7], condition (sup B) and convergence $f(X^n) \xrightarrow{\mathcal{D}} f(X)$ it follows that $(\Gamma^n z, z)\in(T)$ for all $z\in R^d$. In such a way we prove that $(\Pi^n(\varphi))\in(T)$.

The equivalents of the statements 2) and 3) was proved in [6].

Now we shall prove the last implication 3) $\Rightarrow$ 1). At first note that 3) implies $((X^n, \mathbb{F}^n)) \in (AUQ) \cap (T)$ (see for example [5]). So it is enough to prove that for any $T > 0$ and any sequences (δ_n), (τ_n)

$$L(Z^n_{\tau_n}, Z^n_{\tau_n+\delta_n}) \xrightarrow{P} 0 \qquad (3)$$

where $\delta_n \downarrow 0$, $\delta_n \leq 1$ $\forall n \geq 1$, (τ_n) is a sequence of $\mathbb{F}^n$-stopping times $\tau_n \leq T$, L is the Levy - Prokhorov metric on $\mathcal{M}(R^d)$.

Consider the sequence of random elements $((X^n, \sigma_n, Z^n_{\tau_n}, Z^n_{\sigma_n}))$ with values in $D(R^d) \times [0,T+1] \times (\mathcal{M}(R^d))^2$, where $\sigma_n = \tau_n + \delta_n$. From lemma 43.6 in [1] and condition $((X^n, \mathbb{F}^n)) \in (AUQ) \cap (T)$ it follows that the both sequences $(Z^n_{\tau_n})$ and $(Z^n_{\sigma_n})$ are tight. Now it is easy to see that the sequence of measures $(\mu_{(X^n, \sigma_n, Z^n_{\tau_n}, Z^n_{\sigma_n})})$ is relatively compact. Assume that for some subsequence (without restriction we can consider the same sequence), for each $g \in B_{mc}(\Omega \times [0, T+1] \times (\mathcal{M}(R^d))^2)$

$$Eg(X^n, \sigma_n, Z^n_{\tau_n}, Z^n_{\sigma_n}) \to \int g(w,x,\theta,\lambda,\nu)\mu(dw,dx,d\theta,d\lambda,d\nu) \qquad (4)$$

for some measure μ on $(\Omega \times D(R^d) \times [0,T+1] \times (\mathcal{M}(R^d))^2, \mathcal{G} \otimes \mathcal{B}([0,T+1\,) \otimes \mathcal{D}(R^d) \otimes \mathcal{B}((\mathcal{M}(R^d))^2))$.

Note that

$$E(h(X^n)|\mathcal{F}^n) = \int h(w, x) Z^n_\tau(dx) \quad \text{P-a.e.} \qquad (5)$$

for each finite $\mathbb{F}^n$-stopping time τ and any $\mathcal{F}_\tau \otimes \mathcal{D}(R^d)$-measurable real function h such that $Eh(X^n) < \infty$.

Suppose that $f \in B_{mc}(\Omega \times D(R^d) \times [0,T+1] \times \mathcal{M}(R^d))$. Then (4), (5) imply

$$Ef(X^n, \sigma_n, Z^n_{\tau_n}) = E\int f(w, x, \sigma_n, Z^n_{\tau_n}) Z^n_{\tau_n}(dx)$$

$$\to \iint f(w, x, \theta, \lambda)\, \lambda(dx)\, \mu(dw, D(R^d), d\theta, d\lambda, \mathcal{M}(R^d))$$

$$Ef(X^n, \sigma_n, Z^n_{\sigma_n}) = E\int f(w, x, \sigma_n, Z^n_{\sigma_n})\, Z^n_{\sigma_n}(dx)$$

$$\to \iint f(w, x, \theta, \nu)\, \nu(dx)\, \mu(dw, D(R^d), d\theta, \mathcal{M}(R^d), d\nu).$$

Denote

$$\hat{p}_t(\varphi) = \sum_{s \le t} \varphi(\Delta x_s),\ \varphi \in \tilde{C}_+(R^d_0),$$

$$\hat{M}_t(w, x) = f_t(x) - \hat{\alpha}_t(w, x).$$

Now we shall prove that for $\mu(dw, dx, d\theta, d\lambda, \mathcal{M}(R^d))$ - a.e. (w, x, θ, λ) ($\mu(dw, dx, d\theta, \mathcal{M}(R^d), d\nu)$-a.e. (w, x, θ, ν)) $(\hat{M}_t(w,\cdot))_{t \ge \theta}$, $((\hat{M}_t(w,\cdot), z)^2 - (\hat{\Gamma}_t(x,\cdot)z, z))_{t \ge \theta}$, $(\hat{p}_t(\varphi) - \hat{\Pi}_t(w,\cdot,\varphi))_{t \ge \theta}$ are $(\lambda, \mathbb{D}_+)$ - local martingales $((\nu, \mathbb{D}_+)$ - local martingales) and

$$\lambda(x_t \in dy) = \varepsilon_{x_t}(dy),\ t < \theta \tag{6}$$

$$\nu(x_t \in dy) = \varepsilon_{x_t}(dy),\ t < \theta.$$

Then from the conditions $(MP_s)_{s \ge 0}$ it will follow that $\lambda = \nu$ $\mu(dw, dx, d\theta, d\lambda, d\nu)$-a.e. So (3) will hold.

Let

$$T_m(w, x) = \inf\{t : |\hat{M}_t(w, x)| > m\}$$

$$S_n = \inf\{t : |f_t(X^n) - \alpha^n_t| > m+4\} \wedge T_m(\bar{X}^n)$$

and m be such that $P(T_{m-0}(\bar{X}) \ne T_m(\bar{X})) = 0$.

If $s, t \in J(X)$, $s < t$, $g \in B_{mc}(\Omega \times [0,T+1] \times \mathcal{M}(R^d))$, k is $\mathcal{D}_s$-measurable bounded and $\mu(\Omega, dx, [0,T+1], \mathcal{M}(R^d), \mathcal{M}(R^d))$-a.e. continuous function, then

$$E((M^n_{(t \vee \tau_n) \wedge S_n} - M^n_{(s \vee \tau_n) \wedge S_n})k(X^n_{\cdot \vee \tau_n})g(\sigma_n, Z^n_{\tau_n}))$$

$$\to \iint (\hat{M}_{(t \vee \theta) \wedge T_m(w,x)}(w, x)\ \hat{M}_{(s \vee \theta) \wedge T_m(w,x)}(w, x))k(x_{\cdot \vee \theta})x$$

$$g(w,\theta,\lambda)\ \lambda(dx)\ \mu(dw,\ D(R^d),\ d\theta,\ d\lambda,\ \mathcal{M}(R^d))=0.$$

Since s, t, k, g are arbitrary, then the process $(\hat{M}_t(w,\ x))_{t\geq\theta}$ is $(\lambda,\mathbb{D}_+)$-local martingale for $\mu(dw,\ D(R^d),\ d\theta,\ d\lambda,\ \mathcal{M}(R^d))$-a.e. (w,θ,λ).Other cases are proved similarly.

if $h\in C(R)$, $g\in C(R^d)$, then

$$h(0)=Eh(\int g(y_{s\wedge\tau_n})Z^n_{\tau_n}(dy)-g(X^n_{s\wedge\tau_n}))$$

$$\to\iint h(\int g(y_{s\wedge\theta})\lambda(dy)-g(x_{s\wedge\theta}))\ \mu(dw,\ dx\ ,\ d\theta,\ d\lambda,\ d\nu)\ .$$

Since h and g are arbitrary, then (6) is hold.

References

1. D.Aldous, Weak convergence of stochastic processes for processes viewed in the Strasbourg manner, preprint, 1978.
2. A.Butov, On a problem of weak convergence of a sequence of semimartingales to diffusion type process. Uspeki Mat.Nauk, 38, 5, 181-182, 1983.
3. J.Jacod, J.Memin, Sur un type de convergence intermédiaire entre la convergence en loi et la convergence en probabilité. Lect. Notes in Math., 850, 529-546, 1981.
4. K.Kubilius, R.Mikulevičius, On necessary and sufficient conditions for the convergence of semimartingales and point processes. I, II. Lietuvos matem.rink., 24, 3, 139-147, 1984 ; 4, 99-115, 1984 .
5. K.Kubilius, R.Mikulevičius, On necessary and sufficient conditions for the convergence of semimartingales. Lect.Notes in Math., 1021, 339-351, 1983.
6. K.Kubilius, R.Mikulevičius, On necessary and sufficient conditions

for the convergence to non-quasicontinuous semimartingales. Lect. Notes in Control and Inf.Sc., to appear, 1985.

7. Liptser R, A.Shiryaev, On necessary and sufficient conditions in functional central limit theorem for semimartingales. Theory Probab. Appl., 26, 1, 132-137, 1981.

8. R.Liptser, A.Shiryaev, Weak convergence of a sequence of semimartingales to diffusion type process. Math.Sbornik, 121, 2, 176-200, 1983.

9. L.Slominski, Necessary and sufficient conditions for extended convergence of semimartingales, preprint, 1985.

Institute of Mathematics and Cybernetics

Academy of Sciences of the Lithuanian SSR

LIMIT THEOREMS FOR STOCHASTIC DIFFERENTIAL EQUATIONS AND STOCHASTIC FLOWS OF DIFFEOMORPHISMS

Hiroshi Kunita
Department of Applied Science
Kyushu University 36
Fukuoka 812, Japan

The approximation theories and limit theorems related to stochastic differential equations have been studied by many authors from various motivations. The purpose of this report is to look at the problems from the points of the views of the stochastic flow of diffeomorphisms, and present a unified method for a large class of problems.

In Section 1, we survey three limit theorems related to the diffusion processes and stochastic flows. The first is the approximation of the stochastic differential equation developed by Wong-Zakai [27], Ikeda-Nakao-Yamato [7], Ikeda-Watanabe [8], Malliavin [20], Bismut [2], Shu [25] etc. Here, the Brownian motions defining the stochastic differential equation are approximated by sequences of processes with piecewise smooth paths. Polygonal approximations and the approximations by the mollifiers are widely used. The second is the limit theorem for suitable stochastic ordinary differential equations with the small parameter ε, studied by Khasminskii [12], Papanicolaou-Kohler [22], Borodin [4], Kesten-Papanicolaou [9] etc. Under various conditions, they proved that, after a suitable change of the scale of the time, the solutions converge weakly to a diffusion process as $\varepsilon \to 0$. The third is the limit theorem studied by Papanicolaou-Stroock-Varadhan [23], concerning the driving processes and driven processes.

In order to discuss these limit theorems rigorously in a unified method, the recent results on stochastic differential equations and stochastic flows are needed. We discuss these facts in Section 2 following partly to Le Jan [18], Le Jan-Watanabe [19] and Fujiwara-Kunita [5].

In Section 3, we shall formulate the limit theorems and state three theorems. The first (Theorem 3.1) is a rather abstract theorem. Assumptions are stated in the language of the conditional expectations and martingales. Then we consider the two special cases. Theorem 3.2 discusses the case of the mixing property and Theorem 3.3 deals with the case of the ergodic property. Then we check how the limit theorems stated in Section 1 are derived from these theorems. In Section 4, we apply these theorems to limits for stochastic partial differential equations.

1. SURVEYS TO SOME LIMIT THEOREMS RELATED TO STOCHASTIC FLOWS

1.1 We first consider the problem of approximating the solution of the stochastic differential equation. Consider the stochastic differential equation in R^d:

$$(1.1)\qquad dx_t = \sum_{k=1}^{r} F_k(t,x_t)\circ dB_t^k + G(t,x_t)dt, \qquad t\in[0,T],$$

where $B_t = (B_t^1, \ldots, B_t^r)$ is a standard Brownian motion and $F_k(t,x)$, $k = 1, \ldots, r$ are R^d-valued functions having bounded continuous derivatives[1)] of all orders with respect to x, and $\circ dB_t^k$ denotes the Stratonovich integral. Let $B_t^\varepsilon = (B_t^{\varepsilon,1}, \ldots, B_t^{\varepsilon,r})$, $\varepsilon > 0$ be a system of the stochastic processes, piecewise smooth in t, converging uniformly to B_t in $L^2(P)$- sense as $\varepsilon \to 0$. For each ε, we consider the stochastic ordinary differential equation

$$\frac{dx_t^\varepsilon}{dt} = \sum_{k=1}^{r} F_k(t, x_t^\varepsilon)\dot{B}_t^{\varepsilon,k} + G(t, x_t^\varepsilon) \tag{1.2}$$

where $\dot{B}_t^\varepsilon = \frac{d}{dt} B_t^\varepsilon$. Let $\phi_t(x)$ and $\phi_t^\varepsilon(x)$ be the corresponding solutions of equations (1.1) and (1.2) starting at x at time 0. The question we are concerned is whether $\phi_t^\varepsilon(x)$, $\varepsilon > 0$ converges to $\phi_t(x)$ or not as $\varepsilon \to 0$. Perhaps, Wong-Zakai [27] was the first paper studying the problem. They showed that for each x, $\phi_t^\varepsilon(x)$ converges to $\phi_t(x)$ uniformly in t in $L^2(P)$-sense, provided that B_t^ε, $\varepsilon > 0$ is a polygonal approximation of B_t, i.e.,

$$\dot{B}_t^\varepsilon = \frac{1}{\varepsilon}(B_{(n+1)\varepsilon} - B_{n\varepsilon}), \quad \text{if} \quad n\varepsilon < t < (n+1)\varepsilon.$$

Now the map $\phi_t^\varepsilon(\cdot, \omega) : R^d \to R^d$ is a diffeomorphism, since for each ω, it is the solution of the ordinary differential equation. Thus ϕ_t^ε can be regarded as a continuous process with values in G^∞ = Diffeo(R^d) = the topological group consisting of C^∞-diffeomorphisms of R^d. It is called a <u>stochastic flow of diffeomorphisms.</u> The similar property is valid to ϕ_t, but the proof is by no means simple though several proofs are known. Here we shall sketch a proof by the method of the approximation. By Blagoveschenskii-Freidlin [3], the solution $\phi_t(x)$ of equation (1.1) has a modification such that for almost all ω, $\phi_t(x, \omega)$ is smooth in x and derivatives

$$D^\alpha \phi_t(x, \omega) = \left(\frac{\partial}{\partial x_1}\right)^{\alpha_1} \cdots \left(\frac{\partial}{\partial x_d}\right)^{\alpha_d} \phi_t(x, \omega)$$

are continuous in (t,x). Thus for each t, $\phi_t(\cdot, \omega) : R^d \to R^d$ is a smooth map, so that ϕ_t is a stochastic process with values in $C = C^\infty(R^d; R^d)$. Now Ikeda-Watanabe [8] and Bismut [9] have shown that both $D^\alpha \phi_t^\varepsilon$, $\varepsilon > 0$ and $D^\alpha(\phi_t^\varepsilon)^{-1}$, $\varepsilon > 0$ converge uniformly on compact sets in $L^2(P)$-sense, and the former limit coincides with $D^\alpha \phi_t$. As a consequence, we see that the map $\phi_t(\cdot, \omega) : R^d \to R^d$ is also a diffeomorphism a.s. Thus ϕ_t is a stochastic flow of diffeomorphims. Further, since $\phi_{t_{i+1}} \circ \phi_{t_i}^{-1}$, $i = 0, \ldots, n-1$ are independent for any $0 \leq t_0 < t_1 < \ldots < t_n \leq T$, it is often called a <u>Brownian motion with values in G^∞.</u>

Another interesting approximation of B_t is made by the mollifier. Let $c(t)$, $-\infty < t < \infty$ be a smooth nonnegative function supported by $[0,1]$ such that $\int c(t)\,dt = 1$. Define

---

1) Derivatives are continuous in (t,x).

$$B_t^\varepsilon = \frac{1}{\varepsilon}\int c(\frac{t-s}{\varepsilon})B_s ds.$$

It is smooth in t and converges uniformly to B_t as $\varepsilon \to 0$. Malliavin [20] and Shu [25] showed that the approximation theorem similar to the above is valid for the mollifier approximation, too. A more general approximation including the above two was studied by Ikeda-Nakao-Yamato [7]. In their work, the limit $\phi_t(x)$ may not satisfy (1.1), but it satisfies a modified equation with the "correction term $c(t,x)$" i.e.,

$$dx_t = \sum_{k=1}^{r} F_k(t,x_t)\circ dB_t^k + G(t,x_t)dt + c(t,x_t)dt.$$

<u>1.2</u> We shall next look at the limit theorems for stochastic ordinary differential equations, studied by Khasminskii [12], Papanicolaou-Kohler [22], Borodin [4], Kesten-Papanicolaou [9] and others. Consider a stochastic ordinary differential equation with parameter ε.

$$(1.3) \qquad \frac{dx}{dt} = \varepsilon f(x,t,\varepsilon^2 t,\omega) + \varepsilon^2 g(x,t,\varepsilon^2 t,\omega), \qquad t\in[0,\infty)$$

where $f(x,t,u,\omega)$ and $g(x,t,u,\omega)$ are R^d-valued random fields having bounded continuous derivatives of all orders with respect to x a.s. Let $\psi_t^\varepsilon(x,\omega)$ be the solution of equation (1.3) starting at x at time 0. Since the right hand side of (1.3) converges to 0 as $\varepsilon \to 0$, $\psi_t^\varepsilon(x)$, $\varepsilon > 0$ converges to $\psi_t^0(x) \equiv x$ as $\varepsilon \to 0$. Next, we change the scale of the time and consider $\phi_t^\varepsilon(x) \equiv \psi_{t/\varepsilon}^\varepsilon$. Then it satisfies

$$(1.4) \qquad \frac{d\phi_t^\varepsilon}{dt} = f(\phi_t^\varepsilon,\frac{t}{\varepsilon},\varepsilon t) + \varepsilon g(\phi_t^\varepsilon,\frac{t}{\varepsilon},\varepsilon t).$$

Khasiminskii [11] showed that under an ergodic assumption to f, ϕ_t^ε converges to a deterministic curve $\overline{\phi}_t$ a.s., which is a solution of the equation

$$(1.5) \qquad \frac{dx}{dt} = \overline{f}(x),$$

where $\overline{f}(x) = E[f(x,0,0)]$. This may be regarded as a law of the large number for the process ψ_t^ε.

We shall consider the case $\overline{f} = 0$ further. Obviously $\overline{\phi}_t(x) \equiv x$ is satisfied for any t. Then we again change the scale of the time and consider $\phi_t^\varepsilon = \psi_{t/\varepsilon^2}^\varepsilon$. It satisfies

$$(1.6) \qquad \frac{d\phi_t^\varepsilon}{dt} = \frac{1}{\varepsilon}f(\phi_t^\varepsilon,\frac{t}{\varepsilon^2},t) + g(\phi_t^\varepsilon,\frac{t}{\varepsilon^2},t).$$

Under certain assumptions to f and g, the above authors showed that for each x $\phi_t^\varepsilon(x)$, $\varepsilon > 0$ converges weakly to a diffusion process $\phi_t(x)$. It can be regarded as a central limit theorem for $\phi_t^\varepsilon(x)$, $\varepsilon > 0$.

Recently, the author [14], [15] proved the weak convergence of ϕ_t^ε, $\varepsilon > 0$ as stochastic flows under hypotheses that are close to Kesten-Papanicolaou [9]. Set

$$(1.7) \qquad \underline{\underline{G}}_{s,t} = \sigma(f(x,u,v),g(x,u,v)\ ;\ s \leq u \leq t,\ \ 0 \leq v < \infty)$$

and define the strong mixing rate $\beta(t)$ by

$$(1.8)\qquad \beta(t) = \sup_s \sup_{A\in \underline{\underline{G}}_{0,s},\, B\in \underline{\underline{G}}_{s+t,\infty}} |P(A\cap B) - P(A)P(B)|.$$

We assume that

$$(1.9)\qquad \int_0^\infty \beta(\tau)^{\frac{1}{3+3\vee d}}\, d\tau < \infty .$$

Assume further that for each $M > 0$ and multi-index β there is a positive constant $C = C_{M,\beta}$ satisfying

$$(1.10)\qquad E[\sup_{|x|\leq M} |D^\beta f(x,u,v)|^6]^{\frac{1}{3}} \leq C, \qquad \forall u,v$$

and the following limits exist uniformly on compact sets :

$$(1.11)\qquad A_{ij}(x,y,t) = \lim_{\varepsilon\to 0} \frac{1}{\varepsilon^3}\int_t^{t+\varepsilon}\int_t^\tau E[f_i(x,\frac{\sigma}{\varepsilon^2},\sigma)f_j(y,\frac{\tau}{\varepsilon^2},\tau)]\,d\sigma d\tau,$$

$$(1.12)\qquad b_i(x,t) = \lim_{\varepsilon\to 0} \frac{1}{\varepsilon}\int_t^{t+\varepsilon} E[g_i(x,\frac{\sigma}{\varepsilon^2},\sigma)]\,d\sigma,$$

$$(1.13)\qquad c_{ij}(x,t) = \lim_{\varepsilon\to 0} \frac{1}{\varepsilon^3}\int_t^{t+\varepsilon}\int_t^\tau E[f_j(x,\frac{\sigma}{\varepsilon^2},\sigma)\frac{\partial}{\partial x^j} f_i(x,\frac{\tau}{\varepsilon^2},\tau)]\,d\sigma d\tau,$$

where A_{ij}, b_i, c_{ij} have bounded continuous spacial derivatives of all orders. Set

$$a_{ij}(x,y,t) = A_{ij}(x,y,t) + A_{ji}(y,x,t),$$

$$c_i(x,t) = \sum_{j=1}^d c_{ij}(x,t).$$

Then we have the following assertions (i) - (iii).

(i) $\phi_t^\varepsilon(x)$, $\varepsilon > 0$ *converges weakly to* $\phi_t(x)$ *as stochastic flows, i.e.,* $D^\alpha\phi_t^\varepsilon(x)$, $\varepsilon > 0$ *and* $D^\alpha(\phi_t^\varepsilon)^{-1}(x)$, $\varepsilon > 0$ *converge uniformly on compact sets in the sense of the laws for any* α. *(See Section 3.1). The limit* ϕ_t *is a Brownian motion with values in* $G^\infty = \text{Diffeo}(R^d)$ *and satisfies*

$$a_{ij}(x,y,t) = \lim_{h\to 0}\frac{1}{h}E[(\phi_{t+h}\circ\phi_t^{-1}(x) - x)_i(\phi_{t+h}\circ\phi_t^{-1}(y) - y)_j],$$

$$b_i(x,t) + c_i(x,t) = \lim_{h\to 0}\frac{1}{h}E[(\phi_{t+h}\circ\phi_t^{-1}(x) - x)_i].$$

(ii) The integral of the right hand side of (1.6):

$$(1.14)\qquad X_t^\varepsilon(x) = \frac{1}{\varepsilon}\int_0^t f(x,\frac{\tau}{\varepsilon^2},\tau)\,d\tau + \int_0^t g(x,\frac{\tau}{\varepsilon^2},\tau)\,d\tau, \qquad \varepsilon > 0$$

and its derivatives $D^\alpha X_t^\varepsilon(x)$, $\varepsilon > 0$ *converges uniformly on compact sets in the sense of the laws for any* α. *Furthermore, the limit* $X_t = (X_t(x),\ x\in R^d)$ *is a Brownian motion with values in* $C^\infty = C^\infty(R^d;R^d)$ *with the mean* $\int_0^t b_i(x,\tau)d\tau$ *and the covariance* $\int_0^t a_{ij}(x,y,\tau)d\tau$.

(iii) The limit $\phi_t(x)$ *and* $X_t(x) \equiv X(x,t)$ *are linked by the generalized stochastic*

differential equation

$$(1.15) \qquad d\phi_t = X(\phi_t, dt) + c(\phi_t, t)dt.$$

The triple (a,b,c) is called the local characteristics of the Brownian motion (ϕ_t, X_t). It is known that the law of (ϕ_t, X_t) is uniquely determined by the local characteristics. See Fujiwara-Kunita [5].

The precise definition of the generalized stochastic integral at the right hand side of (1.15) will be given at the next section. Here we compare it with the classical one. Suppose that $X_t^\varepsilon(x)$ is defined by

$$(1.16) \qquad X_t^\varepsilon(x) = \sum_{k=1}^{r} \int_0^t F_k(\tau, x)\dot{B}_\tau^{\varepsilon,k} d\tau + \int_0^t G(\tau, x) d\tau.$$

as in the case of 1.1. Then the limit $X_t(x)$ is given by,

$$(1.17) \qquad X_t(x) = \sum_{k=1}^{r} \int_0^t F_k(\tau, x) dB_\tau^k + \int_0^t G(\tau, x) d\tau,$$

which is a C-valued Brownian motion. The corresponding generalized stochastic differential equation (1.15) coincides with the equation (1.1). Generally, any C-valued Brownian motion is expressed as

$$X_t(x) = \sum_{k=1}^{\infty} \int_0^t F_k(\tau, x) dB_\tau^k + \int_0^t G(\tau, x) d\tau,$$

making use of infinite numbers of independent one dimensional Brownian motions B_t^k, $k = 1, 2, \ldots$. Hence the equation (1.15) is a generalization of Itô's stochastic differential equation (1.1).

1.3 We shall next consider the limit theorems studied by Papanicolaou-Stroock-Varadhan [23]. It is formulated as follows. Let $(x^\varepsilon(t), y^\varepsilon(t))$ be the diffusion process in $R^n \times R^m$ defined by the following system of the stochastic differential equations:

$$(1.18) \qquad dx^\varepsilon(t) = [\tfrac{1}{\varepsilon}F^{(1)}(x^\varepsilon(t), y^\varepsilon(t)) + G^{(1)}(x^\varepsilon(t), y^\varepsilon(t))]dt + \sum_{j=1}^{p} \sigma_{\cdot j}^{(1)}(x^\varepsilon(t), y^\varepsilon(t)) d\beta_j^{(1)}(t),$$

$$(1.19) \qquad dy^\varepsilon(t) = [\tfrac{1}{\varepsilon^2}F^{(2)}(x^\varepsilon(t), y^\varepsilon(t)) + \tfrac{1}{\varepsilon}G^{(2)}(x^\varepsilon(t), y^\varepsilon(t)) + H^{(2)}(x^\varepsilon(t), y^\varepsilon(t))]dt + \sum_{\ell=1}^{q} [\tfrac{1}{\varepsilon}\sigma_{\cdot\ell}^{(2)}(x^\varepsilon(t), y^\varepsilon(t)) + \sigma_{\cdot\ell}^{(3)}(x^\varepsilon(t), y^\varepsilon(t))] d\beta_\ell^{(2)}(t),$$

where $(\beta_1^{(1)}(t), \ldots, \beta_p^{(1)}(t))$ and $(\beta_1^{(2)}(t), \ldots, \beta_q^{(2)}(t))$ are p-dimensional and q-dimensional Brownian motions, respectively and are independent each other. The coefficients are assumed to have bounded continuous derivatives of all orders. As $\varepsilon \to 0$, the right hand side of (1.19) diverges and hence the system of solutions $y^\varepsilon(t)$, $\varepsilon \geq 0$ does not converge. On the other hand, the first component $x^\varepsilon(t)$ varies slowly compared with $y^\varepsilon(t)$. Papanicolaou-Stroock-Varadhan have shown that under some conditions for coefficients, $x^\varepsilon(t)$ converges weakly to a diffusion process.

Now let $(\phi_t^\varepsilon(x,y), \psi_t^\varepsilon(x,y))$ be the solution of (1.18) and (1.19) starting at (x,y) at time 0. The pair defines a stochastic flow of diffeomorphisms in $R^n \times R^m$, but the first component $\phi_t^{\varepsilon,y} \equiv \phi_t^\varepsilon(\cdot,y)$, y being fixed, does not in general. However if coefficients $F^{(2)}$, $G^{(2)}$, $H^{(2)}$, $\sigma_{\cdot\ell}^{(2)}$, $\sigma_{\cdot\ell}^{(3)}$ do not depend on x and hence (1.19) defines a closed system, $\psi_t^\varepsilon(x,y)$ does not depend on x. In this case, the mapping $\phi_t^{\varepsilon,y}$; $R^n \to R^n$ becomes a diffeomorphism for each y. Recently, Matsumoto [21] showed that for each y, $\phi_t^{\varepsilon,y}$, $\varepsilon > 0$ and the following X_t^ε, $\varepsilon > 0$

(1.20) $$X_t^\varepsilon(x) = \int_0^t \{\frac{1}{\varepsilon} F^{(1)}(x,y^\varepsilon(t)) + G^{(1)}(x,y^\varepsilon(t))\}dt + \sum_{j=1}^n \int_0^t \sigma_{\cdot j}^{(1)}(x,y^\varepsilon(t))d\beta_j^{(1)}(t)$$

converge weakly as $\varepsilon \to 0$ as stochastic flows, if the transition probabilities $P(t,y,\cdot)$ of the diffusion process $\tilde{y}^0(t)$ determined by

$$d\tilde{y}^0(t) = F^{(2)}(\tilde{y}^0(t))dt + \sum_{\ell=1}^q \sigma_{\cdot\ell}^{(2)}(\tilde{y}(t))d\beta_\ell^{(2)}$$

satisfies the following three conditions (1.21) - (1.23).

(1.21) There exists a unique invariant measure μ on R^n such that $\lim_{t\uparrow\infty} P(t,y,A) = \mu(A)$ exists uniformly on compact sets.

(1.22) The potential kernel

$$\psi(y,A) = \int_0^\infty (P(t,y,A) - \mu(A))dt$$

exists and transforms $C_b^\infty(R^n)$ into itself.

(1,23) $$\int F^{(1)}(x,y)\mu(dy) = 0, \qquad \forall x \in R^m.$$

(Cf. Matsumoto [21]). Under the above condition, the pair $(\phi_t^\varepsilon, X_t^\varepsilon)$ *converges weakly as stochastic flows and the limit* (ϕ_t, X_t) *is a Brownian motion with the local characteristics* (a,b,c) *given by*

(1.24) $$a_{ij}(x_1,x_2) = \int \{\sum_{k=1}^p \sigma_{ik}^{(1)}(x_1,y)\sigma_{jk}^{(1)}(x_2,y)\}\mu(dy) + \int \{\psi(F_i^{(1)})(x_1,y)F_j^{(1)}(x_2,y) + \psi(F_j^{(1)})(x_2,y)F_i^{(1)}(x_1,y)\}\mu(dy),$$

(1.25) $$b_i(x) = \int G^{(1)}(x,y)\mu(dy),$$

(1.26) $$c_i(x) = \sum_j \int \psi(\frac{\partial F_i^{(1)}}{\partial x_j})(x,y)F_j^{(1)}(x,y)\mu(dy)$$

where

(1.27) $$\psi(F_i^{(1)})(x,y) \equiv \int F_i^{(1)}(x,z)\psi(y,dz).$$

1.4 In Sections 1.1, 1.2 and 1.3 we looked at three types of limit theorems concerning the stochastic differential equations and stochastic flows. A common feature for these cases is that $X_\tau^\varepsilon(x)$ defined by (1.14), (1.16) and (1.20) are continuous semimartingales with values in $C^\infty = C^\infty(R^d; R^d)$ and that the solution $\phi_t^\varepsilon(x)$ for the corresponding stochastic (ordinary) differential equation is given by

$$\phi_t^\varepsilon(x) - x = \int_0^t X^\varepsilon(\phi_\tau^\varepsilon(x), d\tau).$$

Here in case of (1.14) and (1.16), $X_t^\varepsilon(x)$ is piecewise smooth in t (with no martingale part) and the right hand side of the above is the usual integral $\int_0^t \dot{X}_\tau^\varepsilon(\phi_\tau^\varepsilon)d\tau$, where $\dot{X}_t^\varepsilon = \frac{d}{dt}X_t^\varepsilon$. In case of (1.20), $X_t^\varepsilon(x)$ contains a martingale part and the right hand side of the above is the generalized Itô (or Stratonovich) integral.

In any case, $X_t^\varepsilon(x)$, $\varepsilon > 0$ converges weakly (or strongly) to a Brownian motion with values in C^∞ and ϕ_t^ε, $\varepsilon > 0$ converges weakly (or strongly) to a stochastic flow. Further, these two are related by

$$\phi_t(x) - x = \int_0^t X(\phi_\tau, d\tau) + \int_0^t c(\phi_\tau, \tau)d\tau$$

with a correction term $c(x,t)$. In the next section, we shall present a unified method, taking account of these common features.

2. STOCHASTIC DIFFERENTIAL EQUATIONS AND STOCHASTIC FLOWS

2.1 C-valued continuous semimartingales. Let $(\Omega, \underline{\underline{F}}, P)$ be a probability space equipped with the filteration $\underline{\underline{F}}_t$, $t \in [0,T]$. Let $X(x,t)$, $t \in [0,T]$ be a continuous random field with values in R^d. Setting $X_t = X(\cdot, t)$, we may consider that X_t is a continuous C-valued process, where $C = C(R^d; R^d)$. The X_t is called a C-valued martingale if it is $\underline{\underline{F}}_t$-adapted, integrable and satisfies $E[X(x,t)|\underline{\underline{F}}_s] = X(x,s)$ holds a.s. for any $x \in R^d$ and $s < t$. If the quadratic variation $\langle X^i(x,t), X^j(y,t)\rangle$ is absolutely continuous with respect to t a.s. for any x, y, the density function $a_{ij}(x,y,t,\omega) = d\langle X^i(x,t), X^j(y,t)\rangle/dt$ is called the characteristic of the C-valued martingale. An $\underline{\underline{F}}_t$-adapted continuous C-valued process $X(x,t)$ is called a C-valued semimartingale with the characteristics $(a(x,y,t), b(x,t))$ if it is written as

$$(2.1) \qquad X(x,t) = \int_0^t b(x,\tau)d\tau + M(x,t),$$

where $b(x,t) = b(x,t,\omega)$ is a measurable random field such that for each x it is $\underline{\underline{F}}_t$-adapted, and $M(x,t)$ is a C-valued continuous martingale with the characteristics $a(x,y,t)$. In the following, we only consider the C-valued semimartingale whose characteristics $a(x,y,t)$ and $b(y,t)$ are continuous in x, y, t a.s. The following proposition is easily verified.

Proposition 2.1 *Let* $X(x,t)$ *be a continuous C-semimartingale with continuous characteristics* (a,b). *Then we have for any* x, y, t,

$$\lim_{h\to 0} \frac{1}{h} E[X(x,t+h) - X(x,t)|\underline{\underline{F}}_t] = b(x,t) \qquad \text{a.s.},$$

$$\lim_{h\to 0} \frac{1}{h} E[(X^i(x,t+h) - X^i(x,t))(X^j(y,t+h) - X^j(y,t)) | \underline{\underline{F}}_t] = a_{ij}(x,y,t) \qquad \text{a.s.}$$

The next proposition is a martingale characterization of a C-valued Brownian motion. It is an analogue of Lévy's characterization of one dimensional Brownian motion by means of the martingale.

Proposition 2.2 *Suppose that the characteristics* (a,b) *of continuous C-valued semimartingale* X_t *do not depend on* ω *a.s. Then* X_t *is a C-valued Brownian motion: for any* $t_1 < t_2 < \dots < t_n$, $X_{t_{i+1}}(x) - X_{t_i}(x)$, $i = 1,\dots,n$ *are independent Gaussian random variables with the means* $\int_{t_i}^{t_{i+1}} b(x,\tau)d\tau$ *and the covariances* $\int_{t_i}^{t_{i+1}} a_{ij}(x,y,\tau)d\tau$.

2.2 Stochastic integrals by C-valued semimartingales. Let $M(x,t)$ be a continuous C-martingale with the characteristic a adapted to $\{\underline{\underline{F}}_t\}$. Let $f_t(\omega)$ be a continuous R^d-valued process adapted to $\{\underline{\underline{F}}_t\}$. We shall define the stochastic integral of f_t based on $M(x,t)$. Let $\Delta = \{0 = t_0 < t_1 < \dots < t_n = T\}$ be a partition. Set

$$y_t^\Delta = \sum_{i=0}^{n-1} \{M(f_{t_i\wedge t}, t_{i+1}\wedge t) - M(f_{t_i\wedge t}, t_i\wedge t)\}.$$

Then it is a continuous R^d-valued (local) martingale and converges uniformly in t in probability as $|\Delta| \to 0$. The limit is denoted by $\int_0^t M(f_r, dr)$. It is a continuous R^d-valued (local) martingale. The quadratic variation satisfies

$$(2.2) \qquad \langle \int_0^t M^i(f_r, dr), \int_0^t M^j(g_r, dr) \rangle = \int_0^t a_{ij}(f_r, g_r, r)dr.$$

See Le Jan [18] and Le Jan-Watanabe [19].

Next, let $X(x,t)$ be a C-semimartingale decomposed as $\int_0^t b(x,\tau)d\tau + M(x,t)$. The stochastic integral of f_t by $X(x,t)$ is defined by

$$\int_0^t X(f_r, dr) \equiv \int_0^t b(f_r, r)dr + \int_0^t M(f_r, dr).$$

2.3 Stochastic differential equation. Given a continuous C-semimartingale $X(x,t)$ with the characteristics (a,b), we shall consider Itô's stochastic differential equation. A continuous $\underline{\underline{F}}_t$-adapted R^d-valued process ϕ_t is called the solution of the stochastic differential equation

$$(2.3) \qquad dx_t = X(x_t, dt),$$

starting at x at time s if it satisfies

$$\phi_t = x + \int_s^t X(\phi_r, dr), \qquad \forall t > s.$$

The equation has a unique solution if the characteristics (a,b) satisfy the Lipschitz conditions below: There is a positive constant L such that

$$|\mathrm{Tr}\{a(x,y,t)\}| \leq L(1+|x|)(1+|y|),$$

$$|\mathrm{Tr}\{a(x,x,t) - 2a(x,y,t) + a(y,y,t)\}| \leq L|x-y|^2,$$

$$|b(x,t)| \leq L(1+|x|),$$

$$|b(x,t) - b(y,t)| \leq L|x-y|.$$

Now let us compare the above stochastic differential equation with the standard Itô's stochastic differential equation. Suppose that $X(x,t)$ is a C-valued Brownian motion of the form

$$X(x,t) = \sum_{k=1}^{r} \int_0^t F_k(\tau,x)dB_\tau^k + \int_0^t G(\tau,x)d\tau, \tag{2.4}$$

where $(B_t^1,\ldots,B_t^r)$ is a standard Brownian motion. Then equation (2.3) is noting but the standard stochastic differential equation

$$dx_t = \sum_{k=1}^{r} F_k(t,x_t)dB_t^k + G(t,x_t)dt.$$

Note that the local characteristics of (2.4) is $(\Sigma_{k=1}^r F_k^i(\tau,x)F_k^j(\tau,x), G(\tau,x))$. Then the Lipschitz condition for $a(x,y,\tau)$ is written by

$$\left|\sum_{i=1}^{d}\sum_{k=1}^{r} F_k^i(\tau,x)F_k^i(\tau,y)\right| \leq L(1+|x|)(1+|y|),$$

$$\sum_{i=1}^{d}\sum_{k=1}^{r} |F_k^i(\tau,x) - F_k^i(\tau,y)|^2 \leq L|x-y|^2,$$

which are obviously satisfied if $F_k(\tau,x)$, $k = 1,\ldots,r$ satisfy the Lipschitz condition in the usual sense. Therefore, the stochastic differential equation (2.3) is a natural generalization of the standard Itô's stochastic differential equation.

We can prove the following theorems similarly as in the case of the standard Itô's stochastic differential equation. See Kunita [13].

Theorem 2.1 *Suppose that the characteristics of the continuous C-semimartingale* X_t *satisfy the Lipschitz condition. Then there is a modification* $\phi_{s,t}(x,\omega)$ *of the solution satisfying the following properties.*

(ϕ.1) $\phi_{s,t}(x,\omega)$ *is an* R^d*-valued continuous random field.*

(ϕ.2) $\phi_{s,u}(x,\omega) = \phi_{t,u}(\phi_{s,t}(x,\omega),\omega)$ *holds for any* $s < t < u$ *and* $x \in R^d$ *a.s.*

(ϕ.3) *The map* $\phi_{s,t}(\cdot,\omega)$; $R^d \to R^d$ *is a homeomorphism for any* $s < t$ *a.s.*

In particular, if the characteristics (a,b) *do not depend on* ω *a.s.,* $\phi_{s,t}$ *has independent increments, i.e., for any* $t_1 < t_2 < \ldots < t_n$, $\phi_{t_i,t_{i+1}}$, $i = 1,\ldots,n-1$ *are independent each other.*

Theorem 2.2 *Suppose that the characteristics* $(a(x,y,t),b(x,t))$ *have bounded and continuous spacial derivatives of orders less than or equal to* k *and there is a positive constant* L *such that*

(2.5) $\quad |D_x^\alpha D_y^\beta a(x,y,t)| \le L, \qquad 1 \le \forall |\alpha|, |\beta| \le k,$

(2.6) $\quad |D^\alpha b(x,t)| \le L, \qquad 1 \le \forall |\alpha| \le k.$

Then there is a modification of the solution satisfying $(\phi.2)$ *and the following* $(\phi.1)_{k-1}$ *and* $(\phi.3)_{k-1}$.

$(\phi.1)_{k-1}$ $\phi_{s,t}(x,\omega)$ *is (k-1)-times continuously differentiable with respect to* x.

$(\phi.3)_{k-1}$ *The map* $\phi_{s,t}(\cdot,\omega)\ ;\ R^d \to R^d$ *is a* C^{k-1}*-diffeomorphisms.*

<u>Remark</u> Let G^m be the totality of C^m-diffeomorphisms of R^d. It is a group by the composition of maps $\phi\circ\psi$, $\phi, \psi \in G$. Let ρ be the compact uniform metric of $C(R^d:R^d)$. Set

$$d_m(\phi,\psi) = \sum_{|\alpha|\le m} \rho(D^\alpha\phi, D^\alpha\psi) + \rho(D^\alpha\phi^{-1}, D^\alpha\psi^{-1}).$$

Then G^m is a complete topological group. Now $\phi_t \equiv \phi_{0,t}$ of Theorem 2.2 can be regarded as a continuous G^{k-1}-valued process. It holds $\phi_{s,t} = \phi_t\circ\phi_s^{-1}$ if $s < t$. In particular, if $\phi_{s,t}$ has independent increments, it is called a G^{k-1}-valued Brownian motion. The process ϕ_t is often called the <u>stochastic flow of diffeomorphisms generated by the continuous C-valued semimartingale</u> X_t.

3. LIMIT THEOREMS

<u>3.1</u> <u>Weak and strong convergence of stochastic flows.</u> Suppose we are given on the probability space $(\Omega, \underline{\underline{F}}, P)$ a system of the filterations $\{\underline{\underline{F}}_t^\varepsilon, t\in[0,T]\}$, $\varepsilon > 0$ and a system of $\underline{\underline{F}}_t^\varepsilon$-adapted continuous C-semimartingales X_t^ε, $\varepsilon > 0$ with the characteristics $(a_{ij}^\varepsilon(x,y,t), b_i^\varepsilon(x,t,\omega))$, $\varepsilon > 0$ satisfying the Lipschitz conditions. Let $\phi_t^\varepsilon(x)$ be the stochastic flow of homeomorphisms generated by X_t^ε. The problem we are concerned in this section is the weak convergence of $(\phi_\tau^\varepsilon, X_\tau^\varepsilon)$ as stochastic flows.

We shall first give the definition of the weak convergence. Let $G^m = \mathrm{Diffeo}^m(R^d)$ and $C^m = C^m(R^d;R^d)$ as before. Let $\hat{W}_m = C([0,T];G^m)$ (or $\tilde{W}_m = C([0,T];C^m)$) be the totality of maps from $[0,T]$ into G^m (resp. C^m). Denote the element of $\hat{W}_m$ (or $\tilde{W}_m$) by ϕ (resp. X) and the value at $t\in[0,T]$ by ϕ_t (resp. X_t). The space $\hat{W}_m$ (or $\tilde{W}_m$) is a complete separable metric space by the metric

(3.1) $$\hat{d}_m(\phi,\psi) = \sup_t d_m(\phi_t,\psi_t)$$

(or $\tilde{\rho}_m(X,Y) = \sup_t \rho_m(X_t,Y_t)$, where $\rho_m(\phi,\psi) = \sum_{|\alpha|\le m} \rho(D^\alpha\phi, D^\alpha\psi)$). Set $W_m = \hat{W}_m \times \tilde{W}_m$ and let $\underline{\underline{B}}(W_m)$ be its topological Borel field.

Now, for almost all ω, the pair $(\phi_t^\varepsilon(x,\omega), X_t^\varepsilon(y,\omega))$ may be regarded as an element of W_m. Hence the law P_ε of the pair $(\phi^\varepsilon, X^\varepsilon)$ is defined as a probability measure on $(W_m, \underline{\underline{B}}(W_m))$ such that

(3.2) $$P_\varepsilon(A) = P\{\omega;\ (\phi^\varepsilon(\omega), X^\varepsilon(\omega)) \in A\}.$$

We will say that system $(\phi_t^\varepsilon, X_t^\varepsilon)$, $\varepsilon > 0$ <u>converges weakly to</u> (ϕ_t, X_t) <u>as stochastic</u>

flows if the corresponding laws P_ε, $\varepsilon > 0$ converges weakly. We will say that X_t^ε, $\varepsilon > 0$ (or ϕ_t^ε, $\varepsilon > 0$) converges strongly to X_t (or to ϕ_t, respectively) if $\tilde{\rho}_m(X^\varepsilon, X)$ (or $\hat{d}_m(\phi^m, \phi)$) converges to 0 in probability as $\varepsilon \to 0$.

3.2 Main theorem. Martingale formulation. We introduce hypotheses for the system of local characteristics $(a^\varepsilon, b^\varepsilon)$, $\varepsilon > 0$ that will allow the weak convergence of the measures P_ε, $\varepsilon > 0$. Set

$$\bar{b}^\varepsilon(x,t) = E[b^\varepsilon(x,t)], \qquad \tilde{b}^\varepsilon(x,t) = b^\varepsilon(x,t) - \bar{b}^\varepsilon(x,t). \tag{3.3}$$

We shall assume that a^ε and $\bar{b}^\varepsilon$ converge to $\bar{a}$ and $\bar{b}$ moderately, but $\tilde{b}^\varepsilon$ does not. The assumption will be made so that the integral $\int_0^t \tilde{b}^\varepsilon(\cdot,\tau)d\tau$ converges to a martingale with the local characteristic $\tilde{a}$. The precise statement of the assumption is as follows.

$(H.1)_k$ $a^\varepsilon(x,y,t,\omega)$ and $b^\varepsilon(x,t,\omega)$ are k-times continuously differentiable with respect to x, y a.s. Further, $b^\varepsilon(x,t,\omega)$ is $\underline{\underline{F}}_0$-adapted.

$(H.2)_{k,r}$ For each $M > 0$ there is a positive constant $K = K_M$ such that

$$E[\sup_{|x|,|y|\le M} |D_x^\alpha D_y^\alpha a_{ij}^\varepsilon(x,y,t)|^r] \le K, \qquad \forall |\alpha| \le k,\ \forall t \in [0,T],\ \forall \varepsilon > 0 \tag{3.4}$$

$$\sup_{|x|\le M} |D^\alpha \bar{b}^\varepsilon(x,t)| \le K, \qquad \forall t \in [0,T],\ \forall \varepsilon > 0 \tag{3.5}$$

$$E[\sup_{|x|\le M} |E[\int_s^t D^\alpha \tilde{b}^\varepsilon(x,\tau)d\tau|\underline{\underline{G}}_s^\varepsilon]|^r |D^\alpha \tilde{b}^\varepsilon(x,s)|^r] \le K, \qquad \forall |\alpha| \le k,\ \forall \varepsilon > 0 \tag{3.6}$$

where

$$\underline{\underline{G}}_s^\varepsilon = \sigma(X_u^\varepsilon(x), a^\varepsilon(x,u), b^\varepsilon(x,u)\ ;\ 0 \le u \le s,\ x \in R^d). \tag{3.7}$$

Furthermore,

$$E[\sup_{|x|\le M} |E[\int_s^t \tilde{b}^\varepsilon(x,\tau)d\tau|\underline{\underline{G}}_s^\varepsilon]|^2] \xrightarrow[\varepsilon \to 0]{} 0. \tag{3.8}$$

$(H.3)_k$ There are functions $a_{ij}(x,y,t)$, $\bar{b}_i(x,t)$, $\tilde{A}_{ij}(x,y,t)$ and $c_{ij}(x,t)$ having bounded continuous spacial derivatives of orders less than or equal to k and the followings are satisfied for each $M > 0$:

$$E[\sup_{|x|,|y|\le M} |\int_0^t a_{ij}^\varepsilon(x,y,\tau)d\tau - \int_0^t a_{ij}(x,y,\tau)d\tau|] \xrightarrow[\varepsilon \to 0]{} 0, \tag{3.9}$$

$$\sup_{|x|\le M} |\int_0^t \bar{b}_i^\varepsilon(x,\tau)d\tau - \int_0^t \bar{b}_i(x,\tau)d\tau| \xrightarrow[\varepsilon \to 0]{} 0, \tag{3.10}$$

$$E[\sup_{|x|,|y|\le M} |E[\int_s^t \tilde{b}_i^\varepsilon(x,\tau)d\tau \int_s^\tau \tilde{b}_j^\varepsilon(y,\sigma)d\sigma|\underline{\underline{G}}_s^\varepsilon] - \int_s^t \tilde{A}_{ij}(x,y,\tau)d\tau|] \xrightarrow[\varepsilon \to 0]{} 0, \tag{3.11}$$

$$E[\sup_{|x|\le M} |E[\int_s^t \frac{\partial}{\partial x_j} \tilde{b}_i^\varepsilon(x,\tau)d\tau \int_s^\tau \tilde{b}_j^\varepsilon(x,\sigma)d\sigma|\underline{\underline{G}}_s^\varepsilon] - \int_s^t c_{ij}(x,\tau)d\tau|] \xrightarrow[\varepsilon \to 0]{} 0. \tag{3.12}$$

Theorem 3.1 *Suppose* $(H,1)_k$, $(H.2)_{k,r}$ *and* $(H.3)_k$ *for some* $k \geq 2$ *and* $r > 1$. *Let* P_ε *be the law of* $(\phi_t^\varepsilon, X_t^\varepsilon)$ *defined over* W_{k-2}. *Then the measures* P_ε, $\varepsilon > 0$ *converge weakly as* $\varepsilon \to 0$. *Furthermore, the limiting measure* P_0 *satisfies the following properties*

(i) $X_t(x) = X(x,t)$ *is a* C^{k-2}*-valued Brownian motion with the characteristics* $(a+\tilde{a}, \bar{b})$, where

$$\tilde{a}_{ij}(x,y,t) = \tilde{A}_{ij}(x,y,t) + \tilde{A}_{ji}(y,x,t). \tag{3.13}$$

(ii) ϕ_t *is a* G^{k-2}*-valued Brownian motion generated by* $X_t + \int_0^t c(x,\tau)d\tau$, *where*

$$c_i(x,t) = \sum_j c_{ij}(x,t). \tag{3.14}$$

Moreover, if $X^\varepsilon(x,t)$, $\varepsilon > 0$ *converge strongly as* $\varepsilon \to 0$, *then the stochastic flows* ϕ_t^ε, $\varepsilon > 0$ *converge strongly to* ϕ_t.

Definition The pair (ϕ_t, X_t) is called a $G^{k-2} \times C^{k-2}$-valued Brownian motion with the local characteristics $(a+\tilde{a}, \bar{b}, c)$.

3.3 Mixing case. As applications of Theorem 3.1, we shall consider the case where the system of continuous C-valued semimartingales saitsfies a suitable mixing condition. Let $X^\varepsilon(x,t)$, $\varepsilon > 0$ be a system of $\underline{\underline{F}}_t^\varepsilon$-adapted continuous C-semimartingales with the characteristics $(a^\varepsilon, b^\varepsilon)$. Set

$$\underline{\underline{G}}_{s,t}^\varepsilon = \sigma(X^\varepsilon(x,u) - X^\varepsilon(x,v), a^\varepsilon(x,u), b^\varepsilon(x,u)\ ;\ s \leq u,\ v \leq t) \tag{3.15}$$

and denote by $\beta^\varepsilon(t)$ the strong mixing rate of $\underline{\underline{G}}_{s,t}^\varepsilon$ (defined by (1.8)). We shall introduce assumptions so that $\beta^\varepsilon(t) \to 0$, $E[|\tilde{b}^\varepsilon|^r] \to \infty$ as $\varepsilon \to 0$ and these rates of the convergence and the divergence are balanced. Then it will imply that $\tilde{b}^\varepsilon(x,t)$ converges to the white noise or $\int_0^t \tilde{b}^\varepsilon(x,\tau)d\tau$ converges to a Brownian motion. The precise assumptions are given as follows.

$(H.2)'_{k,r}$ The mixing rate $\beta^\varepsilon(t)$ satisfies the following properties.

$$\lim_{\varepsilon \to 0} \int_0^T \beta^\varepsilon(\tau)^{\frac{1}{3+3\vee d}} d\tau = 0, \tag{3.16}$$

$$\frac{1}{c(\varepsilon)} \int_0^{c(\varepsilon)} d\sigma \int_\sigma^T d\tau\, \beta^\varepsilon(\tau)^{\frac{1}{3+3\vee d}} = o\left(\int_0^T \beta^\varepsilon(\tau)^{\frac{1}{3+3\vee d}} d\tau\right) \tag{3.17}$$

as $\varepsilon \to 0$. Here and in the following, $c(\varepsilon)$ is a positive increasing function such that $\lim_{\varepsilon\to 0} c(\varepsilon) = 0$. Furthermore, for each $M > 0$ there is a positive constant $K = K_M$ satisfying (3.4), (3.5) and

$$E[\sup_{|x| \leq M} |D^\beta \tilde{b}^\varepsilon(x,\sigma)|^6]^{\frac{1}{3}} \int_0^T \beta^\varepsilon(\tau)^{\frac{1}{3+3\vee d}} d\tau \leq K, \qquad \forall |\beta| \leq k,\quad \forall \varepsilon > 0. \tag{3.18}$$

Instead of $(H.3)_k$, we assume the following.

$(H.3)'_k$ There are functions $a_{ij}(x,y,t)$, $\bar{b}_i(x,t)$, $\tilde{A}_{ij}(x,y,t)$ and $c_{ij}(x,t)$ having bounded continuous spacial derivatives of orders less than or equal to k and satisfy (3.9), (3.10) and

$$(3.19)\qquad |A_{ij}(x,y,t) - \frac{1}{c(\varepsilon)}\int_t^{t+c(\varepsilon)} d\tau \int_t^{\tau} d\sigma\, E[b_i^{\varepsilon}(x,\tau) b_j^{\varepsilon}(y,\sigma)]| \xrightarrow[\varepsilon\to 0]{} 0,$$

$$(3.20)\qquad |c_{ij}(x,t) - \frac{1}{c(\varepsilon)}\int_t^{t+c(\varepsilon)} d\tau \int_t^{\tau} d\sigma\, E[b_j^{\varepsilon}(x,\sigma)\frac{\partial}{\partial x_j} b_i^{\varepsilon}(x,\tau)]| \xrightarrow[\varepsilon\to 0]{} 0,$$

uniformly on compact sets.

Theorem 3.2 *Suppose* $(H.1)_k$, $(H.2)'_{k,r}$ *and* $(H.3)'_k$ *for* $k \geq 2$. *Then the assertions of Theorem 3.1 are valid.*

Remark The above theorem is a generalization of Theorem 1 in Kunita [16], where the case $a_{ij}^{\varepsilon} \equiv 0$ is discussed. It is shown in [16] that the approximation theorem in Section 1.1 follows from Theorem 1.1. Here we shall check that the above theorem implies the assertion of Section 1.2. Note that the characteristics of the C-valued process defined by (1.14) are $(0, \frac{1}{\varepsilon}f(x,t/\varepsilon^2,t) + g(x,t/\varepsilon^2,t))$ and the mixing rate is $\beta^{\varepsilon}(t) = \beta(t/\varepsilon^2)$. Hence we have

$$\int_0^{\infty} \beta^{\varepsilon}(\tau)^{\frac{1}{3+3\sqrt{d}}}\, d\tau \leq \varepsilon^2 \int_0^{\infty} \beta(\tau)^{\frac{1}{3+3\sqrt{d}}}\, d\tau .$$

Then (3.16) and (3.17) are obviously satisfied. Since $\tilde{b}^{\varepsilon} = \frac{1}{\varepsilon}f$, inequality (3.18) follows from the above inequality and (1.10). Hence hypothesis $(H.2)'_{k,r}$ is satisfied. Hypothesis $(H.3)'_k$ follows from (1.11) - (1.13). Therefore the assertion of Section 1.2 is valid.

3.4 Ergodic case. Let $(\Omega,\underline{\underline{F}},P)$ be a probability space equipped with the filterations $\underline{\underline{F}}_t$, $t \geq 0$, where Ω is a complete metric space and $\underline{\underline{F}}$ is the topological σ-field. Let $\underline{\underline{G}}$ be a sub σ-field of $\underline{\underline{F}}_0$. Suppose that for each $\varepsilon \geq 0$ we are given a one parameter family of bi-measurable transformations θ_t^{ε}, $t\in[0,\infty)$ of $(\Omega,\underline{\underline{G}})$, that is, $\theta_t^{\varepsilon} ; \Omega \to \Omega$ is one to one, onto, bi-measurable with respect to $\underline{\underline{G}}$, and $\theta_{t+s}^{\varepsilon} = \theta_t^{\varepsilon}\cdot\theta_s^{\varepsilon}$. We suppose further that as $\varepsilon \to 0$, $\theta_t^{\varepsilon}\omega$ converges to $\theta_t^0\omega$ in probability, and the limit θ_t^0 is a measure preserving ergodic transformation, that is, $P((\theta_t^0)^{-1}A) = P(A)$ is satisfied for all $A\in\underline{\underline{G}}$ and $P(A) = 0$ or 1 if $\theta_t^0 A = A$, $A\in\underline{\underline{G}}$. A sub σ-field $\underline{\underline{G}}_0$ is called a generator of $(\theta_t^0,\underline{\underline{G}})$ if

$$\underline{\underline{G}} = \sigma(\theta_t A \ ; A\in\underline{\underline{G}}_0,\ t\in[0,\infty)).$$

In the following, we fix the generator $\underline{\underline{G}}_0$.

Now suppose we are given for each $\varepsilon > 0$, an $\underline{\underline{F}}_t$-adapted continuous C-semimartingale of the form

$$(3.21)\qquad X(x,t,\varepsilon,\omega) = \varepsilon\int_0^t F(x,\theta_{\tau}^{\varepsilon}\omega)\,d\tau + \varepsilon^2\int_0^t G(x,\theta_{\tau}^{\varepsilon}\omega)\,d\tau + \varepsilon^2 M^{\varepsilon}(x,t,\omega),$$

where $M^{\varepsilon}(x,t)$ is an $\underline{\underline{F}}_t$-adapted C-martingale with the characteristic $a(x,y,\theta_t^{\varepsilon}\omega)$.

Here we assume that F, G, a are continuous in (x,y,ω) and $\underline{\underline{G}}_0$-measurable. Let $\psi_t^\varepsilon(x)$ be the stochastic flow generated by $X(x,t,\varepsilon)$. We are interested in the limit behaviour of $\psi_t^\varepsilon(x)$ as $\varepsilon \to 0$ with $\varepsilon^2 t$ remaining fixed. For this purpose, we shall change the scale of the time. Set $\underline{\underline{F}}_t^\varepsilon = \underline{\underline{F}}_{t/\varepsilon^2}$ and

(3.22) $$\phi_t^\varepsilon = \psi_{t/\varepsilon^2}^\varepsilon, \qquad X^\varepsilon(x,t) = \frac{1}{\varepsilon^2} X(x, \frac{t}{\varepsilon^2}, \varepsilon).$$

Then $X^\varepsilon(x,t)$ is an $\underline{\underline{F}}_t^\varepsilon$-adapted semimartingale with the characteristics

(3.23) $$a^\varepsilon(x,y,t) = a(x,y,\theta_{t/\varepsilon^2}^\varepsilon \omega),$$

(3.24) $$b^\varepsilon(x,t) = \frac{1}{\varepsilon} F(x, \theta_{t/\varepsilon^2}^\varepsilon \omega) + G(x, \theta_{t/\varepsilon^2}^\varepsilon \omega)$$

and ϕ_t^ε is the stochastic flow generated by $X^\varepsilon(x,t)$. We shall introduce the following conditions for a, F and G.

$(H.4)_{k,r}$ (a) $a(x,y,\omega)$, $F(x,\omega)$, $G(x,\omega)$ have bounded continuous spacial derivatives of orders less than or equal to k a.s.

(b) F is of mean 0.

(c) The following limit exists. It is k-times continuously differentiable with respect to x and belongs to $L^{2r}(P)$:

(3.25) $$\widehat{D^\beta F}(x,\omega) = \lim_{\varepsilon \to 0} E[\int_0^{t/\varepsilon^2} D^\beta F(x, \theta_t^\varepsilon \omega)\, dt \,|\underline{\underline{G}}_0], \qquad |\beta| \leq 1.$$

Theorem 3.3 *Suppose* $(H.4)_{k,r}$ *for some* $k \geq 2$ *and* $r > 1$. *Then* $(\phi_t^\varepsilon, X_t^\varepsilon)$ *converges weakly to a* $G_{k-2} \times C_{k-2}$-*valued Brownian motion with the characteristics* $(\bar{a}, b, c)$, *where*

$$\bar{a}_{ij}(x,y) = E[a_{ij}(x,y)] + E[\hat{F}_i(x)F_j(y) + \hat{F}_j(x)F_i(y)],$$

$$b_i(x) = E[G_i(x)],$$

$$c_i(x) = \sum_j E\{\widehat{\frac{\partial F^i}{\partial x_j}}(x) F_j(x)\}.$$

Remark As an application, we shall consider the limit theorem stated in Section 1.3. Associated with (1.19), we shall consider the following stochastic differential equation for each $\varepsilon \geq 0$

$$d\tilde{y}^\varepsilon(t) = [F^{(2)}(\tilde{y}^\varepsilon(t)) + \varepsilon G^{(2)}(\tilde{y}^\varepsilon(t)) + \varepsilon^2 H^{(2)}(\tilde{y}^\varepsilon(t))]\, dt + \sum_{\ell=1}^{q} [\sigma_{\cdot\ell}^{(2)}(\tilde{y}^\varepsilon(t)) + \varepsilon \sigma_{\cdot\ell}^{(3)}(\tilde{y}^\varepsilon(t))]\, d\tilde{\beta}_\ell^{(2)}(t),$$

where $\tilde{\beta}_\ell^{(2)}(t)$, $\ell = 1,\dots,q$ are independent standard Brownian motions. Let $\tilde{\theta}_t^\varepsilon ; \Omega \to \Omega$ be the shift operator defined by $\tilde{y}^\varepsilon(s, \tilde{\theta}_t^\varepsilon \omega) = \tilde{y}^\varepsilon(s+t, \omega)$. Then $\tilde{\theta}_t^\varepsilon \omega$ converges to $\tilde{\theta}_t^0 \omega$ as $\varepsilon \to 0$ since $\tilde{y}^\varepsilon(t)$ converges to $\tilde{y}^0(t)$. Further, taking the initial distribution of $\tilde{y}^0(t)$ as μ, $\tilde{y}^0(t)$, $t \geq 0$ may be regarded as a stationary process, so that $\tilde{\theta}_t^0$ is a measure preserving and ergodic transformation. Let $\tilde{\psi}_t^\varepsilon$ be the stochastic flow generated by

$$\tilde{X}(x,t,\varepsilon) = \varepsilon\int_0^t F^{(1)}(x,\tilde{y}^\varepsilon(\tau))d\tau + \varepsilon^2\int_0^t G^{(1)}(x,\tilde{y}^\varepsilon(\tau))d\tau$$
$$+ \varepsilon^2 \sum_{j=1}^{p} \int_0^t \sigma_{\cdot j}^{(1)}(x,\tilde{y}^\varepsilon(\tau))d\tilde{\beta}_j^{(1)}(\tau),$$

where $(\tilde{\beta}_j^{(1)})$ is a standard Brownian motion independent of $(\tilde{\beta}_j^{(2)})$. Define $\tilde{\phi}_t^\varepsilon$ and $\tilde{X}^\varepsilon(x,t)$ by (3.22). Then the law of $(\tilde{\phi}_t^\varepsilon, \tilde{X}_t^\varepsilon)$ coincides with that of $(\phi_\tau^\varepsilon, X_\tau^\varepsilon)$ in Section 1.2. Then the weak convergence of $(\phi_\tau^\varepsilon, X_t^\varepsilon)$ follows from Theorem 3.3, since hypothesis $(H.4)_{k,r}$ follows from (1.22) and (1.23).

4. LIMIT THEOREMS FOR STOCHASTIC PARTIAL DIFFERENTIAL EQUATIONS

4.1 Limit theorems for stochastic flows discussed in the previous section can be applied to limit theorems for suitable deterministic and stochastic partial differential equations. As a survey, we shall first look at the averaging problem for the partial differential equation. Let $a_{ij}(x,t)$ and $b_i(x,t)$ be bounded smooth functions of (x,t), periodic relative to t with period 2π. The matrix $a(x,t) = (a_{ij}(x,t))$ is assumed to be symmetric and positive definite. Consider the parabolic partial differential equation with the parameter ε:

$$(4.1)\qquad \begin{cases} \dfrac{\partial}{\partial t} u_t^\varepsilon = \dfrac{1}{2}\sum_{i,j} a_{ij}(x,\tfrac{t}{\varepsilon})\dfrac{\partial^2}{\partial x_i \partial x_j} u_t^\varepsilon + \sum_i b_i(x,\tfrac{t}{\varepsilon})\dfrac{\partial}{\partial x_i} u_t^\varepsilon \\ u_0^\varepsilon = f \end{cases}$$

Khasminskii [10] showed that the solution $u_t^\varepsilon(x)$ converges uniformly in (t,x) as $\varepsilon \to 0$ and the limiting function $u_t(x)$ is the solution of the parabolic partial differential equation:

$$(4.2)\qquad \frac{\partial}{\partial t} u_t = \frac{1}{2}\sum_{i,j} \bar{a}_{ij}(x)\frac{\partial^2}{\partial x_i \partial x_j} u_t + \sum_i \bar{b}_i(x)\frac{\partial}{\partial x_i} u_t,$$

where

$$(4.3)\qquad \bar{a}_{ij}(x) = \frac{1}{2\pi}\int_0^{2\pi} a_{ij}(x,\tau)\,d\tau, \qquad \bar{b}_i(x) = \frac{1}{2\pi}\int_0^{2\pi} b_i(x,\tau)\,d\tau.$$

We shall apply the limit theorems of the stochastic flows to the above averaging problem. In order to construct the solution of equation (4.1) probabilistically, it is appropriate to consider the backward stochastic differential equations and backward flows instead of the forward ones discussed in the preceding sections.

The definition of the backward integral is as follows. Given a C-valued Brownian motion $X_t(x)$, let $\underline{\underline{F}}_{s,t}$ be the sub σ-field generated by $X_u(x) - X_v(x)$; $s \leq u$, $v \leq t$, $x \in R^d$. For a moment we fix $t > 0$. Let f_s, $0 \leq s \leq t$ be a continuous R^d-valued process adapted to $\underline{\underline{F}}_{s,t}$. The backward Itô integral is defined by

$$\int_s^t X(f_r, \hat{d}r) \equiv \lim_{|\Delta|\to 0} \sum_{i=0}^{n-1} \{X(f_{t_{i+1}}, t_{i+1}) - X(f_{t_{i+1}}, t_i)\}$$

and the backward Stratonovich integral is defined by

$$\int_s^t \circ X(f_r, \hat{d}r) = \lim_{|\Delta|\to 0} \frac{1}{2}\Big[\sum_{i=1}^{n-1} \{X(f_{t_{i+1}}, t_{i+1}) - X(f_{t_{i+1}}, t_i)\}$$

$$+ \sum_{i=0}^{n-1} \{X(f_{t_i}, t_{i+1}) - X(f_{t_i}, t_i)\}\Big],$$

where $\Delta = \{s = t_0 < t_1 < \ldots < t_n = t\}$. The Storatonovich integral is well defined if $X_t(x)$ is a C^1-valued Brownian motion and f_s is a continuous semimartingale. The integrals are related by

$$\int_s^t \circ X(f_r, \hat{d}r) = \int_s^t X(f_r, \hat{d}r) + \frac{1}{2}\sum_{i=1}^{d} \langle \int_s^t (\frac{\partial}{\partial x_i} X)(f_r, \hat{d}r), f_r^i \rangle.$$

Now if the characteristics (a,b) of the C-valued Brownian motion $X_t(x)$ satisfies the Lipschitz condition, it generates the backward stochastic flow $\hat{\phi}_{s,t}(x,\omega)$, $0 \leq s \leq t < \infty$, namely,

$$\hat{\phi}_{s,t}(x) = x + \int_s^t X(\hat{\phi}_{r,t}(x), \hat{d}r).$$

Then the function $\hat{u}_t(x) \equiv E[f(\hat{\phi}_{0,t}(x))]$ is the solution of the parabolic equation

$$(4.4) \qquad \frac{\partial}{\partial t}\hat{u}_t = L_t \hat{u}_t, \qquad \hat{u}_0 = f$$

where

$$(4.5) \qquad L_t = \frac{1}{2}\sum_{i,j} a_{ij}(x,x,t)\frac{\partial^2}{\partial x_i \partial x_j} + \sum_i b_i(x,t)\frac{\partial}{\partial x_i}.$$

See Kunita [13].

Let us return to the averaging problem. Suppose that there are bounded smooth functions $\sigma(x,t) = (\sigma_{ij}(x,t))$, $i = 1, \ldots, d$, $j = 1, \ldots, m$ such that $\sigma\sigma^* = a$. Let $B_t = (B_t^1, \ldots, B_t^m)$ be a standard Brownian motion. Set

$$X_t^\varepsilon(x) = \sum_{j=1}^{m} \int_0^t \sigma_{\cdot j}(x, \frac{t}{\varepsilon}) dB_\tau^j.$$

Let $\hat{\phi}_{s,t}^\varepsilon(x)$ be the backward flow generated by $X_t^\varepsilon(x)$. Then $u_t^\varepsilon(x) = E[f(\hat{\phi}_{0,t}^\varepsilon(x))]$ is the solution of (4.1). H. Matsumoto [21] showed that the family of stochastic flows $\{\hat{\phi}_{0,t}^\varepsilon(x)\}$ converges weakly as $\varepsilon \to 0$ to a stochastic flow with the local characteristics

$$\bar{a}(x,y) = \frac{1}{2\pi}\int_0^t \sigma(x,\tau)\sigma(y,\tau)^* d\tau, \qquad \bar{b}(x) = \frac{1}{2\pi}\int_0^{2\pi} b(x,\tau)\, d\tau$$

and that $D^\alpha u_t^\varepsilon(x)$, $\varepsilon > 0$ converges uniformly to $D^\alpha u_t(x)$ in (t,x) for any α.

4.2 We shall next consider the limit theorems for parabolic partial differential equations with the random drift terms: Consider a family of equations with parameter ε:

(4.6) $$\frac{\partial}{\partial t}u_t^\varepsilon(x) = L_t u_t^\varepsilon(x) + \sum f_i^\varepsilon(x,t,\omega)\frac{\partial}{\partial x_i}u_t^\varepsilon(x), \qquad u_0^\varepsilon = f$$

where L_t is the operator of (4.5) and $f^\varepsilon(x,t,\omega)$ is a random field similar to the right hand side of (1.6). Pardoux-Boux [24] and Kushner-Huang [17] studied the limit of u_t^ε as $\varepsilon \to 0$ when $Y_t^\varepsilon(x) = \int_0^t f^\varepsilon(x,\tau)d\tau$ converges weakly. Concerning this problem, we have the following.

Theorem 4.1. Suppose that $f^\varepsilon(x,t,\omega)$ *is k-times continuously differentiable in* x *and satisfy* $(H.2)_k'$, $(H.3)_{k,r}'$ *for some* $k \geq 2$ *by setting* $b^\varepsilon(x,t) = f^\varepsilon(x,t)$ *and* $\underline{\underline{G}}_{s,t}^\varepsilon = \sigma(f^\varepsilon(x,u)\ ;\ s \leq u \leq t,\ x \in R^d)$. *Then the family of* C^k*-valued processes* $Y_t^\varepsilon(x) = \int_0^t f^\varepsilon(x,\tau)d\tau$ *converges weakly to a* C^{k-2}*-Brownian motion* $Y_t(x)$. *Furthermore, the solutions* $u_t^\varepsilon(x)$, $\varepsilon > 0$ *of (4.6) together with derivatives* $D^\alpha u_t^\varepsilon(x)$, $|\alpha| \leq k-2$ *converges weakly to the solution* $u_t(x)$ *and its derivatives of the following stochastic partial differential equation*

(4.7) $$u_t(x) = f(x) + \int_0^t L_\tau u_\tau(x)d\tau + \sum_i \int_0^t \frac{\partial}{\partial x_i}u_\tau(x)\circ Y^i(x,d\tau)$$
$$+ \sum_i \int_0^t \frac{\partial}{\partial x_i}u_\tau(x)h^i(x,\tau)d\tau,$$

where

(4.8) $$h(x,t) = c(x,t) - \frac{1}{2}\sum_i \frac{\partial}{\partial y^i}\tilde{a}^{i\cdot}(x,y)\big|_{y=x}.$$

Here $c(x,t)$ *is the function of (3.20) and* $\tilde{a}$ *is the characteristic of* Y_t.

We shall sketch briefly how the limit theorems of the stochastic flows can be applied to the above theorem. Let $(W, \underline{\underline{B}}_W, Q)$ be another probability space where a C^k-valued Brownian motion $X_t(x,w)$ with the characteristics (a,b) is given. Let $(\Omega\times W, \underline{\underline{F}}\otimes\underline{\underline{B}}_W, P\otimes Q)$ be the product probability space. Let $\underline{\underline{F}}_{s,t}^\varepsilon$ be the sub σ-field generated by $X_u(x) - X_v(x)$, $f^\varepsilon(x,u)$; $s \leq u,\ v \leq t,\ x \in R^d$, and set

(4.9) $$\tilde{X}_t^\varepsilon(x) = X_t(x) + \int_0^t f^\varepsilon(x,\tau)d\tau.$$

It is a backward C^k-semimartingale adapted to $\underline{\underline{F}}_{s,t}^\varepsilon$. Let $\tilde{\phi}_{s,t}^\varepsilon(x,\omega,w)$ be the backward flow generated by $\tilde{X}_t^\varepsilon(x)$, and define

$$u_t^\varepsilon(x,\omega) = E_Q[\tilde{\phi}_{0,t}^\varepsilon(x,\omega,\cdot)].$$

It satisfies (4.6) similarly as in 4.1.

Now the family of $\tilde{X}_t^\varepsilon$, $\varepsilon > 0$ satisfies $(H.1)_k$, $(H.2)_k'$ and $(H.3)_k'$. Then the family $(\tilde{\phi}_t^\varepsilon, \tilde{X}_t^\varepsilon)$, $\varepsilon > 0$ converges weakly by Theorem 3.2. By Skorohod's embedding, we may assume that it converges strongly. Let $(\phi_{s,t}, \tilde{X}_t)$ be its limit. Then $\tilde{X}_t(x)$ is the sum of two independent C^{k-2}-Brownian motions $X_t(x,w)$ and $Y_t(x,\omega)$. The pair is related by

$$(4.10)\qquad \tilde{\phi}_{s,t}(x) = x + \int_s^t Y(\tilde{\phi}_{r,t}(x), \hat{dr}) + \int_s^t c(\tilde{\phi}_{r,t}(x))\,dr + \int_s^t X(\phi_{r,t}(x), \hat{dr}).$$

Using the Stratonovich backward integral, it is written by

$$\tilde{\phi}_{s,t}(x) = x + \int_s^t \circ Y(\tilde{\phi}_{r,t}(x), \hat{dr}) + \int_s^t h(\tilde{\phi}_{r,t}(x), r)\,dr + \int_s^t X(\phi_{r,t}(x), \hat{dr}).$$

Define now

$$u_t(x,\omega) = E_Q[f(\hat{\phi}_{0,t}(x,\omega,\cdot)].$$

We can prove similarly as in [14] that it satisfies the equation (4.7).

Remark Our characterization of the limiting process $u_t(x)$ is more direct than Pardoux-Boux and Kushner-Huang's. Indeed, in the latters $u_t(x)$ is characterized as a solution of a suitable martingale problem.

The details of the proof of theorems will be discussed elsewhere.

Reference

[1] P. Billingsley: Convergence of probability measures, John Willey and Sons, New York, 1968.

[2] J. Bismut: Mécanique Aléatoire, Lecture Notes in Math. 866, Springer-Verlag, Berlin, Heidelberg, New York, 1981.

[3] Yu.N. Blagoveschenskii-M.I. Freidlin: Certain properties of diffusion processes depending on a parameter, Soviet Math. Dokl. 2 (1961), 633-636.

[4] A.N. Brodin: A limit theorem for solutions of differential equations with random right hand side, Theory Probab. Appl. 22 (1977), 482-497.

[5] T. Fujiwara-H. Kunita: Stochastic differential equations of jump type and Lévy processes in diffeomorphisms group, Kyoto Math. J. 25 (1985), 71-106.

[6] I.A. Ibragimov-Yu.V. Linnik: Independent and stationary sequences of random variables, Groningen: Wolters-Noordhoff, 1971.

[7] N. Ikeda-S. Nakao-Y. Yamato: A class of approximations of Brownian motion, Publ. RIMS Kyoto Univ. 13 (1977), 285-300.

[8] N. Ikeda-S. Watanabe: Stochastic differential equations and diffusion processes North-Holland-Kodansha, 1981.

[9] H. Kesten-G.C. Papanicolaou: A limit theorem for turbulent diffusion, Commun. Math. Phys. 65 (1979), 97-128.

[10] R.Z. Kahsminskii: Principle of averaging for parabolic and elliptic differential equations and for Markov processes with small diffusion, Theory Probab. Appl. 8 (1963), 1-21.

[11] R.Z. Khasminskii: On stochastic processes defined by differential equations with a small parameter, Theory Probab. Appl. 11 (1966), 211-228.

[12] R.Z. Khasminskii: A limit theorem for solutions of differential equations with random right hand sides, Theory Probab. Appl. 11 (1966), 390-406.

[13] H. Kunita: Stochastic differential equations and stochastic flows of diffeomorphisms, Lecture Notes in Math. 1097 (1984), 144-303.

[14] H. Kunita: Stochastic partial differential equations connected with non-linear filtering, Lecture Notes in Math. 972 (1981), 100-168.

[15] H. Kunita: On the convergence of solutions of stochastic ordinary differential equations as stochastic flows of diffeomorphisms, Osaka J. Math. 21 (1984), 883-911.

[16] H. Kunita: Convergence of stochastic flows connected with stochastic ordinary differential equations, submitted to Stochastics.

[17] H. Kushner-H. Huang: Limits for parabolic partial differential equations with wide band stochastic coefficients and its application to filtering theory,

Stochastics, 14 (1985), 115-148.

[18] Y.Le Jan: Flots de diffusions dans R^d, C.R. Acad. Sci. Paris 294 (1982), Serie I, 697-699.

[19] Y. Le Jan-S. Watanabe: Stochastic flows of diffeomorphisms, Taniguchi Symp. S A Katata, 1982, 307-332.

[20] P. Malliavin: Stochastic calculus of variations and hypoelliptic operators, Proc. of Intern. Symp. SDE Kyoto 1976, Kinokyniya, Tokyo, 1978.

[21] H. Matsumoto: Convergence of driven flows of diffeomorphisms, submitted to Stochastics.

[22] G.C. Papanicolaou-W. Kohler: Asymptotic theory of mixing stochastic ordinary differential equations, Comm. Pure Appl. Math. 27 (1974), 641-668.

[23] G.C. Papanicolaou-D.W. Stroock-S.R.S. Varadhan: Martingale approach to some limit theorems, 1976 Duke Turbulence Conf., Duke Univ. Math. Series III, 1977.

[24] E. Pardoux - R. Boux: PDE with random coefficients: Asymptotic expansion for the moments, Lecture Notes in Control and Inf. Science 42, Ed Fleming and Gorostiza 1982, 276-289.

[25] J.G. Shu: On the mollifier approximation for solutions of stochastic differential equations, J. Math. Kyoto Univ. 22 (1982), 243-254.

[26] H. Watanabe: A note on the weak convergence of solutions of certain stochastic ordinary differential equations, Proc. fourth Japan-USSR Symp. Probab. Theory, Lecture Notes in Math. 1021 (1983), 690-698.

[27] E. Wong-M. Zakai: On the relation between ordinary and stochastic differential equations, Intern. J. Engng. Sci. 3 (1965), 213-229.

WEAK CONVERGENCE AND APPROXIMATIONS FOR PARTIAL DIFFERENTIAL EQUATIONS WITH RANDOM PROCESS COEFFICIENTS

Harold J. Kushner
Lefschetz Center for Dynamical Systems
Division of Applied Mathematics
Brown University
Providence, Rhode Island 02912

ABSTRACT

For a parabolic equation with wide bandwidth coefficients, it is shown that the solution converges weakly to that of a stochastic PDE driven by an infinite dimensional Wiener process as the bandwidth tends to infinity. The treatment is novel and purely probabilistic. The solution to the "wide band" coefficient system is represented as a conditional expectation of a functional of a certain diffusion. By a weak convergence argument, the conditional expectation (and its mean square derivatives) converges weakly to a conditional expectation of a functional of a "limit" diffusion. It is then shown that this "limit" functional satisfies the appropriate stochastic PDE. The infinite dimensional Wiener process is represented explicitly in terms of the original system noise. No coercivity or strict ellipticity conditions are required. The result provides a partial justification for the use of infinite dimensional Wiener processes in distributed systems. Since the method is based on weak convergence arguments for Itô-type equations with wide bandwidth coefficients and "PDE methods" are avoided, it is likely that the technique will find greater use in the analysis of infinite dimensional stochastic systems. The methods have already proved to be very useful in studying approximations to nonlinear filtering problems with wide bandwidth observation noise [10].

1. INTRODUCTION

Let $z(\cdot)$ denote a bounded stationary stochastic process, to be further specified below. For each $\varepsilon > 0$, let u^ε solve the PDE

$$u_t^\varepsilon = Au^\varepsilon + \frac{1}{\varepsilon}\sum_1^n h_i^\varepsilon u_{x_i}^\varepsilon + \frac{1}{\varepsilon} h_{n+1}^\varepsilon u^\varepsilon + \frac{1}{\varepsilon} h_0^\varepsilon, \quad u^\varepsilon(x,0) = u_0(x) \, , \tag{1.1}$$

where $h_i^\varepsilon(x,t) = h_i(x,z^\varepsilon(t)), z^\varepsilon(t) = z(t/\varepsilon^2)$, and

$$Au(x,t) = \frac{1}{2}\sum_{i,j=1}^n a_{ij}(x)u_{x_i x_j}(x,t) + \sum_{i=1}^n c_i(x)u_{x_i}(x,t) + c_{n+1}(x)u(x,t) + c_0(x).$$

The $h_i(x,z^\varepsilon(t))/\varepsilon$ terms are effectively wide bandwidth noise processes. The scaling $z(t/\varepsilon^2)/\varepsilon$ is a frequently used method of obtaining such a wide bandwidth process. Interesting weak convergence methods for the sequence $\{u^\varepsilon\}$ were developed in [1], [2], [3]. The sequences $\{u^\varepsilon\}$ converged weakly (in an appropriate space) to a process $u^\varepsilon(\cdot)$ which satisfied a stochastic partial differential equation driven by a cylindrical Wiener process. Here, we take a different approach, which has the advantage of being more intuitive, and gives substantial insight into the nature of the processes which are involved. PDE methods are avoided entirely. We exploit a representation of u^ε as a functional of a particular diffusion $X^\varepsilon(\cdot)$ and obtain the limits of $\{u^\varepsilon\}$ by weak convergence arguments concerning $\{X^\varepsilon(\cdot)\}$. This stochastic differential equation perspective gives more physical intuition into the processes and should be quite useful in analysis and applications. As for the case of ordinary stochastic differential equations, wide bandwidth noise is often more realistic than white noise or a cylindrical Wiener process. Since the analysis with the latter process is substantially simpler, limit theorems for systems with wide bandwidth noise are desirable. Fuller details are in [7]. The work which led to this paper was done jointly by the author and Dr. Huang Hai of Bell Telephone Labs.

Section 2 states some assumptions and gives our convenient representation of u^ε as a functional of a stochastic differential equation with wide bandwidth processes in the coefficients. Section 3 deals with weak convergence of the solution processes to these stochastic differential equations, and a representation of the weak limit of $u^\varepsilon(x,t)$ as an expectation of a functional $u(x,t)$ of the limit process (the solution to the limit stochastic differential equation), conditioned on one of the driving Wiener processes. Section 4 develops a 'Taylor' formula in terms of mean square derivatives and states the PDE which the limit $u(\cdot,\cdot)$ satisfies - namely, the PDE which is satisfied by $u(\cdot,\cdot)$ and its mean square derivatives. In Section 5, we briefly discuss the situation when the assumption (A3) is weakened.

2. ASSUMPTIONS AND A REPRESENTATION FOR u^ε

(A1) and (A2) below are the two basic assumptions. (A2) simplifies a few of the calculations, but the full power of the mixing condition is never used.

A1. $\{a_{ij}(x)\} = \sigma(x)\sigma'(x)$. The $h_i(\cdot,z), c_i(\cdot), \sigma(\cdot)$, and $u_0(\cdot)$ are bounded and have continuous and bounded partial x-derivatives up to order three for $c_i(\cdot)$ and $\sigma(\cdot)$, and up to order five for the $h_i(\cdot)$.

A2. $z(\cdot)$ is bounded and stationary on $(-\infty,\infty)$ and $E\, h_i(x,z(t)) \equiv 0$. $z(\cdot)$ is ϕ-mixing in both the forward and backward directions with $\int_0^\infty \phi^{1/2}(u)du < \infty$ [8]. Also there is a version of $z(\cdot)$ which is right continuous, and one which is left continuous.

We write $\xi(\cdot)$ for the reverse time process $\xi(t) = z(T_1-t)$. For purposes of analysis we set (w.l.o.g.) $T_1 = 0$. An example of (A2) is an ergodic finite state Markov chain. The analysis is easier under (A3). Later we indicate the method when

(A3) is violated (see also [7]).

A3. <u>For possible vector valued</u> h_i, z, <u>let</u> $h_i(x,z) = h_i(x)z$.

Fix $T > 0$. Define $v^\varepsilon(x,t) = u^\varepsilon(x,T-t)$ and $\xi^\varepsilon(t) = z(\frac{T-t}{\varepsilon^2})$. Then

$$v_t^\varepsilon + Av^\varepsilon + \frac{1}{\varepsilon}\sum_{i=1}^{n} h_i(x,\xi^\varepsilon(t))v_{x_j}^\varepsilon + \frac{1}{\varepsilon}h_{n+1}(x,\xi^\varepsilon(t))v^\varepsilon + \frac{1}{\varepsilon}h_0(x,\xi^\varepsilon(t)) = 0 ,$$
$$v^\varepsilon(x,T) = u_0(x). \tag{2.1}$$

We now define a stochastic differential equation, and then represent v^ε as a functional of the solution. This representation is the key to the analysis.

Let $w^\varepsilon(\cdot)$ denote a standard vector-valued Wiener process which is independent of $z(\cdot)$ and define the processes $x_i^\varepsilon(\cdot)$ by $(x^\varepsilon = (x_1^\varepsilon,\ldots,x_n^\varepsilon))$

$$dx_0^\varepsilon = [c_0(x^\varepsilon) + h_0(x^\varepsilon,\xi^\varepsilon(t))/\varepsilon]x_{n+1}^\varepsilon dt$$
$$dx_i^\varepsilon = [c_i(x^\varepsilon) + h_i(x^\varepsilon,\xi^\varepsilon(t))/\varepsilon]dt + \sum_j \sigma_{ij}(x^\varepsilon)dw_j^\varepsilon, \quad 1 \le i \le n, \tag{2.2}$$
$$dx_{n+1}^\varepsilon = [c_{n+1}(x^\varepsilon) + h_{n+1}(x^\varepsilon,\xi^\varepsilon(t))/\varepsilon]x_{n+1}^\varepsilon dt.$$

In order to get a more convenient representation for (2.2), define $X = (x_0,x,x_{n+1})$, $X^\varepsilon = (x_0^\varepsilon,x^\varepsilon,x_{n+1}^\varepsilon)$ and define the vectors $H(x,\xi) = \{h_i(x,\xi),\ 0 \le i \le n+1\}'$, $C(x) = \{c_i(x),\ 0 \le i \le n+1\}'$. Let $J(x_{n+1})$ denote the $(n+2)$-dimensional diagonal matrix with diagonal entries $(x_{n+1},1,\ldots,1,x_{n+1})$. Define $\overline{H}(X,\xi) = J(x_{n+1})H(x,\xi)$ and $\overline{C}(X) = J(x_{n+1})C(x)$. Define the matrix $\overline{\sigma}(x)$ by

$$\overline{\sigma}_{ij}(x) = \sigma_{ij}(x) \quad \text{for } 1 \le i \le n \text{ and } \overline{\sigma}_{ij} = 0 \text{ if } i \text{ is } 0 \text{ or } n+1.$$

Now, rewrite (2.2) as

$$dX^\varepsilon = [\overline{C}(X^\varepsilon) + \overline{H}(X^\varepsilon,\xi^\varepsilon)/\varepsilon]dt + \overline{\sigma}(x^\varepsilon)dw^\varepsilon, \tag{2.3}$$

Let $E_{x,t}^\xi$ (respectively, $E_{X,t}^\xi$) denote expectation conditioned on the initial condition $x^\varepsilon(t) = x$ (respectively, $X^\varepsilon(t) = X$) and on $\{\xi^\varepsilon(\lambda),\ -\infty < \lambda < \infty\}$. Using the definition of $x^\varepsilon(\cdot)$, we can write ([9],Chapter 8) the solution to (2.2) as (with $x_0(t) = 0$, $x_{n+1}(t) = 1$)

$$v^\varepsilon(x,t) = E_{x,t}^\varepsilon u_0(x^\varepsilon(T))\exp\int_t^T ds[c_{n+1}(x^\varepsilon(s)) + h_{n+1}(x^\varepsilon(s),\xi^\varepsilon(s))/\varepsilon]ds$$
$$+ E_{x,t}^\xi\int_t^T ds[c_0(x^\varepsilon(s)) + h_0(x^\varepsilon(s),\xi^\varepsilon(s))/\varepsilon]\exp\int_t^s d\lambda[c_{n+1}(x^\varepsilon(\lambda)) \tag{2.4a}$$
$$+ h_{n+1}(x^\varepsilon(\lambda),\xi^\varepsilon(\lambda))/\varepsilon]$$
$$= E_{x,t}^\xi[x_{n+1}^\varepsilon(T)u_0(x^\varepsilon(T)) + x_0^\varepsilon(T)] \equiv E_{x,t}^\xi F(X^\varepsilon(T)) .$$

The solution to (1.1) or (2.1) is not necessarily unique. We suppose that (2.4a) is <u>the</u> desired solution.

For arbitrary values $x_0(t) = x_0$ and $x_{n+1}(t) = x_{n+1}$, we define $\bar{v}^\varepsilon$ by

$$\bar{v}^\varepsilon(X,t) = E^\xi_{X,t}[x^\varepsilon_{n+1}(T)u_0(x^\varepsilon(T)) + x^\varepsilon_0(T)] \ . \tag{2.4b}$$

Then $\bar{v}^\varepsilon(0,x,1,t) = v^\varepsilon(x,t)$. For most of the analysis, we let x_0 and x_{n+1} be arbitrary. In the sequel, we exploit the representation (2.4) and the weak convergence properties of (2.3) and of its mean square (with respect to the components of x) derivatives, in order to obtain and characterize the limit $u(\cdot)$ of $\{u^\varepsilon(\cdot)\}$.

3. WEAK CONVERGENCE OF $\{X^\varepsilon(\cdot)\}$.

The symbol $\Rightarrow$ denotes weak convergence. Define the process $B^\varepsilon(t) = \int_0^t \xi^\varepsilon(s)ds/\varepsilon$, and let $X^\varepsilon_X(\cdot)$, $X^\varepsilon_{XX}(\cdot)$ and $X^\varepsilon_{XXX}(\cdot)$ denote the processes (of the appropriate dimension) which are the (first, second and third order) mean-square derivatives of $X^\varepsilon(\cdot)$, with respect to the initial condition X. We will use (w.l.o.g.) $\xi(s) = z(-s)$. Then $E\xi(s)\xi'(0) = Ez(0)z'(s)$. Define $\bar{R} = \int_{-\infty}^{\infty} E\xi(s)\xi'(0)ds$.

The mixing condition (A2) and smoothness (A1) yield the weak convergence result Theorem 3.1. We let the initial time be zero (w.l.o.g.).

Define the matrix $\bar{\Sigma}$ and vector $\bar{Q}$ by (well defined by (A2))

$$\bar{\Sigma}(X) = \int_{-\infty}^{\infty} E\ \bar{H}(X,\xi(s))\bar{H}'(X,\xi(0))ds \tag{3.1}$$

$$\bar{Q}(X) = \int_0^{\infty} E\ \bar{H}'_X(X,\xi(s))\bar{H}(X,\xi(0))ds \ . \tag{3.2}$$

Where $\bar{H}_X$ is the matrix whose columns are the X-gradients of the $(n+2)$ components of $\bar{H}$. Under (A3), $\bar{\Sigma}(X) = \bar{H}(X)\bar{R}\bar{H}'(X)$.

For future use, write $\bar{Q} = (q_0,\dots,q_{n+1})'$. The components of (3.2) are

$$q_i(X) = \int_0^{\infty} E[\sum_{j=1}^{n} h_{i,x_j}(x,\xi(s))h_j(x,\xi(0))]ds, \ n \geq i \geq 1,$$

$$q_0(X) = \int_0^{\infty} E[\sum_1^n h_{0,x_j}(x,\xi(s))h_j(x,\xi(0)) + h_0(x,\xi(s))h_{n+1}(x,\xi(0))]x_{n+1}ds$$

$$q_{n+1}(X) = \int_0^{\infty} E[\sum_1^n h_{n+1,x_j}(x,\xi(s))h_j(s,\xi(0)) + h_{n+1}(x,\xi(s))h_{n+1}(x,\xi(0))]x_{n+1}ds \ .$$

The following Theorem can be proved by the methods of [5].

Theorem 3.1. Assume (A1) and (A2). For any compact set R_1, the sequence $\{X^\varepsilon(\cdot), B^\varepsilon(\cdot), w^\varepsilon(\cdot); X^\varepsilon(0) \in R_1, \varepsilon>0\}$ is tight in $D^k[0,\infty)$ (for the appropriate k). If $X^\varepsilon(0) \to X(0)$ weakly, then the sequence converges weakly and the unique limit $(X(\cdot), B(\cdot), w(\cdot))$ of $\{X^\varepsilon(\cdot), B^\varepsilon(\cdot), w^\varepsilon(\cdot), \varepsilon>0\}$ solves the martingale problem with operator $\mathscr{A}$ given by (3.3).

$$\mathscr{A}f(X,B,w) = \int_0^\infty ds\ E[f_X'(X,B,w)\overline{H}(X,\xi(s)) + f_B'(X,B,w)\xi(s)]_X'\overline{H}(X,\xi(0))$$
$$+ \int_0^\infty ds\ E[f_X'(X,B,w)\overline{H}(X,\xi(s)) + f_B'(X,B,w)\xi(s)]_B'\xi(0) \tag{3.3}$$
$$+ f_X'(X,B,w)\overline{C}(X) + \sum_{i,j} f_{x_i w_j}(X,B,w)\overline{\sigma}_{ij}(x)$$
$$+ \frac{1}{2}\sum_i f_{w_i w_i}(X,B,w) + \frac{1}{2}\sum_{i,j=1}^{n} f_{x_i x_j}(X,B,w)a_{ij}(x)$$

$w(\cdot)$ is a standard Wiener process and $\operatorname{cov} B(t) = t\,\overline{R}$. If $\overline{\sum}(X)$ has a continuous square root $\overline{\sum}^{1/2}(X)$, then there is a standard Wiener process $\overline{B}(\cdot)$ independent of $w(\cdot)$ such that

$$dX = [\overline{C}(X) + \overline{Q}(X)]dt + \overline{\sigma}(x)dw + \overline{\sum}^{1/2}(X)d\overline{B}\ . \tag{3.4}$$

Also if $X^\varepsilon(0) \to X$, $\{X^\varepsilon(\cdot),X_X^\varepsilon(\cdot),X_{XX}^\varepsilon(\cdot),B^\varepsilon(\cdot),w^\varepsilon(\cdot)\}$ converges weakly to a diffusion $(X(\cdot),X_1(\cdot),X_2(\cdot),B(\cdot),w(\cdot))$, where $B(\cdot)$ and $w(\cdot)$ are as above. Under (A3), $\overline{\sum}^{1/2}(X)d\overline{B}$ in (3.4) can be replaced by $\overline{H}(X)dB$ and we have $X_1(\cdot) = X_X(\cdot), X_2(\cdot) = X_{XX}(\cdot)$.

Define ($F(\cdot)$ is defined in (2.4a))

$$\overline{v}(X,t) = E_{X,t}^B(X(T))\ ,$$

where $E_{X,t}^B$ denotes conditioning on the initial condition $X(t) = X$ and on $\{B(\lambda) - B(t),\ \lambda \geq t\}$.

The following theorem is one of the key results. For each initial pair (X,t), it characterizes the limit as a conditional expectation. See [7] for the proof.

Theorem 3.2. Let F_0 be bounded and continuous and assume (A1) to (A3). Then

$$\overline{v}_0^\varepsilon(X,t) \equiv E_{X,t}^\xi F_0(X^\varepsilon(T)) \to \overline{v}_0(X,t) \equiv E_{X,t}^B F_0(X(T)) \tag{3.5}$$

in distribution as $\varepsilon \to 0$, for each X and t. The function $\overline{v}_0^\varepsilon(\cdot,\cdot)$ is continuous in (X,t) for each ω and ε, and $\overline{v}_0(\cdot,\cdot)$ is stochastically continuous in (X,t). It has a version which is separable and measurable as a function of X,t and ω (the "X and t σ-algebras being Borel").

Let $\{X(\cdot),X_X(\cdot),X_{XX}(\cdot),B(\cdot),w(\cdot)\}$ denote the weak limit of $\{X^\varepsilon(\cdot),X_X^\varepsilon(\cdot),X_{XX}^\varepsilon(\cdot),B^\varepsilon(\cdot),w^\varepsilon(\cdot)\}$ and assume Skorohod imbedding so that the convergence is w.p.1. Then for any compact set R_1 and any $q > 0$,

$$\sup_{\substack{x\in R_1\\ t\leq T}} E_{X,t}|\overline{v}_0^\varepsilon(X,t) - \overline{v}_0(X,t)|^q \overset{\xi}{\to} 0. \tag{3.6}$$

The above results hold if $X(\cdot)$ and $X^\varepsilon(\cdot)$ in (3.5) and (3.6) are replaced by $X(\cdot),X^\varepsilon(\cdot)$ and their mean square derivatives (in X) up to order two.

Our function $F(\cdot)$ is not bounded. The following result, proved by a perturbed Liapunov function method, enables us to carry Theorem 3.2 over.

Theorem 3.3. Assume (A1) and (A2). Then for any integer $q > 0$ and compact R_1 and $X^\varepsilon(t) = X$,

$$\sup_{X\in R_1} \sup_{\substack{t\le\tau\le T\\ \varepsilon>0,\, t\le T}} [E|X^\varepsilon(\tau)|^{2q} + E|X_X^\varepsilon(\tau)|^{2q} + E|X_{XX}^\varepsilon(\tau)|^{2q} + E|X_{XXX}^\varepsilon(\tau)|^{2q}] < \infty .$$

We now have

Theorem 3.4. Assume (A1) to (A3). Then (3.5) and (3.6) hold for F replacing F_0, and in particular for any continuous function F_0 which is bounded by (for any $q > 0$), $K(1 + |X^\varepsilon(T)|^q + \ldots + |X_{XX}(T)|^q)$. Also, the convergences (3.5) and (3.6) hold for any such F_0 if the $\bar{v}$ and $\bar{v}^\varepsilon$ are replaced by their mean-square derivatives in X (for smooth enough F_0), up to order two.

Theorem 3.5. Assume (A1) to (A3). For any x,k and $t_1,\ldots,t_k,\ldots,\tau_1,\ldots,\tau_k$,

$$\{v^\varepsilon(x,t_i),\ i \le k\} \Rightarrow \{v(x,t_i),\ i \le k\}$$

$$\{u^\varepsilon(x,\tau_i),\ i \le k\} \Rightarrow \{u^\varepsilon(x,\tau_i),\ i \le k\}$$

The above theorems give weak convergence results for $v^\varepsilon(x,t)$ and characterize the limits of the solutions to (1.1), (2.1), for each t, or at best, for a finite set $\{t_i\}$. They need to be extended to hold for the functions $v^\varepsilon(\cdot,t)$, $v^\varepsilon(x,\cdot)$ or $v^\varepsilon(\cdot,\cdot)$, and we state a result for them.

Tightness and weak convergence of $\{u^\varepsilon(x,\cdot)\}$. We next deal with tightness of $\{u^\varepsilon(x,\cdot)$ in $D[0,\infty)$ to help show that $u^\varepsilon(x,\cdot) = v^\varepsilon(x,T-\cdot)$ converges weakly to $u(x,\cdot) = v(x,T-\cdot)$, an improvement over the pointwise convergence result of Theorem 3.4. The proof [7] uses the perturbed test function method of [5].

Theorem 3.6. Assume (A1) and (A2). Then for each x, $\{u^\varepsilon(x,\cdot)\}$ is tight in $D[0,\infty)$. Under the additional condition (A3), for each x, there is a continuous version of $u(x,\cdot) = v(x,T-\cdot)$ and $u^\varepsilon(x,\cdot) \Rightarrow u(x,\cdot)$.

4. THE LIMIT PARTIAL DIFFERENTIAL EQUATION

In this section we let $\bar{\sigma}(\cdot)$, $\bar{H}(\cdot)$ and $\bar{b}(\cdot)$ be bounded and continuous functions, together with their partial derivatives up to order three. Let (real-valued) $F(\cdot)$ and its partial derivatives (up to order three) be continuous and bounded by a polynomial. Define $X(\cdot)$ by

$$dX = \bar{b}(X)dt + \bar{H}(X)dB + \bar{\sigma}(X)dw ,$$

where $B(\cdot)$ and $w(\cdot)$ are mutually independent Wiener Processes, $w(\cdot)$ is standard and $\operatorname{cov} B(t) = t\bar{R}$.

Define

$$\bar{v}(X,t) = E^B_{X,t} F(X(T)) .$$

A consequence [7, Lemma 4.1] is that

$$\begin{aligned}
\bar{v}_{x_i}(X,t) &= E^B_{X,t} \sum_k F_{x_k}(X(T)) x_{k,x_i}(T) \\
\bar{v}_{x_i x_j}(X,t) &= E^B_{X,t} \sum_{k,\ell} F_{x_k x_\ell}(X(T)) x_{k,x_i}(T) x_{\ell,x_j}(T) \\
&\quad + E^B_{X,t} \sum_k F_{x_k}(X(T)) x_{k,x_i x_j}(T) ,
\end{aligned} \tag{4.1}$$

$$E|\bar{v}_{x_i x_j}(X+y,t) - \bar{v}_{x_i x_j}(X,t)|^2 \le K_1(X)|y|(1+|y|) , \tag{4.2}$$

where $K_1(X)$ is bounded on each compact x-set.

The following lemma, akin to a truncated Taylor expansion will be used to obtain the PDE which $v(\cdot,\cdot)$ satisfies.

<u>Lemma 4.1. For all</u> X,y,t w.p.1 ($\bar{v}_X, \bar{v}_{XX}$ <u>are the mean-square derivatives)</u>

$$\begin{aligned}
\bar{v}(X+y,t) - \bar{v}(X,t) &= \int_0^1 \bar{v}_X(X+sy,t)y\, dt \\
&= \bar{v}_X(X,t)y + \int_0^1 s\, ds \int_0^1 d\tau\; y'\bar{v}_{XX}(X+s\tau y,t)y \\
&= \bar{v}_X'(X,t)y + \frac{1}{2} y'\bar{v}_{XX}(X,t)y + y'Qy
\end{aligned} \tag{4.3}$$

$$Q = \int_0^1 s\, ds \int_0^1 d\tau [\bar{v}_{XX}(X+s\tau y,t) - \bar{v}_{XX}(X,t)],$$

$$E|Q|^2 \le K_2(X)(|y|^2 + |y|) ,$$

<u>where</u> $K_2(\cdot)$ <u>is bounded on each compact x-set. Equation (4.3) holds if</u> y <u>is replaced by a random variable which is independent of</u> $\{B(\lambda) - B(t),\ w(\lambda) - w(t),\ \lambda \ge t\}$.

The next theorem gives the stochastic partial differential equation which $\bar{v}(\cdot,\cdot)$ satisfies.

<u>Theorem 4.2. W.p.1, for each</u> X,t, (<u>use</u> $\bar{a} = \bar{\sigma}\,\bar{\sigma}'$ <u>and</u> $\{\bar{R}_{ij}\} = \bar{R} = \text{cov } B(1)$)

$$\begin{aligned}
\bar{v}(X,t) - \bar{v}(X,T) &= \frac{1}{2} \int_t^T \sum_{i,j} \bar{v}_{x_i x_j}(X,s)\bar{a}_{ij}(X)ds \\
&\quad + \frac{1}{2} \int_t^T \text{trace } \bar{v}_{XX}(X,s) \cdot \bar{H}(X)\bar{R}\,\bar{H}'(X)ds \\
&\quad + \int_t^T \bar{v}_X(X,s)\bar{H}(X)dB(s) + \int_t^T \bar{v}_X'(X,s)\bar{b}(X)ds ,
\end{aligned} \tag{4.4}$$

<u>By using the separable and measurable (parameter</u> X,t) <u>versions of the terms in (4.4), (4.4) holds for all</u> $X,t \le T$, <u>w.p.1.</u>

Remark. The stochastic integral in (4.4) is a "backwards" integral. It is defined as the limit of the sum $\sum \bar{v}_X'(X,i\Delta+\Delta)\bar{H}(X)[B(i\Delta+\Delta) - B(i\Delta)]$, where $\bar{v}_X(X,i\Delta+\Delta)$ is independent of (the "prior" increment) $[B(i\Delta+\Delta) - B(i\Delta)]$. Recall that $\bar{v}(X,i\Delta+\Delta)$ is a function of $B(i\Delta+\Delta+\lambda) - B(i\Delta+\Delta),\ \lambda \geq 0\}$, so the stochastic integral is well defined. "Backwards" stochastic integrals were used by Pardoux in a filtering problem [6].

We use Theorem 4.2 to obtain a stochastic PDE for $v(\cdot,\cdot)$ itself.

Theorem 4.3. Assume (A1) to (A3). Then

$$v(x,t) - v(x,T) = \frac{1}{2}\int_t^T \sum_{i,j=1}^n v_{x_i x_j}(x,s)[a_{ij}(x) + \tilde{a}_{ij}(x)]ds + \tag{4.5}$$
$$+ \int_t^T \sum_1^n v_{x_i}(x,s)[c_i(x)ds + q_i(x)ds + h_i(x)dB(s)]$$
$$+ \int_t^T v(x,s)[c_{n+1}(x)ds + q_{n+1}(x)ds + h_{n+1}(x)dB(s)]$$
$$+ \int_t^T [c_0(x)ds + q_0(x)ds + h_0(x)dB(s)]$$
$$+ \int_t^T \sum_1^n v_{x_i}(x,s)\tilde{a}_{n+1,i}(x)ds$$

where

$$\tilde{a}_{ij}(x) = \int_{-\infty}^{\infty} Eh_i(x,\xi(s))h_j(x,\xi(0))ds,\quad 1 \leq i,\ j \leq n+1\ ,$$

and $q_0(x) = q_0(X)$, $q_{n+1}(x) = q_{n+1}(X)$, evaluated at $x_0 = 0$ and $x_{n+1} = 1$. (See above Theorem 3.1 for the definition of $q_i(\cdot)$). By taking a separable and measurable version of all integrals (parameter x,t) we can suppose that (4.5) holds w.p.1 for all x and $t \leq T$.

Remark. With an appropriate identification of terms and a reversal of time, (4.5) agrees with the result in [2], [3].

Proof. Use $\bar{v}(x,t) = v(0,x,1,t)$ and $\bar{b}(X) = \bar{C}(X) + \bar{Q}(X)$, where the components $\{q_i(X)\}$ of $\bar{Q}(X)$ are defined above Theorem 3.1. Since

$$dx_0 = [\text{function of } x]x_{n+1}dt + [\text{function of } x]x_{n+1}dB$$

$$dx_{n+1} = [\text{function of } x]x_{n+1}dt + [\text{function of } x]x_{n+1}dB,$$

we have

$$x_0(T) = x_0(t) + [\text{function of } x(s),\ T \geq s \geq t]x_{n+1}(t)$$

$$x_{n+1}(T) = [\text{function of } x(s),\ T \geq s \geq t]x_{n+1}(t)\ .$$

Thus

$$\frac{\partial \bar{v}(X,t)}{\partial x_0} = 1, \quad \frac{\partial \bar{v}(X,t)}{\partial x_{n+1}} = v(x,t), \quad \frac{\partial^2 \bar{v}(X,t)}{\partial^2 x_{n+1}} = 0$$

In order to obtain (4.5), simply use the above substitutions. The last term in (4.5) is simply

$$\int_t^T \sum_{i=1}^{n+1} v_{x_i x_{n+1}}(x,s)\tilde{a}_{i,n+1}(x)ds \ . \qquad \text{QED}$$

Equation (4.5) is a stochastic PDE, driven by a <u>finite dimensional</u> Wiener process $B(\cdot)$. $B(\cdot)$ is finite dimensional owing to the form (A3). In this case, the covariance operators $K(x,y)$ of the driving noise for the stochastic PDE's obtained in [2] or [3] can be factored into $K_1'(x)K_1(y)$ for some finite dimensional matrix valued function $K_1(\cdot)$, and the driving cylindrical Wiener process appearing in these references reduces to an ordinary finite dimensional Wiener process.

Theorem 3.4 gives a weak convergence result for $\{v^\varepsilon(x,\cdot),\ \varepsilon > 0\}$ for each <u>fixed</u> initial condition x. With little additional work, a function space type of convergence can be obtained. In order to save space here we refer the reader to [7], Section 5, where we treat the limit of $v^\varepsilon(\cdot)$ as a process with values in an L_2-space with the weak topology.

5. <u>GENERAL</u> $h(\cdot,\cdot)$.

When (A3) is dropped, an expansion technique is used, and the infinite dimensional driving Wiener process is constructed directly from the data. Replace (A3) by

<u>A4.</u> <u>The</u> $\{h_i(x,\xi),\ 0 \le i \le n+1\} = H(x,\xi)$ <u>has support in some compact hypercube</u> R_0.

<u>A5.</u> <u>For each</u> s, $\xi(0)$ <u>and</u> $\xi(s)$ <u>are exchangeable in that for each Borel set</u> A,

$$P\{(\xi(0),\xi(s)) \in A\} = P\{(\xi(s),\xi(0)) \in A\} \ .$$

Let D_x^i denote the vector of partial derivatives of order i - arranged in some order. Define $\hat{H}(x,\xi) = \{H'(x,\xi), D_x^1 H'(x,\xi), \ldots, D_x^3 H'(x,\xi)\}'$ and

$$\mathcal{K}(x,y) = \int_{-\infty}^{\infty} E\ \hat{H}(x,\xi(s))\hat{H}'(x,\xi(0))ds \ ,$$

a symmetric and non-negative definite function. Define $\mathcal{K}f(x) = \int \mathcal{K}(x,y)f(y)dy$, $\mathcal{K}$ being an operator from $L_2(R_0)$ to $L_2(R_0)$. Let $\{\lambda_i\}$ and $\{G_i\}$ denote the eigenvalues and eigenvectors, with g_{i0} the vector of the first $(n+2)$ components of G_i. Then $\lambda_i \ge 0$ and $\sum \lambda_i < \infty$. For each integer m and initial time t, define the processes (5.1) to (5.3), where $\{B_i(\cdot), w(\cdot)\}$ are mutually independent and $\operatorname{cov} B_i(t) = \lambda_i t$. Define

$$\xi_i(s) = \int \hat{H}'(x,\xi(s))G_i(x)dx, \quad \xi_i^\varepsilon(s) = \xi_i(s/\varepsilon^2) \ ,$$

$$H^m(x,\xi^\varepsilon(s)) = \sum_1^m \xi_i^\varepsilon(s) g_{i0}(x), \quad \bar{H}^m(X,\xi^\varepsilon(s)) = J(x_{n+1}) H^m(x,\xi^\varepsilon(s)), \quad B_i^\varepsilon(s) = \int_t^s \xi_i^\varepsilon(\lambda)\,d\lambda/\varepsilon.$$

$$dX^{\varepsilon,m} = [\bar{C}(X^{\varepsilon,m}) + \bar{H}^m(X^{\varepsilon,m},\xi^\varepsilon(s))/\varepsilon]dt + \bar{\sigma}(x^{\varepsilon,m})dw \tag{5.1}$$

$$dX^m = [\bar{C}(X^m) + \bar{Q}^m(X^m)]dt + \bar{\sigma}(x^m)dw + \sum_{i=1}^m J(x_{n+1}^m) g_{i0}(x^m) dB_i \tag{5.2}$$

$$dX = [\bar{C}(X) + \bar{Q}(X)]dt + \bar{\sigma}(x)dw + \sum_{i=1}^\infty J(x_{n+1}) g_{i0}(x) dB_i, \quad s \geq t, \tag{5.3}$$

where

$$\bar{Q}^m(X) = \frac{1}{2}\sum_1^m \lambda_i [J(x_{n+1}) g_{i0}(x)]_X' [J(x_{n+1}) g_{i0}(x)] .$$

We have $\{X^{\varepsilon,m}(\cdot), w^\varepsilon(\cdot), B_i^\varepsilon(\cdot), i\leq m\} \Rightarrow \{X^m(\cdot), w(\cdot), B_i(\cdot), i \leq m\}$ (5.3) is the weak limit of $\{X^\varepsilon(\cdot)\}$, (5.2) is the weak limit of (5.1), and (5.1) is used as an approximation to $X^\varepsilon(\cdot)$ in the analysis. Let $E_{X,t}^B$ denote conditioning on the initial condition $X(t) = y$ and on $\{B_i(\lambda) - B_i(t), \lambda \geq t, i < \infty\}$. Then

<u>Theorem 5.1.</u> $v^\varepsilon(x,t) \to E_{x,t}^B F(X(t))$ <u>in distribution as</u> $\varepsilon \to 0$. <u>Also</u> $v(\cdot,\cdot)$ <u>satisfies (4.5) with</u> $H(x)dB(s)$ <u>replaced by</u> $\sum_1^\infty g_{i0}(x) dB_i(s)$.

We have thus constructed the infinite dimensional driving $B(\cdot)$ process directly from the $H(X,\xi(\cdot))$ process. See [7] for details.

REFERENCES

[1] E. Pardoux and R. Bouc, PDE with random coefficients: asymptotic expansion for the moments, Lecture Notes in Control and Inf. Sciences <u>42</u>, ed. Fleming and Gorostiza, 1982, 276-289.

[2] R. Bouc and E. Pardoux, Asymptotic analysis of PDE's with wide band noise disturbances, and expansion of the moments, Preprint, Université de Provence, 1983.

[3] H.J. Kushner and H. Huang, Limits for parabolic partial differential equations with wide band stochastic coefficients, LCDS Report #83-27, Brown University, December 1983, submitted to STOCHASTICS.

[4] H.J. Kushner, Jump-diffusion approximations for ordinary differential equations with wideband random right hand sides, SIAM J. on Control and Optim. 17 (1979), 729-744.

[5] H.J. Kushner, <u>Approximation and weak convergence methods for random processes with applications to stochastic systems theory</u>, M.I.T. Press, Cambridge, Mass. U.S.A., April 1984.

[6] E. Pardoux, Stochastic partial differential equations and the filtering of diffusion processes, Stochastics <u>3</u> (1979), 127-167.

[7] H.J. Kushner and Hai Huang, Weak convergence and approximations for partial differential equations with stochastic coefficients, LCDS Report #84-1, Brown University, submitted to STOCHASTICS.

[8] P. Billingsley, Convergence of probability measures, Wiley, New York, 1968.

[9] I.I. Gihman and A.V. Skorohod, Introduction to the Theory of Random Processes, Saunders, Philadelphia, 1965.

[10] H.J. Kushner and Hai Huang, Approximate and limit results for non-linear filters with wide bandwidth observation noise, LCDS Report #84-36, Brown University, submitted to STOCHASTICS.

OPTIMAL CONTROL OF REFLECTED DIFFUSION PROCESSES : AN EXAMPLE OF STATE CONSTRAINTS.

P.L. Lions
Ceremade, Université Paris-Dauphine
Place de Lattre de Tassigny, 75775 Paris Cedex 16.

I. Introduction.

We want to give one example of optimal control problems of diffusion processes with complete observations and state constraints. We follow the approach of dynamic programming and thus we will derive some fully nonlinear second-order elliptic equations - called the Hamilton-Jacobi-Bellman equations (HJB in short) - with appropriate boundary conditions corresponding to the state constraints. The type of state constraints we consider here is the following : we want the state of the system we control to stay, during the full interval of time considered, in a prescribed region. Associated with such problems are at least three difficulties : i) find reasonable models of such state constraints, ii) derive and justify the boundary conditions for HJB equations, iii) show that the value function is the unique solution of HJB equation with such boundary conditions.

As we mentionned in [14], we know three possible answers to i) ; first, the so-called exit problems where the state of the system is stopped when it leaves the prescribed region ; this corresponds to Dirichlet boundary conditions and we refer to P.L. Lions [15], [16], [17] for the most general results known on questions ii), iii) in that case (see also the references therein). The other possibility consists in restricting ourselves to control processes such that the state never leaves the given region : then, if the processes are nondegenerate at the boundary, this leads to infinite Dirichlet or Neumann boundary conditions (see J.M. Lasry and P.L. Lions [13]) ; while if the processes degenerate at the boundary (as it is the case of course for deterministic problems) the boundary conditions are quite different (see M. Soner [26], I. Capuzzo-Dolcetta and P.L. Lions [10]). Finally, the third possible answer to i) is the use of boundary mechanisms which prevent the process from exiting, the simplest of which being the standard reflection. We wish here to concentrate on this particular situation.

We may now describe the systems we are studying : let (Ω, F, F_t, P) be a usual probability space, B_t some m-dimensional F_t -adapted Brownian motion. Let A be a separable metric space, a control process α_t will be a F_t -progressively measurable process with values in A . Let $\mathcal{O}$ be a smooth open domain in $\mathbb{R}^N$ and let γ be a smooth vector-field satisfying

(1) $$\exists \nu > 0 , \forall x \in \partial\mathcal{O} , \quad (\gamma(x), n(x)) \geq \nu$$

where n denotes the unit outward normal to $\partial\mathcal{O}$. We consider now a stochastic differential equation with reflection along γ on $\partial\mathcal{O}$: we denote by (X_t, A_t) a couple of continuous processes, F_t -adapted such that A_t is nondecreasing, X_t $X_t \in \bar{\mathcal{O}}$ for all $t \geq 0$ and

(2) $$dX_t = \sigma(X_t, \alpha_t) dB_t + b(X_t, \alpha_t) dt - \gamma(X_t) dA_t , \quad X_o = x$$

(3) $$A_t = \int_o^t 1_{\partial\mathcal{O}}(X_s)\, dA_s$$

where x is given in $\bar{\mathcal{O}}$, σ, b are vector-valued functions with smoothness properties that are detailed below.

We now consider a cost function for the initial position x and the control α_t

(4) $$J(x, \alpha_t) = E \int_o^\infty f(X_t, \alpha_t)\, e^{-\lambda t}\, dt$$

where $\lambda > 0$, f is some real-valued function on $\bar{\mathcal{O}} \times A$. We will assume (to simplify) in all that follows that σ_{ij} $(1 \leq i \leq N , 1 \leq j \leq m)$, b_i $(1 \leq i \leq N)$, f are bounded continuous in $(x, \alpha) \in \bar{\mathcal{O}} \times A$, smooth in x uniformly in α (say two bounded derivatives in x , bounded independently of α). Finally, we consider the value function

(5) $$u(x) = \inf_{\alpha_t} J(x, \alpha_t)$$

where the infimum is taken over all controls α_t .

Obviously, we expect in view of the general dynamic programming argument - see W.H. Fleming and R. Rishel [7], A. Bensoussan [1], A. Bensoussan and J.L. Lions [2], [3], N.V. Krylov [12]... - that u solves the (HJB) equation

$$\text{(HJB)} \qquad \sup_{\alpha \in A} \; [A_\alpha u(x) - f_\alpha(x)] = 0 \qquad \text{in} \quad \mathcal{O}$$

where $f_\alpha(x) = f(x,\alpha)$, $A_\alpha = -a_{ij}(x,\alpha)\partial_{ij} - b_i(x,\alpha)\partial_i + \lambda$ and $a(x,\alpha) = \frac{1}{2}\,\sigma(x,\alpha)\cdot$ $\cdot\sigma^T(x,\alpha)$.

The main new question, when passing from the exit problem to such a model problem, lies in the nature of the boundary condition. The standard interpretation of reflected Brownian motion leads to the following heuristic boundary condition

$$(6) \qquad \frac{\partial u}{\partial \gamma} = 0 \qquad \text{on} \quad \Gamma$$

(or at least on some part of the boundary). But, easy arguments show that if σ degenerates on Γ then (6) is much too stringent (consider for instance deterministic problems) and thus one has to find a weaker formulation of the boundary condition.

This problem was solved in [18] for deterministic problems by the use of a "viscosity formulation" of the boundary condition. The notion of viscosity solutions has been introduced by M.G. Crandall and the author [5] in the context of general first-order Hamilton-Jacobi equations (see also M.G. Crandall, L.C. Evans and the author [4]) ; the relations with control problems and the extension to the second-order case (and stochastic control problems) being given in [19], [16]. For stochastic control problems, the notion allows one in particular to justify completely the derivation of (HJB). Let us also mention that recently M.G. Crandall and the author proposed a complete theory of first-order Hamilton-Jacobi equations in infinite dimensional spaces by the use of viscosity solutions ([6]), thus giving some hope that viscosity solutions could help in the treatment of stochastic control problems with partial observations (see Fohlmann [11]). We propose here a notion of viscosity solution of (HJB) with the boundary condition (6) ; it is given in section II below together with some uniqueness result. Section III is devoted to some regularity results.

We would like to mention that the above problem is just an example of our methods and results and that we can treat as well finite horizon problems, optimal stopping, reflection controls with pay-offs, switching costs, impulse control, jump diffusion processes, ergodic problems, and we refer to P.L. Lions and N.S. Trudinger [23], [24], P.L. Lions and B. Perthame [21], F. Gimbert and the author [9], F. Gimbert [6], P.L. Lions [20]...

II. Viscosity formulation and uniqueness results.

We call the reflected problem (R) the combination of the (HJB) equation together with the boundary condition (6) (of oblique type).

Definition : $u \in C(\overline{\mathcal{O}})$ is a viscosity solution of (R) if the following holds for all $\varphi \in C^2(\overline{\mathcal{O}})$

i) let x_o be a local maximum point in $\overline{\mathcal{O}}$ of $u-\varphi$, then

(7) $$\sup_{\alpha \in A} \left\{-a_{ij}(x_o,\alpha)\partial_{ij}\varphi(x_o) - b_i(x_o,\alpha)\partial_i\varphi(x_o) + \lambda u(x_o) - f(x_o,\alpha)\right\} \leqslant 0$$

provided $x_o \in \mathcal{O}$ or $x_o \in \partial\mathcal{O}$ and $\frac{\partial\varphi}{\partial\gamma}(x_o) \geqslant 0$.

ii) let x_o be a local minimum point in $\overline{\mathcal{O}}$ of $u-\varphi$, then

(8) $$\sup_{\alpha \in A} \left\{-a_{ij}(x_o,\alpha)\partial_{ij}\varphi(x_o) - b_i(x_o,\alpha)\partial_i\varphi(x_o) + \lambda u(x_o) - f(x_o,\alpha)\right\} \geqslant 0$$

provided $x_o \in \mathcal{O}$, or $x_o \in \partial\mathcal{O}$ and $\frac{\partial\varphi}{\partial\gamma}(x_o) \leqslant 0$. ■

Then, combining the methods and results of [16], [18] we obtain

Proposition 1 : The value function $u \in C_b(\overline{\mathcal{O}})$ and u is a viscosity solution of (R).

We next wish to give a uniqueness result for viscosity solutions : we need to assume that $\Gamma = \partial\mathcal{O}$ is the disjoint union of three closed, possibly empty subsets Γ_1, Γ_2, Γ_3 and that

(9) $$\exists\, \nu > 0,\ \forall (x,\alpha) \in \Gamma_1 \times A, \quad (a(x,\alpha)) \geqslant \nu I$$

(10) $$\forall (x,\alpha) \in \Gamma_2 \times A, \quad a_{ij}(x,\alpha)n_i n_j = 0, \quad b_i(x,\alpha)n_i - a_{ij}(x,\alpha)\partial_{ij}d(x) \leqslant 0$$

(11) $$\forall (x,\alpha) \in \Gamma_3 \times A, \quad D^\mu\sigma(x,\alpha) = 0 \quad \text{for} \quad |\mu| \leqslant 1,$$

where $d(x) = \text{dist}(x,\partial\mathcal{O})$.

Theorem 2 : Under assumptions (9)-(11), the value function u is the unique viscosity solution of (R) in $C_b(\overline{\mathcal{O}})$.

Remarks : i) Assumptions (9)-(11) respectively mean that the diffusions are non-degenerate on Γ_1, do not hit Γ_2 and become "deterministic" on Γ_3.

ii) Actually, one can prove comparison results for viscosity subsolutions and supersolutions with different data f .

iii) Uniqueness and comparizon results are proved in [18] for first-order Hamilton-Jacobi equations (see also B. Perthame and R. Sanders [25] for a different proof).

Sketch of proof : Let $v \in C(\bar{\mathcal{O}})$ be another viscosity solution of (R). Without loss of generality, we can assume that the value function u is Lipschitz continuous and that $\mathcal{O}$ is bounded. In view of [15], we see that

$$\max_{\bar{\mathcal{O}}} |u-v| = \max_{\Gamma_1 \cup \Gamma_3} |u-v| .$$

Next, using some informations on Γ_1 recalled in the next section and the above definition, ones deduces that

$$\max_{\bar{\mathcal{O}}} |u-v| = \max_{\Gamma_3} |u-v|$$

At this stage, to simplify notations, we take $\Gamma_3 = \Gamma$. We then consider $\xi_\varepsilon \in \mathcal{D}(\mathcal{O})$ satisfying : $\xi_\varepsilon \equiv 1$ if $d(x) \geqslant 2\varepsilon$, $\xi_\varepsilon \equiv 0$ if $d(x) \leqslant \varepsilon$, $0 \leqslant \xi_\varepsilon \leqslant 1$, $|\nabla\xi_\varepsilon| \leqslant \frac{C}{\varepsilon}$. And we build control problems as before just replacing σ by $\sigma_\varepsilon = \xi_\varepsilon\sigma$. Again, we may assume without loss of generality that the corresponding value functions u_ε are uniformly Lipschitz and that $\max_{\bar{\mathcal{O}}} |u_\varepsilon - u| \leqslant C\varepsilon^2$: this automatically holds if λ is large (depending only on $\gamma, \mathcal{O}, \sigma, b$) or in the finite horizon problem. Indeed, observe that in view of (11)

$$E \, [\sup_{0 \leqslant t \leqslant T} |X_t^\varepsilon - X_t|^2] \leqslant C_T \varepsilon^4 , \quad \text{for all } T < \infty$$

(using for example the a priori estimates methods in P.L. Lions and A.S. Sznitman [22]).

Next, we compare v and u_ε on the set $\mathcal{O}_\varepsilon = \{x \in \mathcal{O} , d(x) < \varepsilon\}$. We thus follow the comparison proof in [18] considering the second-order terms in (HJB) as a perturbation and using (11) to deduce ;

$$\max_{\bar{\mathcal{O}}_\varepsilon} |u_\varepsilon - v| \leqslant \max_{\Gamma_\varepsilon} |u_\varepsilon - v| + C\varepsilon^2 , \quad \text{with } \Gamma_\varepsilon = \{x \in \mathcal{O} , d(x) = \varepsilon\}.$$

In conclusion, we find

$$\max_{\bar{\mathcal{O}}} |u-v| = \max_{\Gamma} |u-v| \leqslant \max_{\Gamma_\varepsilon} |u-v| + C\varepsilon^2$$

Now, using once more the results of [16], we see that

$$\max_{\Gamma_\varepsilon} |u-v| \leqslant \max_{\Gamma} |u-v| \sup_{x \in \Gamma_\varepsilon, \alpha_t} E[e^{-\lambda\tau}]$$

where τ is the first exit time from $\mathcal{O}$. To bound this term, we introduce a function $\tilde{d}$ equal to d in a neighborhood of $\partial\mathcal{O}$ but smooth on $\bar{\mathcal{O}}$ and positive on $\mathcal{O}$. We then consider $w(x) = \exp(-\delta\tilde{d})$ and for $\delta > 0$ small enough we see that

$$A_\alpha w(x) \geqslant 0 \qquad \text{for all } x \in \bar{\mathcal{O}}, \quad \alpha \in A$$

and thus for ε small enough

$$\sup_{x \in \Gamma_\varepsilon, \alpha_t} E[e^{-\lambda\tau}] \leqslant \sup_{x \in \Gamma_\varepsilon} w(x) = \exp(-\varepsilon\delta)$$

Combining this with the above inequalities, we obtain

$$\max_{\bar{\mathcal{O}}} |u-v| \leqslant e^{-\delta\varepsilon} \max_{\bar{\mathcal{O}}} |u-v| + C\varepsilon^2$$

and we conclude easily.

III. Some regularity results.

We recall some results taken from [16], [23] and N.S. Trudinger [27].

Theorem 3 : i) There exists a constant λ_o (depending only on $\gamma, \mathcal{O}, \sigma, b$) such that the value function $u \in C^{0,\theta}(\bar{\mathcal{O}})$ with $\theta = \lambda/\lambda_o$ if $\lambda < \lambda_o$, θ arbitrary in $(0,1)$ if $\lambda = \lambda_o$ and $\theta = 1$ if $\lambda > \lambda_o$.

ii) If (9) holds, there exists a neighborhood of Γ_1 (relative to $\bar{\mathcal{O}}$) such that u is of class C^2 on that neighborhood and $\frac{\partial u}{\partial \gamma} = 0$ on Γ_1.

iii) If (9),(10) hold with $\Gamma = \Gamma_1 \cup \Gamma_2$ (i.e. $\Gamma_3 = \emptyset$), then there exists a constant λ_1 (depending only on σ, b) such that for $\lambda > \lambda_1$, $u \in C^{0,1}(\bar{\mathcal{O}})$,

$u(x) - \frac{1}{2} C|x|^2$ is concave for some C , $A_\alpha u \in L^\infty(\mathcal{O})$ for all $\alpha \in A$ and $A_\alpha u$ is bounded in $L^\infty(\mathcal{O})$ independently of α , and finally

(12) $$\sup_{\alpha \in A} [A_\alpha u - f_\alpha] = 0 \qquad \text{a.e. in } \mathcal{O} .$$

iv) If $a(x,\alpha)$ is definite positive uniformly in $(x,\alpha) \in \bar{\mathcal{O}} \times A$, then $u \in C^2(\bar{\mathcal{O}})$ and u is the unique classical solution of (HJB)-(6).

Remark : In the finite horizon case, as usual, the assumptions $\lambda > \lambda_o$ or $\lambda > \lambda_1$ are not needed.

References :

[1] A. Bensoussan : Stochastic control by functional analysis methods. North-Holland, Amsterdam, 1982.

[2] A. Bensoussan and J.L. Lions : Applications des inéquations variationnelles en contrôle stochastique, Dunod, Paris, 1976.

[3] A. Bensoussan and J.L. Lions : Contrôle impulsionnel et inéquations quasi-variationnelles, Dunod, Paris, 1982.

[4] M.G. Crandall, L.C. Evans and P.L. Lions : Some properties of viscosity solutions of Hamilton-Jacobi equations. Trans. Amer. Math. Soc., 282 (1984), p. 487-502.

[5] M.G. Crandall and P.L. Lions : Viscosity solutions of Hamilton-Jacobi equations. Trans. Amer. Math. Soc., 277 (1983), p. 1-42. Announced in C.R. Acad. Sci. Paris, 292 (1981), p. 183-186.

[6] M.G. Crandall and P.L. Lions : Hamilton-Jacobi equations in infinite dimensions. Part I, J. Funct. Anal. (1985) ; Part II, to appear in J. Funct. Anal. Announced in C.R. Acad. Sci. Paris, 300 (1985), p. 67-70.

[7] W.H. Fleming and R. Rishel : Deterministic and stocahstic optimal sontrol, Springer, Berlin, 1975.

[8] F. Gimbert : Problèmes de Neumann quasilinéaires ergodiques. J. Funct. Anal.

[9] F. Gimbert and P.L. Lions : Existence and regularity results for solutions of second-order, elliptic, integrodifferential operators. To appear in Ric. Mat. Napoli.

[10] I. Capuzzo-Dolcetta and P.L. Lions : work in preparation.

[11] M. Kohlmann : Viscosity solutions in partially observed control. Preprint.

[12] N.V. Krylov : Controlled diffusion processes. Springer, Berlin, 1980.

[13] J.M. Lasry and P.L. Lions : work in preparation. Announced in C.R. Acad. Sci. Paris.

[14] P.L. Lions : Optimal stochastic control with state constraints. In "Stochastic Differential Systems", Lecture Notes in Control and Information Sciences, 69, Springer, Berlin, 1985.

[15] P.L. Lions : Optimal control of diffusion processes and Hamilton-Jacobi-Bellman equations, Parts 1, 2, Comm. P.D.E., 8 (1983), p. 1101-1174 and p. 1229-1276 ; Part 3, In Nonlinear Partial Differential equations and their applications. Collège de France Seminar, Vol. V, Pitman, London, 1983.

[16] P.L. Lions : On the Hamilton-Jacobi-Bellman equations. Acta Applicandae, 1 (1983), p. 17-41.

[17] P.L. Lions : Some recent results in the optimal control of diffusion processes. Stochastic analysis, Proc. of the Taniguchi Intern. Symp. on Stochastic Analysis, Katata and Kyoto, 1982 ; Kinokuniga, Tokyo (1984).

[18] P.L. Lions : Neumann type boundary conditions for Hamilton-Jacobi equations. Duke Math. J..

[19] P.L. Lions : Generalized solutions of Hamilton-Jacobi equations. Pitman, London, 1982.

[20] P.L. Lions : Quelques remarques sur les problèmes elliptiques quasilinéaires du second ordre. J. Analyse Math..

[21] P.L. Lions and B. Perthame : Quasi-variational inequalities and ergodic impulse control. SIAM J. Control Optim..

[22] P.L. Lions and A.S. Sznitman : Stochastic differential equations with reflecting boundary conditions. Comm. Pure Appl. Math., 37 (1984), p. 511-537.

[23] P.L. Lions and N.S. Trudinger : Linear oblique derivative problems for the uniformly elliptic Hamilton-Jacobi-Bellman equation.

[24] P.L. Lions and N.S. Trudinger : Optimal control of reflected diffusion processes with optimal stopping. Preprint.

[25] B. Perthame and R. Sanders : The Neumann problem for fully nonlinear second-order singular perturbation problems. M.R.C. report.

[26] M. Soner : Optimal control with state-space constraint. I. Preprint.

[27] N.S. Trudinger : to appear.

ASYMPTOTIC ORDERING OF PROBABILITY DISTRIBUTIONS FOR LINEAR CONTROLLED SYSTEMS WITH QUADRATIC COST

Petr Mandl
Department of Probability and Mathematical Statistics, Charles University
Sokolovská 83, 186 00 Prague 8, Czechoslovakia

1. Autonomous systems

Consider an n-dimensional linear controlled system

(1) $dX_t = f X_t\, dt + g U_t dt + dW_t, \quad t \geq 0, \quad X_0 = x,$

together with the cost functional

$$C_T = \int_0^T (X_t' c X_t + |U_t|^2) dt, \qquad T \geq 0.$$

Prime denotes the transposition, $W = \{W_t, \ t \geq 0\}$ the n-dimensional Wiener process. The control $U = \{U_t, \ t \geq 0\}$ is an m-dimensional process depending in a nonanticipative way on the observation of X. The matrix c is nonzero nonnegatively definite, and the pairs of matrices (f,g) and $(f', \sqrt{c})$ are assumed to be stabilizable. The steady state matrix Riccati equation

(2) $$wf + f'w - w g g' w + c = 0$$

has then a unique nonnegatively definite solution (see e.g. [1]). w yields the optimal stationary control

(3) $$U_t = k X_t, \quad t \geq 0,$$

with

(4) $$k = -g'w.$$

For

(5) $$\theta = \text{trace } w$$

we have under (3)

(6) $$\lim_{T \to \infty} E\, C_T/T = \theta .$$

On the other hand let U be any nonanticipative control, and let under U hold

$$\lim_{t \to \infty} E|X_t|^2 / t = 0.$$

Then

(7) $$\liminf_{T \to \infty} E\, C_T / T \geq \theta.$$

(6) and (7) are easily seen from the following relation

(8) $$C_T - \theta T + X_T' w X_T - x'wx - \int_0^T |U_t - kX_t|^2 dt = 2\int_0^T X_t' w dW_t, \quad T \geq 0.$$

To establish (8) write

$$C_T + X_T' w X_T - x'wx = C_T + \int_0^T d(X'wX) = \int_0^T (X'cX + |U|^2)\, dt +$$
$$+ 2\int_0^T (X'wfX + X'wgU)dt + 2\int_0^T X'w\, dW + \int_0^T \mathrm{tr}\, w\, dt.$$

According to (2)

$$2X'wfX + X'cX = |k\,X|^2 .$$

Hence, with regard to (4) and (5),

$$C_T + X_T' w X_T - x'w\,x = \int_0^T (|kX|^2 + |U|^2 - 2X'k'U)dt +$$
$$+ 2\int_0^T X'w\, dW + \theta T = \int_0^T |U - kX|^2\, dt + 2\int_0^T X'w\, d\, W + \theta\, T.$$

Under the optimal stationary control (3) C_T is subject to the <u>central limit law</u>. Namely,

(9) $$\lim_{T\to\infty} P((C_T - \theta T)/\sqrt{T} \leq y) = \emptyset(y/\sqrt{\Delta}), \qquad y \in (-\infty, \infty),$$

where $\emptyset(z)$ is the distribution function of the standardized normal distribution $N(0,1)$.

(9) follows from our Proposition 2. Let us only sketch the proof. The quadratic variation of the right hand side in (8) is

$$V_T = 4\int_0^T |w\, X_t|^2\, dt, \qquad T \geq 0.$$

Thus we have the representation

(10) $$2\int_0^T X_t'\, w\, d\, W_t = \mathcal{W}_{V_T}, \qquad T \geq 0,$$

where $\mathcal{W} = \{\mathcal{W}_t,\ t \geq 0\}$ is a Wiener process. A relation analogous to (8) holds for V_T, $T \geq 0$, under arbitrary control U. To derive it let v be the unique nonnegatively definite solution of the matrix equation

(11) $$v(f + gk) + (f + gk)'v + 4\, w^2 = 0,$$

and let

$$\Delta = \text{trace } v.$$

Then

(12) $$V_T - \Delta T + X_T' v X_T - x'vx - 2\int_0^T X_t' v g(U_t - kX_t)dt = 2\int_0^T X_t' v dW_t, \quad T \geq 0.$$

In fact, from (11) follows

$$V_T + X_T' v X_T - x' v x = 4 \int_0^T X' w^2 X \, dt + 2 \int_0^T (X' v f X + X' v g U) dt + $$
$$+ 2 \int_0^T X' v \, dW + \int_0^T \mathrm{tr}\, v \, dt = 2 \int_0^T X' v \, g(U - k X) dt + 2 \int_0^T X' v \, dW + \Delta T.$$

Under (3) the strong law of large numbers applies to the martingale on the right hand side of (12). Hence, (12) implies

(13) $$\lim_{T \to \infty} V_T / T = \Delta \qquad \text{a.s.}$$

Further, from (8), (10),

$$(C_T - \theta T) / \sqrt{T} = (x' w x - X_T' w X_T) / \sqrt{T} + \mathcal{W}_{V_T} / \sqrt{T}.$$

The first term on the right is negligible, and the second one is asymptotically $N(0, \Delta)$ in virtue of (13). This yields (9).

(6) and (7) exhibit the optimality of (3) for the <u>average cost criterion.</u> The next proposition states that (9) is also an optimal property.

<u>Proposition 1.</u> Let U be such that

(14) $$\lim_{t \to \infty} E|X_t|^2 / \sqrt{t} = 0.$$

Then

(15) $$\limsup_{T \to \infty} P((C_T - \theta T) / \sqrt{T} \leq y) \leq \Phi(y / \sqrt{\Delta}), \quad y \in (-\infty, \infty).$$

(15) means that $(C_T - \theta T) / \sqrt{T}$, as $T \to \infty$, is <u>asymptotically stochastically larger or equal</u> to a random variable having $N(0, \Delta)$ distribution. The asymptotic lower bound for any α-quantile of the distribution of C_T is

$$\theta T + z_\alpha \sqrt{(\Delta T)} \qquad \text{where} \quad \Phi(z_\alpha) = \alpha .$$

We shall need the following elementary property of the Wiener process.

<u>Lemma 1.</u> For $T \geq 0$, $S > 0$, $b > a$

$$P\Big(\inf_{|t-T| \leq S} \mathcal{W}_t \leq a, \quad \mathcal{W}_T \geq b\Big) \leq 3\, \Phi\Big(\frac{a-b}{\sqrt{S}}\Big).$$

<u>Proof of Proposition 1.</u> It may be assumed that $E \int_0^T |X_t|^2 dt < \infty$, $T \geq 0$. Set

$$A_T = \int_0^T |U - k X|^2 dt, \quad Z_T = \int_0^T |X|^2 dt, \qquad M_T = 2 \int_0^T X' w \, dW.$$

From (14)

$$E\left(2\int_0^T X'v\,dW\right)^2 = 4\int_0^T E|vX|^2 dt = o(T^{3/2}), \qquad T\to\infty ,$$

and

$$\left|2\int_0^T X'v\,g(U-kX)dt\right| \leqq \text{const.}\sqrt{Z_T}\,\sqrt{A_T}, \qquad T\geqq 0.$$

Let ε, δ be small arbitrary, $0<\varepsilon<1$, $\delta>0$. (14) implies

$$P(X_T' w X_T > \varepsilon\sqrt{T}) \leqq \varepsilon , \qquad P(Z_T > \delta^2 T^{3/2}) \leqq \varepsilon$$

for T sufficiently large. Using (12) L large can be found so that

$$P(|V_T - \Delta T| \leqq L(\sqrt{A_T}\sqrt{Z_T} + T^{3/4})) \geqq 1-\varepsilon , \qquad T\geqq 1.$$

There is a Wiener process $\{\mathcal{W}_t, \ t\geqq 0\}$ for which

$$M_T = \mathcal{W}_{V_T} , \qquad T\geqq 0,$$

holds. Hence, for all T sufficiently large,

$$P(C_T - \theta T \leqq y\sqrt{T}) \leqq \varepsilon + P(M_T + A_T - \varepsilon\sqrt{T} \leqq y\sqrt{T}) \leqq$$

$$\leqq 2\varepsilon + P\left(\inf_{|t-\Delta T| \leqq L(\sqrt{A_T}\sqrt{Z_T} + T^{3/4})} \mathcal{W}_t \leqq -A_T + (y+\varepsilon)\sqrt{T}\right) \leqq$$

$$\leqq 2\varepsilon + P(\mathcal{W}_{\Delta T} < (y+2\varepsilon)\sqrt{T}) + \sum_{j=0}^{\infty} P(\mathcal{W}_{\Delta T} \geqq (y+2\varepsilon)\sqrt{T},$$

$$j\sqrt{T} \leqq A_T < (j+1)\sqrt{T}, \quad \inf_{|t-\Delta T| \leqq L(\sqrt{A_T}\sqrt{Z_T} + T^{3/4})} \mathcal{W}_t \leqq -A_T + (y+\varepsilon)\sqrt{T}) \leqq$$

$$\leqq 3\varepsilon + \phi\left(\frac{y+2\varepsilon}{\sqrt{\Delta}}\right) + \sum_{j=0}^{\infty} P(\mathcal{W}_{\Delta T} \geqq (y+2\varepsilon)\sqrt{T} ,$$

$$\inf_{|t-\Delta T| \leqq L(\delta\sqrt{j+1}\,T + T^{3/4})} \mathcal{W}_t \leqq -j\sqrt{T} + (y+\varepsilon)\sqrt{T}).$$

For T large we have

$$L(\delta\sqrt{j+1}\,T + T^{3/4}) \leqq 2\delta L(j+1)\,T, \qquad j=0,1,\ldots$$

Consequently, with regard to Lemma 1,

$$\text{(16)} \quad P(C_T - \theta T \leqq y\sqrt{T}) \leqq 3\varepsilon + \phi\left(\frac{y+2\varepsilon}{\sqrt{\Delta}}\right) + 3\sum_{j=0}^{\infty} \phi\left(\frac{-(j+\varepsilon)\sqrt{T}}{\sqrt{(2\delta L(j+1)T)}}\right).$$

Further,

$$\sum_{j=0}^{\infty} \emptyset\left(\frac{-(j+\varepsilon)}{\sqrt{(2\delta L(j+1))}}\right) \leq \frac{1}{\sqrt{(2\pi)}} \sum_{j=0}^{\infty} \frac{\sqrt{(2\delta L(j+1))}}{j+\varepsilon} \exp\left\{-\frac{(j+\varepsilon)^2}{4\delta L(j+1)}\right\} \leq$$

$$\leq \frac{1}{\sqrt{(2\pi)}} \sum_{j=0}^{\infty} \frac{\sqrt{(2\delta L)}}{\varepsilon} \exp\left\{-\frac{j}{8\delta L}\right\} = \frac{1}{\varepsilon}\sqrt{\frac{\delta L}{\pi}}\left(1-\exp\left\{-\frac{1}{8\delta L}\right\}\right)^{-1}.$$

We conclude that the last term in (16) can be made arbitrarily small by taking δ small. From this (15) follows. □

<u>Proposition 2.</u> Let U be such that (14) holds, and

$$\text{(17)} \qquad \underset{T\to\infty}{\text{p lim}} \frac{1}{\sqrt{T}} \int_0^T |U_t - k X_t|^2 \, dt = 0.$$

Then

$$\text{(18)} \qquad \lim_{T\to\infty} P((C_T - \theta T)/\sqrt{T} \leq y) = \emptyset(y/\sqrt{\Delta}), \quad y \in (-\infty, \infty).$$

Proof. Restate the first paragraph of the proof of Proposition 1, and note that by (17)

$$P(A_T > \delta\sqrt{T}) \leq \varepsilon$$

for large T.

We have for all T s fficiently large

$$P(C_T - \theta T > y\sqrt{T}) \leq P(M_T + A_T + \varepsilon\sqrt{T} > y\sqrt{T}) \leq$$

$$\leq \varepsilon + P\left(\sup_{|t-\Delta T| \leq L(\sqrt{A_T}\sqrt{Z_T} + T^{3/4})} \mathcal{W}_t \geq -A_T + (y-\varepsilon)\sqrt{T}\right) \leq$$

$$\leq 2\varepsilon + P(\mathcal{W}_{\Delta T} > (y-2\varepsilon)\sqrt{T}) + P(\mathcal{W}_{\Delta T} \leq (y-2\varepsilon)\sqrt{T},$$

$$\sup_{|t-\Delta T| \leq L(\delta^2 T + T^{3/4})} \mathcal{W}_t \geq -\delta\sqrt{T} + (y-\varepsilon)\sqrt{T}) \leq$$

$$\leq 2\varepsilon + P(\mathcal{W}_{\Delta T} > (y-2\varepsilon)\sqrt{T}) + 3\,\emptyset\left(\frac{(\delta-\varepsilon)\sqrt{T}}{\sqrt{(L(\delta^2 T + T^{3/4}))}}\right).$$

The last term can be made arbitrarily small as $T\to\infty$ by taking δ sufficiently small. Hence we conclude that

$$\limsup_{T\to\infty} P((C_T - \theta T)/\sqrt{T} > y) \leq 2\varepsilon + 1 - \emptyset\left(\frac{y-2\varepsilon}{\sqrt{\Delta}}\right).$$

Since ε is arbitrary, this together with (15) yields (18). □

2. Nonautonomous systems

The controlled system (1) and the cost functional are specified by the matrices f, g, c. Assume now given a family of such triples of matrices

$$(19)\qquad f[\alpha], \quad g[\alpha], \quad c[\alpha], \quad \alpha \in A.$$

α is, for the sake of simplicity, a one-dimensional parameter ranging in a closed bounded interval A. Let the matrices (19) be continuously differentiable with respect to α.

Nonautonomous systems are introduced by setting

$$\alpha = \beta(t), \quad t \geq 0.$$

We write

$$f[\beta(t)] = f(t), \quad g[\beta(t)] = g(t) \quad \text{etc.}$$

In particular $k[\alpha]$ is the optimal stationary control corresponding to α, and $k(t)$ is obtained by inserting $\alpha = \beta(t)$. Similarly for $\theta(t)$, $w(t)$, $\Delta(t)$, and $v(t)$. The derivative with respect to t is denoted by a dot. We assume $\dot{\beta}(t)$, $t \geq 0$, continous.

The equation for the trajectory is

$$d X_t = f(t) X_t \, dt + g(t) U_t \, dt + d W_t', \quad t \geq 0,$$

and the associated cost equals

$$C_T = \int_0^T (X_t' c(t) X_t + |U_t|^2) dt, \quad T \geq 0.$$

(8) and (12) have the following analogues,

$$(19)\quad C_T - \int_0^T \theta(t) dt + X_T' w(T) X_T - x' w(0) x - \int_0^T |U_t - k(t) X_t|^2 dt = $$
$$= \int_0^T X_t' \dot{w}(t) X_t dt + 2 \int_0^T X_t' w(t) d W_t,$$

$$(20)\quad V_T - \int_0^T \Delta(t) dt + X_T' v(T) X_T - x' w(0) x - 2 \int_0^T X_t' v(t) g(t) (U_t - k(t) X_t) dt =$$
$$= \int_0^T X_t' \dot{v}(t) X_t dt + 2 \int_0^T X_t' v(t) \, d W_t, \quad T \geq 0.$$

(19) and (20) are used to prove the extensions of Propositions 1,2.

Proposition 3. Let

(21) $$\lim_{T\to\infty} \int_0^T |\dot{\beta}(t)| \, dt / \sqrt{T} = 0.$$

Then under each U such that

(22) $$E|X_t|^2 \leq \text{const.}, \quad t \geq 0,$$

it holds

$$\limsup_{T\to\infty} P((C_T - \int_0^T \theta(t)dt)/\sqrt{(\int_0^T \Delta(t)dt)} \leq y) \leq \emptyset(y), \quad y \in (-\infty, \infty).$$

Proposition 4. Let (21) be valid. Then under each U satisfying (22) together with

$$p \lim_{T\to\infty} \frac{1}{\sqrt{T}} \int_0^T |U_t - k(t) X_t|^2 \, dt = 0$$

it holds

$$\lim_{T\to\infty} P((C_T - \int_0^T \theta(t)dt / \sqrt{(\int_0^T \Delta(t)dt)} \leq y) = \emptyset(y), \quad y \in (-\infty, \infty).$$

The case

(23) $$\lim_{t\to\infty} \beta(t) = \alpha_0$$

can be called the occurence of a transient phenomenon. (21) is fulfilled whenever the convergence in (23) is monotonous. Consider α_0 to be a parameter unknown to the controller. A useful class of self-optimizing controls to employ then consists of controls

(24) $$U_t = k[\alpha_t^*] \, X_t, \quad t \geq 0,$$

where α_t^* is a consistent estimate of α_0 based on the observation of X_s, $s \leq t$. See [3] for a study of such controls. The methods presented here were applied to investigate the influence of transient phenomena on the performance of the system under controls (24). For Markov chains the investigation is done in [2].

3. References

[1] V. Kučera: A review of the matrix Riccati equation. Kybernetika (Prague) 9(1973), 42-61.

[2] P. Mandl, G. Hübner : Transient phenomena and self-optimizing control of Markov chains. Acta Univ.Carolinae, Math.et.Phys., 26 (1985), No 1.

[3] B. Pasik - Duncan : On adaptive control. University of Kansas, Lawrence 1985.

ADAPTIVE TRACKING OF DYNAMIC AIRBORNE VEHICLES BASED ON (FLIR) IMAGE PLANE INTENSITY DATA

Peter S. Maybeck

Department of Electrical and Computer Engineering
Air Force Institute of Technology / ENG
Wright-Patterson Air Force Base, Ohio, USA 45432

Abstract

In the recent past, the capability of tracking dynamic targets from forward looking infrared (FLIR) measurements has been improved substantially by replacing standard correlation trackers with adaptive extended Kalman filters or enhanced correlator/Kalman filter combinations. This research investigates a tracker able to handle "multiple hot-spot" targets, in which digital and/or optical signal processing is employed on the FLIR data to identify the underlying target shape. Furthermore, multiple model adaptive filtering is investigated as a means of changing the field-of-view as well as the tracker bandwidth when target acceleration can vary over a wide range. The performance potential of such a tracking algorithm is shown to be substantial.

I. Introduction

This paper addresses the problem of accurately tracking the azimuth and elevation of a close-range, highly maneuverable airborne target, using outputs from a forward-looking infrared (FLIR) sensor as measurements. The shape of the target intensity pattern on the FLIR image focal plane is not assumed to be well known a priori, and it may involve multiple "hot spots" as well as change markedly in time. Consequently, the target shape function must be identified adaptively in real time. Moreover, the target vehicle can exhibit a full gamut of dynamic behaviors, from benign straight-line trajectories to very harsh, high-g turning and jinking maneuvers. It is desired to maintain very precise tracking during the benign phases while also preventing loss-of-lock during maneuver initiation and sustained acceleration. Thus, a capacity to change filter gains and field-of-view rapidly and effectively must be incorporated.

In earlier research, a simple four-state extended Kalman filter [1,2] was developed to track a point source (distant) target with benign dynamics, based on FLIR measurements assumed to be corrupted by temporally and spatially uncorrelated noises [3,4]. This algorithm consistently exhibited an order of magnitude improvement in rms tracking errors over currently used correlation trackers, under nominally assumed conditions; between 0.2 and 0.8 pixel (picture element, 20 μrad on a side) rms errors were attained in various scenarios. This enhanced precision was achieved by allowing the filter to exploit knowledge unused by the usual correlation trackers: size, shape and motion characteristics of the target, and spectral description of atmospheric jitter.

Robustness studies [5,6] demonstrated the performance degradation caused by an accurate portrayal of the tracking problem differing from that assumed in the filter design. Variations in the spread, shape and height of the target intensity pattern in the FLIR image plane and differing target motion characteristics were significant, while changes in rms value of the temporal or spatial correlation of the background noise were of lesser importance for the signal-to-noise ratios under consideration. Design modifications and online adaptation were then incorporated to enable this type of filter to track maneuvering targets with spatially distributed and changing image intensity profiles, against background clutter [5-8]. Alternative target dynamics models were also explored to enhance tracking capabilities [9-11]. Although adaptive gain changing in the filter allowed for maintaining track during gradual acceleration of target image motion, it was not sufficient for the case of harsh maneuver initiation. Residual monitoring provided a means of detecting harsh maneuver onset and responding appropriately. This included immediate gain change, reprocessing of the most recent measurement, and an ad hoc alteration of the state estimate during a period of time following the maneuver detection [6-8]. As experienced by others [12-17], the appropriate adaptation to a changing set of target dynamics was a challenging issue; despite successful tracking in some demanding scenarios, it was still desired to explore alternative adaptation mechanisms.

Up to this point, however, all filter designs were based on the assumption that the target intensity profile in the FLIR image plane would be unimodal and well described by a bivariate Gaussian function, allowing elliptical constant-intensity contours to account for target shape effects. Research was then conducted on ways to handle multiple hot-spot targets, where neither the functional form of the target intensity nor the number of hot spots or their relative spacing could be provided a priori [18-19]. For this situation, online digital or optical signal processing techniques [20-23] would be used to derive a target shape function from the available FLIR sensor information. In one tracker formulation, this shape function is used in the measurement update portion of an extended Kalman filter that is otherwise identical to the previous designs; each is used to estimate target position offsets from the center of the sensor field-of-view, and other pertinent states. In an alternative tracker, the shape function is used as a template for an enhanced correlator, which then provides offset "measurements" to a linear Kalman filter rather than using an extended Kalman filter to process raw FLIR data directly. This latter design is considerably less demanding computationally, and so it is preferable if its performance is adequate. Initial research efforts [18-19] concentrated on demonstrating the feasibility of the adaptive shape function construction and considered only benign target dynamics. Ensuing research [24-25] evaluated performance potential of the two tracker formulations in more highly dynamic close-range scenarios, establishing comparable rms tracking errors; the extended Kalman filter exhibited larger biases but smaller standard deviations than the enhanced correlator/linear Kalman filter algorithm. However, this research also revealed a need for an effective and quickly responding adaptation to large-scale changes in the target dynamics.

Multiple model adaptive estimation [2, 10, 14, 26-34] can be used to provide adaptive expansion and contraction of the effective tracker field-of-view, as well as adaptive selection of an assumed target dynamics model, in order to increase the dynamic range of precision tracking. In a feasibility study [35], two independent filters have been used to generate state estimates from the shared sensor. One is tuned for best performance in the case of benign dynamics and uses a narrow field-of-view; the other is tuned for high-g target maneuvering and uses a wider field-of-view (reduced resolution is accepted in order to provide considerably lower probability of losing lock). Adaptive expansion and contraction of the tracker field-of-view is attained by generating the probabilistically weighted average of the two filter state estimates. A subsequent study [36] has demonstrated the performance benefit of more than two elemental filters.

This paper reviews the earlier filter developments and then concentrates attention on adaptive identification of the target image shape function and the use of multiple model adaptation for varying the effective field-of-view and assumed target dynamics model in real time. Section 2 describes the fundamental filter development, and then a number of such filters are used within the structure of a multiple model adaptive estimator in Section 3. Section 4 evaluates the performance potential of this adaptive filter, and some concluding remarks are made in Section 5.

II. Individual Filter Designs

The FLIR measurement model developed in [3] and [7], the target dynamics models of [7] and [9], and the adaptive target shape identification algorithm of [18] can form the basis for either an extended Kalman filter or a cascade of an enhanced correlator with a linear Kalman filter. This section presents these models and the resulting filter designs. In the next section, a number of such elemental filters will be combined within a multiple model adaptive estimator.

We desire to track the centroid of a spatially distributed dynamic target based on FLIR measurements, in order to provide appropriate inputs to a pointing controller so that the target remains in the center of the field-of-view. This involves determining the pointing errors in the two dimensions of the FLIR image plane (and other states as well), given measurements of average intensity level over each of 64 pixels in an 8-by-8 array ("tracking window") provided as a subset of a larger array by the FLIR at a 30 Hz rate. Letting $x_{peak}(t)$ and $y_{peak}(t)$ locate the centroid of the apparent target intensity function relative to the center of the 8-by-8 array, we can describe that intensity at any point (ξ_x,ξ_y) by the function

$$I_{target}\{[\xi_x-x_{peak}(t)],\ [\xi_y-y_{peak}(t)],\ t\}$$

as depicted in Fig. 1. In earlier research [3-10], this function was assumed to be well modeled as bivariate Gaussian, possibly with some uncertain parameters to be identified. Here the entire I_{target} function is computed adaptively, as discussed later. The apparent centroid location is actually the sum of contributions due to true target dynamics and atmospheric jitter (ignoring vibration effects for a ground-based tracker):

$$x_{peak}(t) = x_d(t) + x_a(t) \tag{1}$$

and similarly for $y_{peak}(t)$. The objective of the tracker is to estimate x_d and y_d accurately so that they can be regulated by closed-loop control.

Even for benign dynamics, it is appropriate to estimate velocity

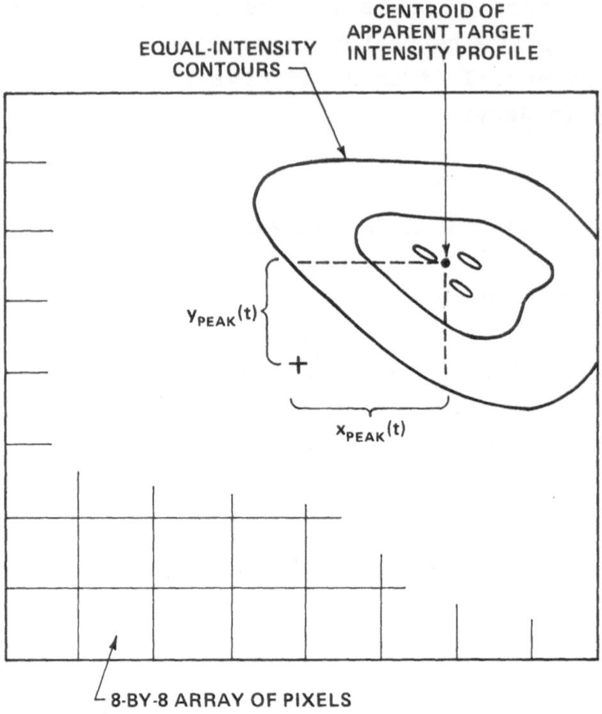

FIG. 1: APPARENT TARGET INTENSITY PATTERN ON IMAGE PLANE

(and perhaps acceleration) as well as position of a close-range target; long range targets may not require as many states to be estimated. Letting $x_d(t)$ and $y_d(t)$ be arrayed in a position vector $\mathbf{p}(t)$, we can write (as an approximation, ignoring the effects of a rotating tracker frame):

$$\dot{\mathbf{p}}(t) = \mathbf{v}(t) \quad ; \quad \dot{\mathbf{v}}(t) = \mathbf{a}(t) \tag{2}$$

Whereas acceleration $\mathbf{a}(t)$ can be modeled as a low-strength white noise for very benign conditions (straight-line flight trajectories, with white noise used for filter tuning), experience in this particular application [6-11, 24, 25] has indicated the performance desirability of two alternatives. First, one can treat acceleration as a first-order Gauss-Markov process,

$$\dot{\mathbf{a}}(t) = -[1/T]\,\mathbf{a}(t) + \mathbf{w}(t) \tag{3}$$

where the correlation time T and strength of the white Gaussian noise $\mathbf{w}(t)$ are treated as design tuning parameters to match a range of characteristics. Secondly, one can invoke a "constant turn-rate" model, very descriptive of many airborne target scenarios:

$$\dot{\mathbf{a}}(t) = -\omega^2 \mathbf{v}(t) + \mathbf{w}(t) \quad ; \quad \omega = \frac{|\mathbf{v}(t) \times \mathbf{a}(t)|}{|\mathbf{v}(t)|^2} \tag{4}$$

where ω is the turn rate. Unlike (3), this is a nonlinear dynamics model, so a tradeoff of performance versus computational loading must be conducted before its use is warranted for online implementation.

Atmospheric disturbances cause wavefront phase distortions, resulting in translational shifts in the FLIR image plane called "jitter". On the basis of spectral properties, atmospheric jitter processes x_a and y_a (see Eq. (1)) were each modeled as outputs of a third order shaping filter [1], described by a transfer function of $K\omega_1\omega_2^2(s+\omega_1)^{-1}(s+\omega_2)^{-2}$, driven by white Gaussian noise [37,38]. Since $\omega_1 << \omega_2$ ($\omega_1 \cong 14$ rad/sec, $\omega_2 \cong 660$ rad/sec) and the lower frequencies are more important, this was well approximated in the filter by the reduced-order model $K\omega_1(s+\omega_1)^{-1}$.

Combining these two states with the six target dynamics states from (2) and either (3) or (4) forms the basis of an eight-state extended Kalman filter [2] propagation algorithm of the form:

$$\dot{\hat{\mathbf{x}}}(t) = \mathbf{f}[\hat{\mathbf{x}}(t), \mathbf{u}(t)] \tag{5}$$

$$\dot{\mathbf{P}}(t) = \mathbf{F}(t)\,\mathbf{P}(t) + \mathbf{P}(t)\,\mathbf{F}^T(t) + \mathbf{Q}(t) \tag{6}$$

where $\mathbf{P}(t)$ is the error covariance and $\mathbf{Q}(t)$ is the white noise strength matrix. If (3) is used, then $\mathbf{f}[\hat{\mathbf{x}}(t), \mathbf{u}(t)] = \mathbf{F}\hat{\mathbf{x}}(t) + \mathbf{B}\mathbf{u}(t)$, i.e., the propagation model is linear and time invariant, and thus is particularly easy to implement. If (4) is used, (5) is nonlinear and $\mathbf{F}(t)$ in (6) is the partial of $\mathbf{f}$ with respect to the state $\mathbf{x}$, evaluated at the current best state estimate.

At each sample time t_i, measurements of the average intensity over each pixel in the 8-by-8 tracking window become available. Let $z_{jk}(t_i)$ denote the scalar measurement corresponding to the j-th row and k-th column of that array, to write:

$$z_{jk}(t_i) = \frac{1}{A_p} \iint_{\substack{\text{REGION OF}\\ \text{JK-TH PIXEL}}} I_{target}\{[\xi_x - x_{peak}(t)],[\xi_y - y_{peak}(t)],t\}\,d\xi_x d\xi_y$$

$$+ \; b_{jk}(t_i) + n_{jk}(t_i) \tag{7}$$

where A_p is the area of one pixel, I_{target} was described previously, $b_{jk}(t_i)$ models the background noise effects on the jk-th pixel and $n_{jk}(t_i)$ the internal FLIR noise effects on that pixel. The 64 scalar measurements of this form at sample time t_i are arrayed in a vector:

$$\mathbf{z}(t_i) = \mathbf{h}[\mathbf{x}(t_i), t_i] + \mathbf{b}(t_i) + \mathbf{n}(t_i) \tag{8}$$

which can be used as the measurement model upon which to base an extended Kalman filter. Online identification of $\mathbf{h}[\mathbf{x}(t_i), t_i]$ will be discussed subsequently. The background and FLIR noises are assumed independent (so that their variances add for a given pixel) and temporally uncorrelated; FLIR noise is assumed spatially uncorrelated and the background noise has a correlation distance of about two pixels (the corresponding noise covariance matrix, $E\{\mathbf{v}(t_i\mathbf{v}^T(t_i)\} = \mathbf{R}(t_i)$, is thus sparse but not diagonal).

An 8-state, 64-measurement extended Kalman filter [2] was designed. To avoid online computation of a 64-by-64 matix inversion, the usual update algorithm was replaced by the algebraically equivalent:

$$\mathbf{P}^{-1}(t_i^+) = \mathbf{P}^{-1}(t_i^-) + \mathbf{H}^T(t_i)\mathbf{R}^{-1}(t_i)\mathbf{H}(t_i) \tag{9}$$

$$\mathbf{P}(t_i^+) = [\mathbf{P}^{-1}(t_i^+)]^{-1} \tag{10}$$

$$\mathbf{K}(t_i) = \mathbf{P}(t_i^+)\mathbf{H}^T(t_i)\mathbf{R}^{-1}(t_i) \tag{11}$$

$$\hat{\mathbf{x}}(t_i^+) = \hat{\mathbf{x}}(t_i^-) + \mathbf{K}(t_i)\,\{\mathbf{z}(t_i) - \mathbf{h}[\hat{\mathbf{x}}(t_i^-),t_i]\} \tag{12}$$

where $\hat{\mathbf{x}}(t_i^+)$ and $\mathbf{P}(t_i^+)$ are the state estimate and error covariance after updating the values $\hat{\mathbf{x}}(t_i^-)$ and $\mathbf{P}(t_i^-)$ with the measurement vector $\mathbf{z}(t_i)$. Here $\mathbf{H}(t_i)$ is the partial $\partial\mathbf{h}/\partial\mathbf{x}$ evaluated at $\hat{\mathbf{x}}(t_i^-)$, and it is to be identified online along with $\mathbf{h}[\hat{\mathbf{x}}(t_i^-),t_i]$. This form only requires two 8-by-8 matrix inversions; $\mathbf{R}^{-1}(t_i)$ is assumed to be generated once offline.

Up to this point, the lower (digital and/or optical) signal processing path of Fig. 2 has been described. Based on the FLIR intensity data $\mathbf{z}(t_i)$ and the identified $\mathbf{h}$ and $\mathbf{H}$ functions, the extended Kalman filter produces state estimates $\hat{\mathbf{x}}(t_i^+)$ and one-step-ahead predictions $\hat{\mathbf{x}}(t_{i+1}^-)$. The latter can be used by a control algorithm to point the center of the field-of-view to where the target is predicted to be located: so that $\hat{x}_d(t_{i+1}^-)$ and $\hat{y}_d(t_{i+1}^-)$ are zeroed. The upper path [18] in Fig. 2 identifies $\mathbf{h}$ and its partial derivative $\mathbf{H}$. It is based on the fact that the actual target image will change rather slowly relative to the sample period of 1/30 sec., while the background noises will typically change more rapidly, especially if a background is being swept behind a moving target. Thus, temporal averaging or filtering of sequential data frames should be exploited in target shape reconstruction; spatial or spatial frequency filtering may also be useful to discriminate between target and background IR intensity patterns [18].

First an FFT is performed on the FLIR data frame (for efficient processing and possible spatial frequency filtering, perhaps optically), and then a negating phase shift is applied to reconstruct the

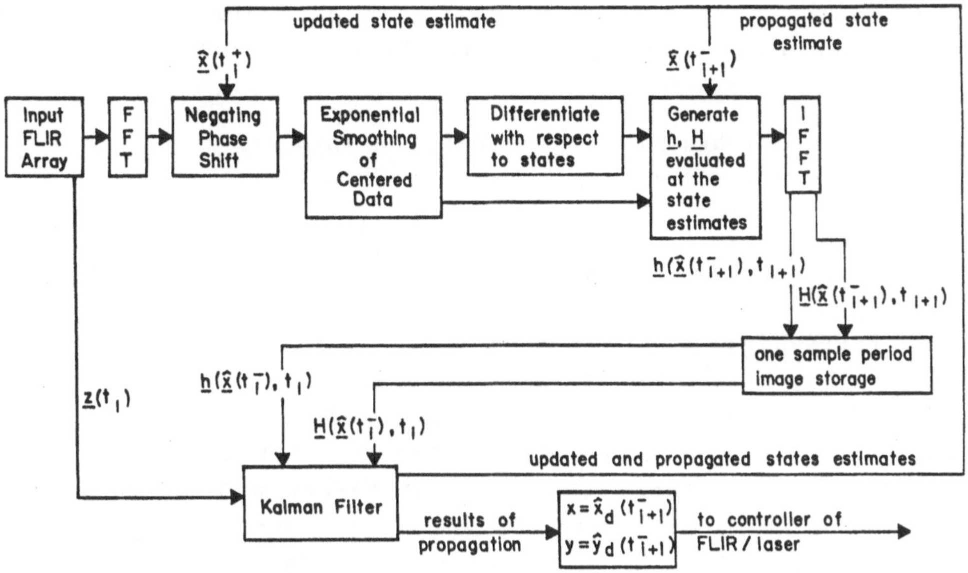

FIG. 2: DATA PROCESSING ALGORITHM

transform of a target as though it were centered in the field-of-view in the original untransformed coordinates (using the shift properties of discrete Fourier transforms [18-20]). The appropriate phase shift for the frame at t_i is based on $\hat{x}_{peak}(t_i^+)$ and $\hat{y}_{peak}(t_i^+)$ derived from Eq. (1) and $\hat{\mathbf{x}}(t_i^+)$ as produced by the filter.

Conceptually, this result can be averaged with the most recent M such centered and transformed data frames, to accentuate the target and attenuate noise. To implement this without requiring explicit storage of these M data sets, finite-memory averaging is approximated by exponential smoothing:

$$\hat{G}(t_i) = \alpha\, G(t_i) + [1-\alpha]\, \hat{G}(t_{i-1}) \qquad (13)$$

where $G(t_i)$ is the current data frame value of G and $\hat{G}(t_i)$ and $\hat{G}(t_{i-1})$ are the current and previous smoothed estimates of G, respectively. Valid α's lie in the range [0,1], with smaller α's corresponding to longer memory; sensitivity studies [18,19] yielded $\alpha = 0.1$. The output of the "Exponential Smoothing of Centered Data" block in Fig. 2 is a representation of the FFT of **h** associated with a centered image.

Again through appropriate phase shifting, $\mathbf{h}[\hat{\mathbf{x}}(t_{i+1}^-),t_{i+1}]$ can be evaluated, assuming that the controller successfully zeroed out the estimated target dynamics position states in the sample period between t_i and t_{i+1}. The spatial derivative $\mathbf{H}[\hat{\mathbf{x}}(t_{i+1}^-),t_{i+1}]$ is readily produced by simple multiplication, using the derivative property of Fourier transforms: if f_x and f_y are spatial frequencies and $\mathcal{F}$ denotes Fourier transform,

$$\begin{aligned} \mathcal{F}[\partial \mathbf{h}(x,y)/\partial x] &= j2\pi f_x \cdot \mathcal{F}[\mathbf{h}(x,y)] \\ \mathcal{F}[\partial \mathbf{h}(x,y)/\partial y] &= j2\pi f_y \cdot \mathcal{F}[\mathbf{h}(x,y)] \end{aligned} \tag{14}$$

Finally, the inverse FFT of both results yields $\mathbf{h}[\hat{\mathbf{x}}(t_{i+1}^-),t_{i+1}]$ and $\mathbf{H}[\hat{\mathbf{x}}(t_{i+1}^-),t_{i+1}]$ for use in the Kalman filter at the <u>next</u> sample time, i.e., at t_{i+1}.

An alternative tracker was also described in [18]. The upper path of Fig. 2 is left intact except that $\mathbf{H}$ no longer requires evaluation. However, $\mathbf{h}[\hat{\mathbf{x}}(t_i^-),t_{i+1}]$ is now used as a template by an enhanced correlator that operates on the raw FLIR data to generate estimated offsets between the apparent target and field-of-view center. These are provided as two scalar measurements to a <u>linear</u> Kalman filter (assuming Eq. (3) is used rather than (4) for the basis of time propagation of estimates). Thresholding of the correlation function to suppress low noise-induced peaks and computing centroid summations are used to approximate correlation function peak detection with lower computational loading and lower sensitivity to multiple peak problems than other peak detection methods; a threshold of one half the maximum correlation value has been adopted based on empirical performance evaluations. The 2-by-2 measurement noise covariance matrix $\mathbf{R}$ for this filter is determined by statistical analysis of errors produced by the proposed correlator under controlled conditions.

Thus, four possible configurations have been described. Two involve an extended Kalman filter that processes the raw FLIR data directly (64-dimensional measurements, nonlinearly modeled); one has a linear propagation cycle based on Eqs. (2) and (3), and the other has a nonlinear propagation based on (2) and (4). In contrast are the two "alternative" trackers based on an enhanced correlator feeding two-dimensional offset "measurements" (modeled linearly) to a Kalman filter; the filter is totally linear if the propagation cycle is based on Eqs. (2) and (3), and it is an extended Kalman filter if based on the nonlinear dynamics of (2) and (4).

Any of these configurations could be "tuned" [1,2] for best performance under benign target conditions by appropriate choice of noise strength $\mathbf{Q}$ in Eq. (6) (and correlation time T in Eq. (3) if it is the basis of the filter propagation cycle). If it is desired to tune for best performance under heavy target maneuvering, not only would $\mathbf{Q}$ and T be changed, but it would also be appropriate to adopt a larger field-of-view (this is also true for target acquisition). For the original filter formulation, in which raw FLIR data is processed directly, the algorithm can be left fundamentally unchanged, except that each scalar measurement is now the average intensity over an N-by-N array of pixels instead of a single pixel; the dynamics model must be rescaled accordingly. For feasibility studies [35,36], N has been set to three, so that the wider field-of-view is 24 pixels by 24 pixels (rather than 8-by-8), assumed to be centered on the same location as the original 8-by-8 array. Since this amount of data was assumed available at each sample time, the narrower field-of-view data is padded with 8 rows and columns of additional noisy data, rather than zeros, in generating the FFT's. Padding with zeros is a common engineering practice to ensure that the inherent periodicity of discrete Fourier transforms will not have seriously degrading effects within the original finite array [20]. However, if significant true target intensity magnitudes exist outside the 8-by-8 array, to pad with zeros would induce an artificial edge in the spatial domain and thus artificially increased magnitudes of high-frequency components in the transform domain. Since the target is more likely to be contained totally within the larger field-of-view, padding with zeros has been adopted for that case. In addition, since the correlation distance of the background noise is two of the original pixels, and the filter "pixels" in the wider field-of-view case are actually 3-by-3 arrays of the original FLIR pixels, **R** is made diagonal in the wider field-of-view filter.

III. Multiple Model Adaptive Filtering

One means of allowing rapid and effective changing of both the bandwidth and field-of-view of the tracker is multiple model adaptive filtering [2, 14, 26-36]. In this approach, two (or more) independent filters are processed in parallel, each based upon a particular model of target dynamics intensity and a corresponding field-of-view. By optimally combining the estimates of these filters at each sample time, an adaptive estimator is produced that will yield desirable high resolution for benign target trajectories while also maintaining lock on highly dynamic targets.

Let **a** denote the vector of uncertain parameters in a given model; here it is composed of the strength of the white noise driving the target acceleration model (3) or (4), and the field-of-view size. In order to make identification of **a** tractable, its continuous range of values is discretized into K representative values. If we define the hypothesis conditional probability $p_k(t_i)$ as the probability that $\mathbf{a}$ assumes the value $\mathbf{a}_k$ (for k = 1, 2,...,K), conditioned on the observed measurement history $\mathbf{Z}(t_i) = [\mathbf{z}^T(t_o), \mathbf{z}^T(t_1), \ldots, \mathbf{z}^T(t_i)]^T$, i.e.,

$$p_k(t_i) = \text{prob}\{\mathbf{a}=\mathbf{a}_k \mid \mathbf{Z}(t_i)=\mathbf{Z}_i\} \tag{15}$$

then it can be shown [2] that $p_k(t_i)$ can be evaluated recursively for all k via:

$$p_k(t_i) = \frac{f_{\mathbf{z}(t_i)|\mathbf{a},\mathbf{Z}(t_{i-1})}(\mathbf{z}_i|\mathbf{a}_k,\mathbf{Z}_{i-1})\cdot p_k(t_{i-1})}{\sum_{j=1}^{K} f_{\mathbf{z}(t_i)|\mathbf{a},\mathbf{Z}(t_{i-1})}(\mathbf{z}_i|\mathbf{a}_j,\mathbf{Z}_{i-1})\cdot p_j(t_{i-1})} \tag{16}$$

and that the Bayesian estimate of the state is the probabilistically weighted average:

$$\hat{\mathbf{x}}(t_i^+) = E\{\mathbf{x}(t_i)|\mathbf{Z}(t_i)=\mathbf{Z}_i\} = \sum_{k=1}^{K} \hat{\mathbf{x}}_k(t_i^+)\cdot p_k(t_i) \tag{17}$$

where $\hat{\mathbf{x}}_k(t_i^+)$ is the state estimate generated by a Kalman filter based on the assumption that the parameter vector equals $\mathbf{a}_k$. Thus, the multiple model filtering algorithm is composed of a bank of K separate Kalman filters, each based on a particular value $\mathbf{a}_1, \ldots, \mathbf{a}_K$ of the parameter vector, as depicted in Fig. 3. When the measurement $\mathbf{z}_i$ becomes available at time t_i, the residuals $\mathbf{r}_1(t_i), \ldots, \mathbf{r}_K(t_i)$ are

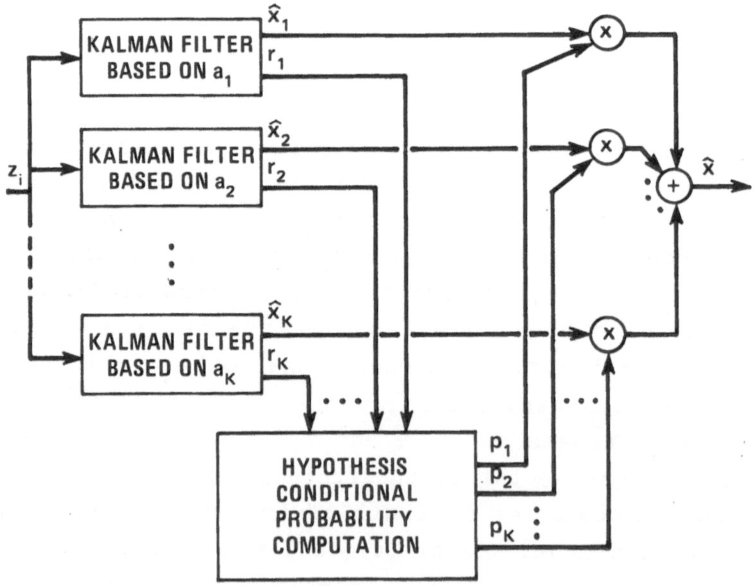

FIG. 3: MULTIPLE MODEL FILTERING ALGORITHM

generated in the K filters and used to compute $p_1(t_i), \ldots, p_K(t_i)$ via Eq. (16). Each numerator density in (16) is Gaussian if linear models are used and approximated as Gaussian if nonlinear models are employed; i.e.,

$$f_{\mathbf{z}(t_i)|\mathbf{a},\mathbf{Z}(t_{i-1})}(\mathbf{z}_i|\mathbf{a}_k,\mathbf{Z}_{i-1}) = \frac{1}{(2\pi)^{m/2}|\mathbf{A}_k(t_i)|^{1/2}} \exp\{\bullet\}$$
$$\{\bullet\} = \{-\tfrac{1}{2}\mathbf{r}_k^T(t_i)\mathbf{A}_k^{-1}(t_i)\mathbf{r}_k(t_i)\} \tag{18}$$

where m is the measurement dimension and $\mathbf{A}_k(t_i)$ is calculated in the k-th Kalman filter as:

$$\mathbf{A}_k(t_i) = \mathbf{H}_k(t_i)\mathbf{P}_k(t_i^-)\mathbf{H}_k^T(t_i) + \mathbf{R}_k(t_i) \tag{19}$$

The denominator in Eq. (16) is simply the sum of all the computed numerator terms and thus is the scale factor required to ensure that the $p_k(t_i)$'s sum to one.

One expects that the residuals of the Kalman filter based upon the "best" model will have mean squared values most in consonance with its own computed $\mathbf{A}_k(t_i)$, while "mismatched" filters will have larger residuals than anticipated through this matrix. Therefore, Eqs. (16)-(19) will most heavily weight the filter based upon the most correct assumed parameter value. However, the performance of the algorithm depends on there being significant differences in the characteristics of residuals in the "correct" and "mismatched" filters. In fact, lack of substantial differences specifically hampered an earlier study on multiple model estimation for this application [9,10]. Therefore, each filter should be tuned for best performance when the true values of the uncertain parameters are identical to its model for these parameters. One should specifically avoid the often-used "conservative" philosophy of adding considerable dynamics pseudonoise to open the filter bandwidth, since this tends to mask the difference between good and bad models. In one study [35], K has been chosen as 2 and the narrow field-of-view filter has been tuned for best performance for constant-velocity trajectories in inertial space, and the wider field-of-view filter has been tuned for best performance on a 20-g pull-up maneuver while still being able to track a constant-velocity target. The latter constraint was added to allow the wide field-of-view filter to be used in initial acquisition, typically of long-range targets with benign dynamics; an alternative would be to add an elemental filter to the bank that is tuned for acquisition. Table 1 displays the tuning values for the various filters, as established in [24, 25, 35]; Q is the strength of the white dynamics noise in Eq. (3) or (4), and T is the correlation time in (3). Only the linear

TABLE 1: FILTER TUNING VALUES

FILTER	NARROW FIELD-OF-VIEW Q(pixel²/sec⁵)	NARROW FIELD-OF-VIEW T(sec)	WIDE FIELD-OF-VIEW Q(pixel²/sec⁵)	WIDE FIELD-OF-VIEW T(sec)
EXTENDED KF (GAUSS-MARKOV ACC.)	300	1.5	5000	1.3
EXTENDED KF (CONST. TURN RATE)	300	-	5000	-
CORRELATOR/ LINEAR KF	150	3.5	2000	1.5

dynamics model was used in conjunction with the correlator / Kalman filter architecture.

To allow the algorithm to adapt to a changing parameter value, computed p_k's were artificially bounded below by a small value, 0.01 [2, 26, 35]. (Other means are also available, as in [2, 14, 30].) Without such bounding, the p_k for a "mismatched" filter could converge to (essentially) zero, precluding appropriate response of the recursion in Eq. (16) to subsequent parameter changes. After lower bounding, the resultant probabilities are rescaled so that their sum remains equal to one.

For the extended Kalman filter formulations, the measurement dimension is 64, and so (18) entails a 64-by-64 inversion for each filter in the bank at every sample time. This was approximated by using only the diagonal terms, requiring only 64 divisions instead. (Other possibilities were also explored, including a full inversion of only those elements associated with the center 4-by-4 region of the field-of-view, and application of the matrix inversion lemma [1] to generate an equivalent expression in terms of 8-by-8 inversions where 8 is the state dimension in this application.) Moreover, the coefficient premultiplying the exponential in (18) was ignored, since its magnitude is independent of the "correctness" of the assumed parameter value, thereby reducing computational loading. In contrast, the correlator/linear filter algorithm is based on 2-dimensional "measurements," and so it poses no such computational burden problems or need for approximations.

Finally, recall Fig. 2. When the multiple model filtering

algorithm is used to feed estimates to the controller, the overall $\hat{x}_d(t_{i+1}^-)$ and $\hat{y}_d(t_{i+1}^-)$ are zeroed out: the FLIR is pointed at the adaptive filter's best estimate of target location. Therefore, for the k-th filter, the phase shift just before the IFFT in Fig. 2 corresponds to

$$[\hat{x}_d(t_{i+1}^-) + \hat{x}_a(t_{i+1}^-)]_{\substack{k\text{-th}\\ \text{filter}}} - [\hat{x}_d(t_{i+1}^-)]_{\substack{\text{adaptive}\\ \text{filter}}}$$

rather than $\hat{x}_a(t_{i+1}^-)$ as in the case of only a single filter.

For very severe target maneuvers, the difference between the position estimates of the small field-of-view filter and those of the multiple model filter can become so large that a cylindrical shift [18-20, 35] of the image approaches a complete cycle. At this point, the small field-of-view filter will diverge; due to the lower bounding on the p_k's, this will eventually cause the entire algorithm to diverge. To preclude this, any time the shift magnitude exceeds 3 pixels, the states of the small field-of-view filter are set to appropriately rescaled values of the large field-of-view filter estimates, and its error covariance is reset correspondingly.

IV. Representative Performance

Monte Carlo analyses involving ten runs each have been used to test the previously described trackers against realistic scenarios. Sample means ± 1 standard deviation are plotted versus time to demonstrate transient characteristics, or temporally averaged over an interval of interest to allow compact comparisons. A three-hot-spot target of dimensions appropriate for a single-place, two-engine aircraft was simulated in 3-space and then projected onto the FLIR image plane, using as trajectories:

(1) straight-and-level flight at 1 km/sec for 5 sec, starting at about 20.5 km range and reaching minimum range of 20 km (with the trajectory in the plane orthogonal to the line of sight) at 5 sec;

(2) same as (1) for 2 sec, then conducting a pull-up maneuver (at 2g, 10g, or 20g levels), using an unrealistically harsh step change in pitch rate to tax the algorithm severely;

(3) same as (2) for 3.5 sec of simulation time, but then returning to straight-line flight, again using a step change in pitch rate;

(4) same as (3) except that, at 3.5 sec, the target turns in toward the tracker, causing more dramatic changes in the target shape in the FLIR image plane.

TABLE 2: PERFORMANCE ANALYSIS - MEAN ± STANDARD DEVIATION OF ERRORS IN $\hat{y}_d(t_i^-)$ AND $\hat{y}_d(t_i^+)$

TEST CASE	CORRELATOR/ LINEAR KF	EXTENDED KF GAUSS-MARKOV ACC	EXTENDED KF CONST TURN-RATE
STRAIGHT TRAJECTORY 1			
MULTIPLE MODEL	-.001 ± .063 -.005 ± .157	-.241 ± .114 -.246 ± .102	-.208 ± .121 -.210 ± .109
NARROW FOV	-.015 ± .170 -.020 ± .152	-.266 ± .118 -.270 ± .106	-.209 ± .123 -.210 ± .112
WIDE FOV	-.017 ± .289 -.020 ± .246	.015 ± .162 .013 ± .128	.013 ± .166 .012 ± .129
PULL-UP TRAJECTORY 2			
2g	-.157 ± .225 .118 ± .209	-.802 ± .129 -.737 ± .119	-.435 ± .145 -.432 ± .134
10g	-.089 ± .747 .084 ± .740	-.232 ± .171 -.107 ± .138	-.203 ± .167 -.221 ± .139
20g	-.243 ± .450 -.026 ± .436	-.923 ± .181 -.710 ± .155	-.285 ± .160 -.300 ± .129
20g WITH WIDE FOV	-.264 ± .250 .036 ± .214	-.884 ± .204 -.674 ± .181	-.117 ± .194 -.123 ± .162

The ratio of target spot maximum intensity to background and FLIR noise rms value is 20, and the hot spots have squared glint dispersions (spreads) of 2.0 pixel². Atmospheric jitter is modeled with mean squared value of 0.2 pixel² and correlation time of 0.07 sec.

Table 2 presents the sample mean ± one standard deviation of the errors committed in estimating the target dynamics elevation position state y_d, averaged over 1.5 seconds of time after initial transients have died out. These are presented for the correlator/linear Kalman filter, the extended Kalman filter based on a Gauss-Markov acceleration model, and the extended Kalman filter based on a constant turn-rate dynamics model. Errors in $\hat{y}_d(t_i^-)$ are tabulated, followed by errors in $\hat{y}_d(t_i^+)$: the former indicates the accuracy of the command sent to the pointing controller and the latter portrays the best precision in state estimation, so these are the most pertinent statistics. The azimuth position statistics are not included because they provide no additional insight into performance and because the maneuvers being performed are predominantly in the y direction.

First consider the table entries for the straight-and-level trajectory 1; the corresponding time histories of statistics for the three forms of multiple model adaptive filter can be viewed in the first 2 seconds of results in Figs. 4-6. (As indicated by the initial

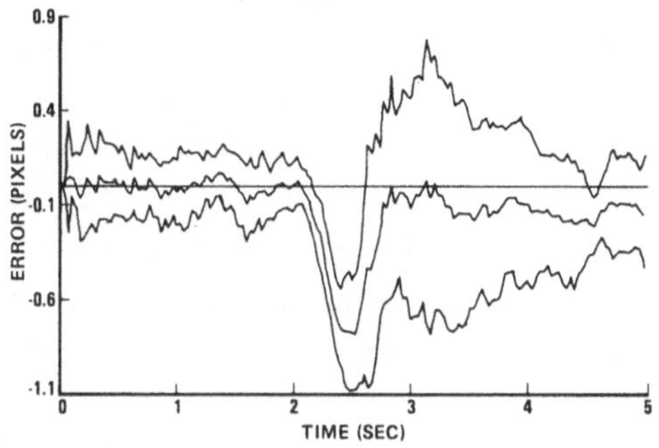

FIG. 4: $\hat{y}(t_i^+)$ ERROR MEAN ± ONE STD. DEV. FOR CORRELATOR/KF ON 2G PULL-UP

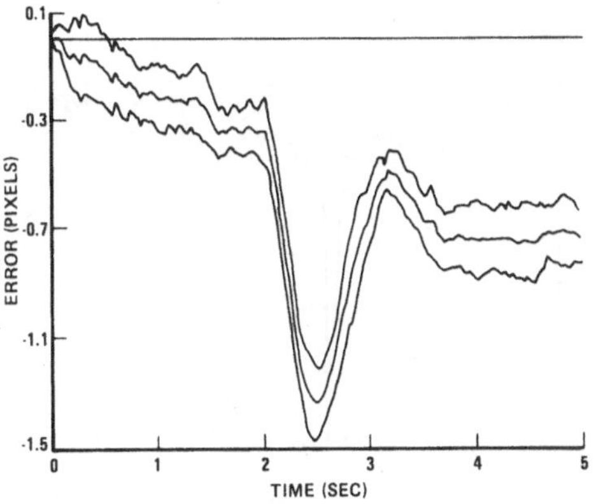

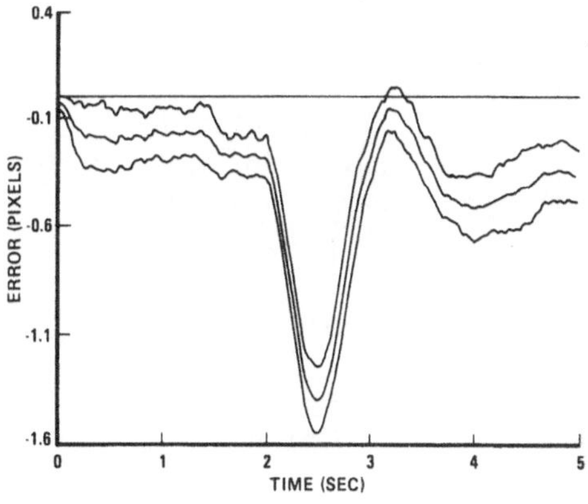

FIGS. 5 AND 6: $\hat{y}(t_i^+)$ ERROR MEAN ± ONE STD. DEV. FOR EKF ON 2G PULL-UP (5: BASED ON GAUSS-MARKOV ACC. MODEL: 6: BASED ON CONST. TURN-RATE MODEL)

transient in the plots, the filters were started with artificial knowledge of the true states; initial acquisition was investigated separately and did not cause difficulties.) From the table, it can be seen that the multiple model filters have performance very similar to the narrow field-of-view single filters that are specifically tuned for this benign trajectory, indicating that the adaptation is appropriately weighting the narrow field-of-view filter very heavily. In the first column, as anticipated, the larger field-of-view filter has poorer performance on the benign trajectory. For the extended Kalman filters, the standard deviations are worse but the biases are reduced for the wide field-of-view case. This is due to tuning of the narrow field-of-view filters; larger assumed dynamics noise strength would reduce these biases, but this was not incorporated in order to retain very distinguishing characteristics in the residuals of the different filters of the multiple model tracker.

The bottom of Table 2 pertains to performance achieved by the multiple model filter against pull-up maneuvers at 2g, 10g, and 20g levels; the 2g case time histories are plotted in Figs. 4-6, while 10g case characteristics are exhibited in the first 3.5 sec of Figs. 7 and 8. The correlator/linear filter generally has smaller biases but larger standard deviations than the extended Kalman filters. Its rms errors are smaller for low-g maneuvers (again due to tuning of the narrow field-of-view extended filters), but the superiority of the extended Kalman filter based on constant turn-rate dynamics is clearly evident for harsher maneuvers. There the extended Kalman filters have

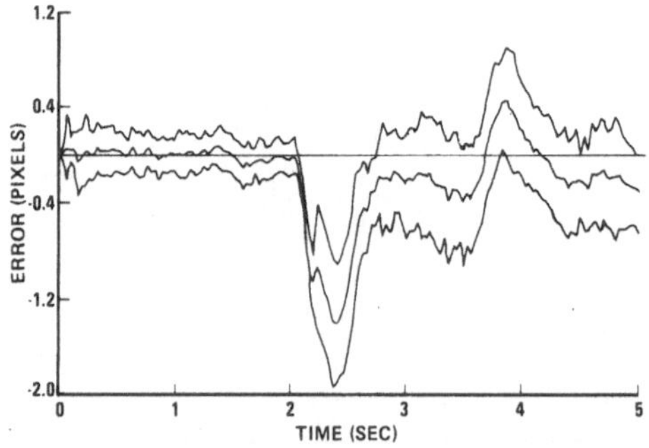

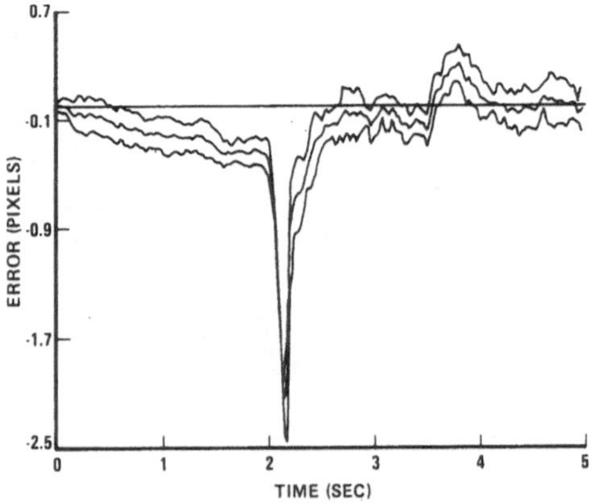

FIGS 7 AND 8: $\hat{y}(t_i^+)$ ERROR MEAN $\pm$ ONE STD. DEV. ON 10G TRAJECTORY 3: (7: CORRELATOR/KF; 8: EXTENDED KF)

much smaller rms errors and shorter transients, and the bias is considerably smaller when the filter is based on constant turn-rate dynamics rather than a Gauss-Markov acceleration model. At low g's, the adaptive filter performance resembles that of the narrow field-of-view filter, but quickly converges to that of the wide field-of-view filter after maneuver initiation. Note that the performance of the correlator/filter is actually worse at 10 g's than at 20 g's: there is no filter in the adaptive filter's bank that is expressly tuned for best performance at these conditions. As seen in the last two rows of Table 2, the performance of the adaptive filter is somewhat worse than that of the large field-of-view filter, due to the lower bounding of the conditional probabilities as discussed previously. The bound of 0.01 is the result of a tradeoff: higher values allow quicker reaction to maneuver initiation, but at the expense of an inappropriately heavy weight on a "wrong" model in the steady-state pull-up. No entry is made in the table for the narrow field-of-view filter, since it was unable to maintain lock on 10g or 20g targets.

As displayed in Figs. 7 and 8, the response to the straightening maneuver of trajectory 3 at 3.5 sec. is similar to that of the pull-up initiation at t = 2 sec. (Fig. 8 pertains to the Gauss-Markov acceleration model, but the constant turn-rate model yields very similar results.) Returning desirably heavy weight to the narrow field-of-view is not as rapid or effective as the response to maneuver onset, partly because the residuals in both elemental filters are consistent with the covariance computed in the corresponding filter via Eq. (19). Similar characteristics have been observed elsewhere [28], but the ad hoc modification suggested therein (artificially forcing a "no-maneuver" declaration) has not been investigated.

Trajectory 4 evaluations were very similar to those of trajectory 2, demonstrating the ability to maintain good estimates of a shape function undergoing large changes in time, and thereby to maintain desirable tracking performance. Robustness to signal-to-noise ratio was tested by leaving the filter-assumed SNR at 20 while the real world was simulated with SNR of 10, resulting in about 35% larger rms errors for a 2g pull-up, equivalent rms errors at 10g, and 5% smaller rms errors at 20g; reduction of real world SNR to unity caused divergence in all cases, as noted previously [5-7]. Finally, as a performance bound, the adaptive filter was artificially told when the pull-up maneuver occurred and appropriate weightings were applied immediately. There was no change in steady state behavior and an imperceptible change in the time for the transients to die out; the only substantial difference was the peak value of the mean error excursion, which was reduced between 10 and 45 percent for the different filters and test trajectories.

V. Summary

Algorithms have been developed for tracking dynamic targets in infrared image data, where the target pattern is uncertain a priori and may be composed of multiple hot spots. Digital and/or optical signal processing is used to identify this target shape adaptively in real time. Multiple model filtering has been shown to provide an effective means of changing the field-of-view and bandwidth of the tracker against a wide dynamic range of targets. In this multiple model algorithm, one elemental filter is tuned to benign dynamics and uses a narrow field-of-view and another is tuned to harsher maneuvering and correspondingly uses a wider field-of-view. There are significant computational loading advantages to using a correlator/ linear Kalman filter combination for each of these filters, but extended Kalman filters processing the raw FLIR data and based upon a constant turn-rate dynamics model provide superior tracking capability, particularly for highly dynamic close-range targets.

Research is continuing, to enhance performance in a number of ways: (1) improved filter tuning for particular target characteristics, with elemental filter reinitialization if divergence occurs due to mismatch with current real world behavior; (2) inclusion of more filters in the bank to allow additional target dynamics levels, fields of view, and combinations thereof; (3) outputting the state estimate corresponding to the highest probability $p_k(t_i)$ versus a probabilistically weighted average as in Eq. (17) (i.e., MAP versus Bayesian estimation), trading off quickness of response to real world changes against accentuation of coarse discretization effects; (4) evaluation of performance as a function of sample period, filter tuning / probability lower bounding combinations, and other pertinent parameters; (5) investigation of robustness to variation in target shape function, dynamic trajectories, atmospheric jitter, and background noise at low SNR; and (6) establishing performance sensitivity to combinations of sensor resolution and noise attributes, controller/ actuator dynamics, and other tracking environment characteristics.

References

1. Maybeck, P.S., Stochastic Models, Estimation and Control, Vol. 1, Academic Press, New York, 1979.
2. Maybeck, P.S., Stochastic Models, Estimation and Control, Vol. 2, Academic Press, New York, 1982.
3. Maybeck, P.S., and D.E. Mercier, "A Target Tracker Using Spatially Distributed Infrared Measurements," IEEE Trans. AC, Vol. AC-25, No. 2, pp. 222-225, April 1980.
4. Mercier, D.E., "An Extended Kalman Filter for Use in a Shared Aperture Medium Range Tracker," M.S. thesis, A.F. Inst. of Tech., Wright-Patterson AFB, Ohio, Dec. 1978.
5. Maybeck, P.S., D.A. Harnly and R.L. Jensen, "Robustness of a New Infrared Target Tracker," Proc. IEEE Nat. Aerospace & Elec. Conf., Dayton, Ohio, pp. 639-644, May 1980.
6. Harnly, D.A., and R.L. Jensen, "An Adaptive Distributed-Measurement Extended Kalman Filter for a Short Range Tracker," M.S. thesis, A.F. Inst. of Tech., Wright-Patterson AFB, Ohio, Dec. 1979.
7. Maybeck, P.S., R.L. Jensen and D.A. Harnly, "An Adaptive Extended Kalman Filter for Target Image Tracking," IEEE Trans. AES, Vol. AES-17, pp. 173-180, Mar. 1981.
8. Maybeck, P.S., "Advanced Applications of Kalman Filters and Nonlinear Estimators in Aerospace Systems," Control and Dynamic Systems (C.T. Leondes, ed.), Vol 20, Academic Press, New York, pp. 67-154, 1983.
9. Maybeck, P.S., W.H. Worsley and P.M. Flynn, "Investigation of Constant Turn-Rate Dynamics Models in Filters for Airborne Vehicle Tracking," Proc. IEEE Nat. Aerospace & Elec. Conf., Dayton, Ohio, pp. 896-903, May 1982.
10. Flynn, P.M., "Alternative Dynamics Models and Multiple Model Filtering for a Short Range Tracker," M.S. thesis, A.F. Inst. of Tech., Wright-Patterson AFB, Ohio, Dec. 1981.
11. "Firefly III IFFC Fire Control System," Tech. Rept. 19008 ACS 12004, General Electric Co., Aircraft Equip. Div., Binghamton, N.Y., Dec. 1979 (revised, Jan. 1981).
12. Bakir, E., "Adaptive Kalman Filter for Tracking Maneuvering Targets," AIAA Jour. Guid., Cont. & Dyn., Vol. 6, No. 6, pp. 414-416, Sept. 1983.
13. Berg, R.F., "Estimation and Prediction for Maneuvering Target Trajectories," IEEE Trans. AC, Vol. AC-28, No. 3, pp. 294-304, Mar. 1983.
14. Chang, C.B., and J.A. Tabaczynski, "Application of State Estimation to Target Tracking," IEEE Trans. AC, Vol. AC-29, No. 2, pp. 98-109, Feb. 1984.
15. Demetry, J.S., and H.A. Titus, "Adaptive Tracking of Maneuvering Targets," Tech. Rept. NPS-52DE8041A, Nav. Postgrad. Sch., Monterey, Cal., Apr. 1968.
16. McAuley, R.J., and E. Denlinger, "A Decision-Directed Adaptive Tracker," IEEE Trans. AES, Vol. AES-9, No. 2, pp. 229-236, Mar. 1973.
17. Nahi, N.E., and B.M. Schaefer, "Decision-Directed Adaptive Recursive Estimators: Divergence Prevention," IEEE Trans. AC, Vol. AC-17, No. 1, pp. 61-67, Jan. 1972.
18. Maybeck, P.S., and S.K. Rogers, "Adaptive Tracking of Multiple Hot-Spot Target IR Images," IEEE Trans. AC, Vol. AC-28, No. 10, pp. 937-943, Oct. 1983.
19. Rogers, S.K., "Enhanced Tracking of Airborne Targets Using Forward

21. Casasent, D.P., "Pattern Recognition: A Review," IEEE Spectrum, pp. 28-33, Mar. 1981.
22. Casasent, D.P., J. Jackson and C. Neuman, "Frequency-Multiplexed and Pipelined Iterative Optical Systolic Array Processors," Applied Optics, Vol. 22, pp. 115-124, Jan. 1983.
23. Roemer, W.A., and P.S. Maybeck, "An Optically Implemented Multiple-Stage Kalman Filter Algorithm," Proc. SPIE: Real Time Signal Processing VI, San Diego, Cal., pp. 221-228, Aug. 1983.
24. Kozemchak, M.R., "Enhanced Image Tracking: Analysis of Two Acceleration Models in Tracking Multiple Hot-Spot Images," M.S. thesis, A.F. Inst. of Tech., Wright-Patterson AFB, Ohio, Dec. 1982.
25. Millner, P.P., "Enhanced Tracking of Airborne Targets Using a Correlator / Kalman Filter," M.S. thesis, A.F. Inst. of Tech., Wright-Patterson AFB, Ohio, Dec. 1982.
26. Athans, M., and C.B. Chang, "Adaptive Estimation and Parameter Identification Using Multiple Model Estimation Algorithm," Technical Note 1976-28, ESD-TR-76-184, Lincoln Laboratory, Lexington, Mass., June 1976.
27. Brown, R.G., "A New Look at the Magill Adaptive Estimator as a Practical Means of Multiple Hypothesis Testing," IEEE Trans. Circuits & Sys., Vol. CAS-30, No. 10, pp. 765-768, Oct. 1983.
28. Korn, J., and L. Beean, "Application of Multiple Model Adaptive Estimation Algorithms to Maneuver Detection and Estimation," Tech. Rept. TR-152, Alphatech, Inc., Burlington, Mass., June 1983.
29. Magill, D.T., "Optimal Adaptive Estimation of Sampled Stochastic Processes," IEEE Trans AC, Vol. AC-10, No. 5, pp. 434-439, Oct. 1965.
30. Moose, R.L., "An Adaptive State Estimation Solution to the Maneuvering Target Problem," IEEE Trans. AC, Vol. AC-20, No. 3, pp 359-362, June 1975.
31. Moose, R.L., H.F. Van Landingham and D.H. McCabe, "Modeling and Estimation for Tracking Maneuvering Targets," IEEE Trans. AES, Vol. AES-15, No. 3, pp. 448-456, May 1979.
32. Moose, R.L., and P.P. Wang, "An Adaptive Estimator with Learning for a Plant Containing Semi-Markov Switching Parameters," IEEE Trans. Sys., Man, Cyber., pp. 277-281, May 1973.
33. Tenney, R.R., R.S. Hebbert and N.R. Sandell, Jr., "A Tracking Filter for Maneuvering Sources," IEEE Trans. AC, Vol. AC-22, No. 2, pp. 246-261, Mar. 1977.
34. Thorp, J.S., "Optimal Tracking of Maneuvering Targets," IEEE Trans. AES, Vol. AES-9, No. 4, pp. 512-519, July 1973.
35. Suizu, R.I., "Enhanced Tracking of Airborne Targets Using Multiple Model Filtering Techniques for Adaptive Field of View Expansion," M.S. thesis, A.F. Inst. of Tech., Wright-Patterson AFB, Ohio, Dec. 1983.
36. Loving, P.A., "Bayesian vs. MAP Multiple Model Adaptive Estimation for Field of View Expansion in Tracking Airborne Targets," M.S. thesis, A.F. Inst. of Tech., Wright-Patterson AFB, Ohio, Mar. 1985.
37. Hogge, C.B., and R.R. Butts, "Frequency Spectra for the Geometric Representation of Wavefront Distortions Due to Atmospheric Turbulence," IEEE Trans. Antennas and Prop., Vol. AP-24, No. 2. pp. 144-154, Mar. 1976.
38. "Advanced Adaptive Optics Control Techniques," Tech. Rept. TR-966-1, The Analytic Sciences Corp., Reading, Mass. (prepared for A.F. Weapons Laboratory, Kirtland AFB, New Mexico), Jan. 1978.

WIDE BAND LIMIT OF LYAPOUNOV EXPONENTS

E. Pardoux
UER de Mathématiques
Université de Provence
3,Place Victor Hugo
F 13331 Marseille Cedex 3
and INRIA

Abstract : We show that, under appropriate conditions, the Lyapounov exponents of a sequence of solutions of linear differential equations with wide-band noise coefficients, converge to that of its diffusion limit.

§1. Introduction : Let X_t be a stochastic process with values in $\mathbb{R}^d$, solution of one of the linear equations :

$$(1.1) \qquad dX_t = AX_t\, dt + \sum_{i=1}^{k} B_i X_t \circ dW_t^i$$

$$(1.2) \qquad \frac{dX_t}{dt} = A(Y_t)X_t$$

where $W_t=(W_t^1,\ldots,W_t^k)$ is a standard Wiener process defined on a probability space with filtration (Ω,F,F_t,P), o denotes Stratonovitch integral, and Y_t is a diffusion process with values in some smooth manifold M. It is known that in both cases, under appropriate conditions, the a.s. asymptotic behavior (as $t \to \infty$) of the solution is caracterized by a unique Lyapounov exponent (see Arnold-Kliemann-Oeljeklaus [1]and Arnold-Oeljeklaus-Pardoux [2]). Theses exponents are rarely easy to compute in practice, but the "real-noise" case (equation of type (1.2)) is certainly worse than the "white noise" case (equation of type (1.1)) in that respect. On the other hand, it is well-known (see e.g. Kurtz [5], Blankenship-Papanicolaou [3], Kushner [6]) that the solution of (1.1) can be considered as the limit in law of a suitably scaled sequence of solutions of (1.2). In fact, virtually any model involving white noise as input should be considered as an idealisation of a more realistic model with a wide band noise as input, see Kushner [6] for a discussion and many applications. It is therefore of interest, to see whether or not the large time behaviour of the solution of (1.1) is in a certain sense the limit of large time behaviours of approximating wide-band systems of type (1.2). Inded, this involves an intercharge of limits, and such intercharges in the same context are not always valid, see Kushner [7].

We show that under appropriate conditions the Lyapounov exponents do converge. Our result is compared below with related results of Blankenship-Papanicolaou [3].

§2. Lyapounov exponents of linear stochastic differential equations.

In this section, we recall some results from Arnold-Kliemann-Oeljeklaus [1] and Arnold-Oeljeklaus-Pardoux [2] about a.s. Lyapounov exponents of linear stochastic differential equations. Consider first equation (1.1), and suppose that $X_o \neq o$ a.s.. Then the process $U_t = \| X_t \|^{-1} X_t$ solves the following stochastic differential equation, written in Stratonovich form :

$$(2.1) \qquad dU_t = h(A,U_t)dt + \sum_{i=1}^{k} h(B_i,U_t) \circ dW_t^i$$

where $h(C,u) \triangleq Cu - (Cu,u)u$, and moreover :

$$(2.2) \qquad \operatorname{Log} \| X_t \| = \int_o^t q(U_s)ds + \sum_{i=1}^{k} \int_o^t (B_i U_s, U_s) dW_s^i$$

with $q(u) = (Au,u) + \frac{1}{2} \sum_{i=1}^{k} [(B_i^2 u,u) + |B_i u|^2 - 2(B_i u,u)^2]$, the last integral being of Ito-type. Note that U_t takes values in S^{d-1}, but since $h(C,-u) = -h(C,u)$, $h(C,.)$ can be viewed as a vector field over the projective space $\mathbb{P}^{d-1}$, which is obtained from S^{d-1} by identifying u and -u. Then (2.1) can be viewed as a S D E on $\mathbb{P}^{d-1}$, and (2.2) makes perfect sense with U_t considered as taking values in $\mathbb{P}^{d-1}$. We now assume :

$$(2.3) \quad \dim \text{L.A}\, \{h(A,.),\ h(B_1,.), \ldots, h(B_k,.)\}(u) = d-1 \quad \forall u \in \mathbb{P}^{d-1}$$

Where $\text{L.A.}\{h(A,.),h(B_1,.),\ldots,h(B_k,.)\}$ denotes the Lie algebra of vector fields over $\mathbb{P}^{d-1}$, generated by $h(A,.),h(B_1,.),\ldots,h(B_k,.)$.

We then have :

Théorem 2.1 : (see [2]) Assume (2.3). Then the $\mathbb{P}^{d-1}$ valued diffusion process $\{U_t\}$ has a unique invariant measure μ, and for any initial X_o F_o-measurable random vector with values in $\mathbb{R}^d-\{o\}$,

$$\lim_{t\to\infty} \frac{1}{t} \operatorname{Log} \| X_t \| = \lambda \quad \text{a.s.}$$

where $\lambda \triangleq \int_{\mathbb{P}^{d-1}} q(u) d\mu(u)$

□

Let us now turn to equation (1.2). We suppose that $y \to A(y)$ is an analytic function from an analytic connected Riemannian manifold M, into the space of real $d \times d$ matrices. We assume that $\{Y_t\}$ is a stationary ergodic diffusion on M, solution of the following Stratonovich S D E :

$$(2.4) \quad dY_t = f_o(Y_t)dt + \sum_{i=1}^{r} f_i(Y_t) \circ dV_t^i$$

where $f_o, f_1, \ldots, f_r$ are analytic vector fields, $\{(V_t^i), 1 \leqslant i \leqslant r\}$ are mutually independent standard Wiener processes defined on (Ω, F, F_t, P), and:

$$(2.5) \quad \dim \text{L.A.} \{f_1, \ldots, f_1\}(y) = \dim M, \ \forall y \in M$$

Defining again $U_t = \| X_t \|^{-1} X_t$, we obtain :

$$(2.6) \quad \frac{dU_t}{dt} = h(A(Y_t), U_t)$$

$$\operatorname{Log} \| X_t \| = \int_o^t (A(Y_s)U_s, U_s) ds$$

As above, we consider U_t as taking valves in $\mathbb{P}^{d-1}$. We finally assume :

$$(2.7) \quad \dim \text{L.A.} \{h(A(y), .), y \in M\}(u) = d-1, \quad \forall u \in \mathbb{P}^{d-1}$$

We then have :

Theorem 2.2 : (see [1]). Under the above assumptions, in particular (2.5) and (2.7), the process $\{(U_t, Y_t)\}$ has a unique invariant mesure ν on $\mathbb{P}^{d-1} \times M$, and for any initial X_o F_o-measurable random vector with values in $\mathbb{R}^d-\{o\}$, if $\{X_t\}$ denotes the solution of (1.2),

$$\lim_{t\to\infty} \frac{1}{t} \operatorname{Log} \| X_t \| = \lambda \quad \text{a.s.}$$

where $\lambda \triangleq \int_{\mathbb{P}^{d-1} \times M} (A(y)u, u) d\nu(u, y)$

□

§3 The wide-band approximation :

Following the works of Kurtz [5], Papanicolaou [9], Papanicolaou-Stroock-Varadhan [8], Blankenship-Papanicolaou [3] and Kushner [6], let us introduce a sequence of equations of type (1.2), whose solutions converge weakly to that of an equation of type (1.1).

We now add new assumptions concerning the diffusion process $\{Y_t\}$. We first assume (this assumption is not at all necessary, by does simplify the notations) that $\{Y_t\}$ is time-reversible, i.e. $\forall g,h$ measurable and bounded functions from M into $\mathbb{R}$, $\forall t > o$,

$$E[\,g(Y_o)h(Y_t)\,] = E[\,g(Y_t)h(Y_o)\,]$$

We need to add some stringent assumption, in order to be able to obtain bounded solutions to some Poisson equations associated with $\{Y_t\}$. One type of hypotheses would be to assume that M is compact, together with Doeblin's condition. We now formulate another possible set of hypotheses, following Bouc-Pardoux [4], which is supposed to hold throughout the sequel. We assume that $M=\mathbb{R}^\ell$, all coefficients in (2.4) together with their first order derivatives are bounded, and $\sigma\sigma^*(y) \geq \alpha I > o$, $\forall y \in \mathbb{R}^\ell$; where $\sigma(y) \triangleq (f_1(y) \ldots . f_r(y))$. Moreover, if $\tilde{f}_o(Y_t)$ denotes the drift of Y_t (i.e. the coefficient of dt in (2.4) writen in Ito form), let us assume that $\exists$ M and $c > o$ s.t.:

$$(3.1) \qquad (\tilde{f}_o(y),y) \leq -c\,|y|\,, \quad \forall y \in \mathbb{R}^\ell \text{ s.t. } |y| \geq M$$

For a detailed explanation of the consequences of (3.1), see [4]and[12].

For $\varepsilon > o$, define $Y_t^\varepsilon = Y_{t/\varepsilon}$, and let $\{X_t^\varepsilon, t \geq o\}$ be the solution of :

$$(3.2) \qquad \frac{dX_t^\varepsilon}{dt} = A(Y_t^\varepsilon)X_t^\varepsilon + \frac{1}{\sqrt{\varepsilon}}\,B(Y_t^\varepsilon)X_t^\varepsilon$$

There does exist an integer $p \leq d^2$, s.t. (3.2) may be rewritten in the form :

$$(3.3) \qquad \frac{dX^\varepsilon}{dt} = A(Y_t^\varepsilon)X_t^\varepsilon + \frac{1}{\sqrt{\varepsilon}}\sum_{i=1}^{p} g_i(Y_t^\varepsilon)B_i\,X_t^\varepsilon$$

where each $g_i(y)$ is real valued, and the B_i's are $d \times d$ matrices. We suppose in addition to the hypotheses of §2 that $A, g_1, \ldots, g_p$ are bounded, and that :

$$(3.4) \qquad \int g_i(y)\,d\bar{\nu}(y) = o\,, \quad i=1,\ldots,p$$

where $\overline{\nu}$ denotes the unique invariant probability measure of Y_t. We suppose moreover that there exists $N > o$ s.t. :

$$(3.5) \quad \int_N^{+\infty} \operatorname{ess\,sup}_{|y| \geqslant t} |g_i(y)| \, dt < \infty \quad , \ i=1,\ldots,p$$

It follows from the above assumptions (in particular (3.1),(3.4) and (3.5)) that

$$\int_{-\infty}^{+\infty} E[\, g_i(Y_t)/Y_o = y \,] \, dt$$

exists and is a bounded function of y ; $i = 1,\ldots,p$; see [4]. Without loss of generality, we may and will assume that for some $k \leqslant p$,

$$(3.6) \quad \int_{-\infty}^{+\infty} E[\, g_i(Y_o) g_j(Y_t) \,] \, dt = \delta_{ij} \quad ; \ i,j = 1,\ldots,k$$

$$(3.7) \quad \int_{-\infty}^{+\infty} E[\, g_i(Y_o) g_j(Y_t) \,] \, dt = o \ ; \ i=1,\ldots,p \ ; \ j=k+1,\ldots,p$$

where $\delta_{ij} = 1$ if $i=j$, $= o$ otherwise. (3.6) and (3.7) are obtained by an orthogonalization procedure. Note that their left hand sides are symetric in (i,j), and

$$\int_{-\infty}^{+\infty} E \ [g_i(Y_o) g_i(Y_t)] \, dt \geqslant o$$

The above may vanish, without the process $\{g_i(Y_t)\}$ beeing necessarily zero.

Consider finally the Statonovich stochastic differential equation :

$$(3.4) \quad dX_t = A X_t \, dt + \sum_{i=1}^{k} B_i X_t \circ dW_t^i$$

where $A = \int_{\mathbb{R}^\ell} A(y) d\overline{\nu}(y)$; $\{W_t^i\}$, $i=1,\ldots,k$, are mutually independent standard Wiener processes. The following convergence result can be essentially found in the references cited in the begining of the present section (see also [4]):

<u>Theorem 3.1</u> : Suppose $X_o^\varepsilon = x_o$, $\forall \varepsilon > o$, where $x_o \in \mathbb{R}^d - \{o\}$. Then under the above conditions $X^\varepsilon_\cdot \Rightarrow X_\cdot$, where $\{X_t\}$ denotes the unique solution of (3.4) with initial condition $X_o = x_o$ ($\Rightarrow$ stands for the convergence in law in $C(\mathbb{R}_+ ; \mathbb{R}^d)$).

□

§4 Convergence of the Lyapounov exponents :

Let us now reformulate the rank condition (2.7) in the context of (3.3) :

$$(4.1) \quad \dim L.A.\{h(A(y),.)+\frac{1}{\sqrt{\varepsilon}}\sum_{i=1}^{p} g_i(y)h(B_i,.), y\in \mathbb{R}^{\ell}\}(u)=d-1, \forall\, u\in \mathbb{P}^{d-1}$$

It is easily seen that in case $A(y) \equiv A$ and $k = p$, (4.1) is equivalent to :

$$(4.2) \quad \dim L.A.\{h(A,.),h(B_1,.),\dots,h(B_k,.)\}(u)=d-1, \ \forall\, u\in \mathbb{P}^{d-1}$$

It now follows from Theorem 2.2 that under (4.1),

$\frac{1}{t}$ Log $\|X_t^{\varepsilon}\| \to \lambda_{\varepsilon}$ a.s., where

$$\lambda_{\varepsilon} \triangleq \int_{\mathbb{P}^{d-1}\times \mathbb{R}^{\ell}} [(A(y)u,u)+\frac{1}{\sqrt{\varepsilon}}\sum_{i=1}^{p} g_i(y)(B_iu,u)]\,d\nu_{\varepsilon}(u,y)$$

ν_{ε} denoting the unique invariant measures of $(U_t^{\varepsilon},Y_t^{\varepsilon})$, where $U_t^{\varepsilon} = \|X_t^{\varepsilon}\|^{-1} X_t^{\varepsilon}$.

From Theorem 2.1, under (4.2),

$\frac{1}{t}$ Log $\|X_t\| \to \lambda$ a.s., where

$$\lambda \triangleq \int_{\mathbb{P}^{d-1}} q(u)\, d\mu\,(u)$$

q being defined as in §2, and μ denoting the unique invariant measure of $U_t \triangleq \|X_t\|^{-1} X_t$.

We can now state :

Theorem 4.1 : Under the hypotheses of §2 and 3, and (4.1),(4.2), $\lambda_{\varepsilon} \to \lambda$, as $\varepsilon \to 0$

Before proving this result, let us first deduce from it :

Corollary 4.2 : Suppose the S.D.E. (3.4) is a.s. asymptotically exponentially stable (i.e. $\lambda < 0$). Then $\exists\, \varepsilon_0 > 0$ s.t. $\forall\, \varepsilon \in]0,\varepsilon_0[$, (3.3) is a.s. asymptotically exponentially stable (i.e. $\lambda_{\varepsilon} < 0$).

□

Remark 4.3 : A result very similar to Corollary 4.2 can be found in Blankenship-Papanicolaou [3], where the notion of a Lyapounov exponent is not used, and conditions for stability are given in terms of the existence of a Lyapounov function

□

Proof of Theorem 4.1 : We have :

$$\lambda_\varepsilon = \lim_{t\to\infty} \frac{1}{t}\int_o^t [(A(Y_s^\varepsilon)U_s^\varepsilon, U_s^\varepsilon) + \frac{1}{\sqrt{\varepsilon}}\sum_{i=1}^p g_i(Y_s^\varepsilon)(B_i U_s^\varepsilon, U_s^\varepsilon)]\, ds$$

This limits exists a.s. from the ergodic theorem, and equals the limit of the expectations taken with any initial law for $(U_o^\varepsilon, Y_o^\varepsilon)$. We suppose now that $(U_o^\varepsilon, Y_o^\varepsilon)$ is a F_o measurable random vector, whose law is ν_ε, the invariant law of $\{(U_t^\varepsilon, Y_t^\varepsilon), t \geq o\}$. We now have : $\forall\, t > o$,

$$\lambda_\varepsilon = \frac{1}{t} E \int_o^t [(A(Y_s^\varepsilon)U_s^\varepsilon, U_s^\varepsilon) + \frac{1}{\sqrt{\varepsilon}}\sum_{i=1}^p g_i(Y_s^\varepsilon)(B_i U_s^\varepsilon, U_s^\varepsilon)]\, ds$$

Let us define the real valued process :

$$Z_t = \int_o^t [(A(Y_s^\varepsilon)U_s^\varepsilon, U_s^\varepsilon) + \frac{1}{\sqrt{\varepsilon}}\sum_{i=1} g_i(Y_s^\varepsilon)(B_i U_s^\varepsilon, U_s^\varepsilon)]\, ds \ , \ t \geq o$$

and denote by Q_ε the law on $C(\mathbb{R}_+ ; \mathbb{P}^{d-1} \times \mathbb{R})$ of $\{(U_t^\varepsilon, Z_t^\varepsilon), t \geq o\}$. As $\varepsilon \to o$, $Q_\varepsilon \Rightarrow Q$, where Q is the law of the process $\{(U_t, Z_t), t \geq o\}$, where $\{(U_t, Z_t), t \geq o\}$ solves the Stratonovich differential system :

$$dU_t = h(A, U_t)dt + \sum_{i=1}^k h(B_i, U_t) \circ dW_t^i$$

$$dZ_t = (AU_t, U_t)dt + \sum_{i=1}^k (B_i U_t, U_t) \circ dW_t^i$$

the law of U_o being μ, the unique invariant measure; and $Z_o = o$. Here $\{W_t^i\}$, i=1...k, are mutually independent standard Wiener processes.

This convergence combines a result similar to Theorem 3.1 and a Corollary of that result (which here, since tightness is obvious, is relatively easy) which says that $\mu_\varepsilon \Rightarrow \mu$, where :

$$\mu_\varepsilon(.) = \int_{\mathbb{R}^\ell} \nu_\varepsilon(., dy)$$

It is moreover easily checked that for fixed $t > o$, the family of random variables $\{Z_t^\varepsilon, \varepsilon > o\}$ is uniformly integrable. It then follows ($t > o$ is fixed):

$$\lambda_\varepsilon \to \frac{1}{t} E [\int_o^t (A\, U_s, U_s)ds + \sum_{i=1}^k \int_o^t (B_i U_s, U_s) \circ dW_s^i]$$

clearly, this limit equals :

$$\frac{1}{t} E \int_o^t q(U_s) ds = \lim_{t\to\infty} \frac{1}{t} E \int_o^t q(U_s)) ds$$

$$= \lim_{t\to\infty} \frac{1}{t} \int_o^t q(U_s) ds \quad \text{a.s.}$$

$$= \lambda$$

The ergodic theorem has again been used in the last lines.

□

§5 Convergence of invariant measures :

We finally study a question which again was already analized in Blankenship-Papanicolaou [3], but we use a different approach, by exploiting the results from Pardoux-Pignol [10]and [11].

Suppose we are given, in addition, to the data of §3, $F(y)=(F_1(y),\ldots,F_d(y))'$ with :

$$(5.1) \qquad F_i \in L^1(\mathbb{R}^\ell ;\bar{\nu}),\ i=1,..p$$

where again $\bar{\nu}$ denotes the unique invariant measure of $\{Y_t\}$, and $G_1,\ldots,G_p \in \mathbb{R}^d$.

Consider the following differential equation :

$$(5.2) \qquad \frac{dX_t^\varepsilon}{dt} = A(Y_t^\varepsilon)X_t^\varepsilon + F(Y_t^\varepsilon) + \frac{1}{\sqrt{\varepsilon}} \sum_1^p (B_i X_t^\varepsilon + G_i) g_i(Y_t^\varepsilon)$$

where now $p \leqslant d^2 + d$, the g_i's satisfying the same hypotheses as in §4. The following is a particular case of a result in [11]:

Proposition 5.1 : Suppose that the Lyapounov exponent λ_ε of equation (5.2) without forcing terms (i.e. with $F \equiv o$, $G_1 \equiv o,\ldots, G_p \equiv o$)satisfies:

$$\lambda_\varepsilon < o$$

Then (5.2) possess a unique invariant probability γ_ε, which is the law of the random vector :

$$\int_{-\infty}^o \Phi_\varepsilon(t)^{-1}[F(Y_t^\varepsilon) + \frac{1}{\sqrt{\varepsilon}} \sum_1^p G_i\, g_i(Y_t^\varepsilon)]dt$$

where $\{\Phi_\varepsilon(t), t \in \mathbb{R}\}$ is the $d \times d$ matrix valued process which solves :

$$\frac{d}{dt} \Phi_\varepsilon(t) = A(Y_t^\varepsilon)\, \Phi_\varepsilon(t) + \frac{1}{\sqrt{\varepsilon}} \sum_1^p B_i \Phi_\varepsilon(t) g_i(Y_t^\varepsilon)$$

$$\Phi_\varepsilon(o) = I$$

□

Again, $X^{\varepsilon}_{\cdot} \Rightarrow X_{\cdot}$, where $\{X_t\}$ solves the Stratonovich SDE :

$$(5.3) \quad dX_t = (AX_t + F)dt + \sum_{i=1}^{k} (B_i X_t + G_i) \circ dW_t^i$$

where $F = \int_{\mathbb{R}^{\ell}} F(y)\, d\bar{\nu}(y)$, and it follows from the results in [10]:

<u>Proposition 5.2</u> : Suppose that the Lyapounov exponent λ of equation (5.3) without forcing terms (i.e. with $F = o$, $G_1 = o, \ldots, G_k = o$) satisfies:

$$\lambda < o$$

Then (5.3) possess a unique invariant probability γ, which is the law of the random vector :

$$\int_{-\infty}^{o} \Phi(t)^{-1} [F\,dt + \sum_{1}^{k} G_i \circ dW_t^i]$$

where $\{\Phi(t), t \in \mathbb{R}\}$ is the $d \times d$ matrix valued process which solves :

$$d\Phi(t) = A\,\Phi(t) + \sum_{i=1}^{k} B_i\,\Phi(t) \circ dW_t^i$$

$$\Phi(o) = I$$

□

We can finally prove (λ, γ and $\lambda_{\varepsilon}, \gamma_{\varepsilon}$ are defined in Proposition 5.2 and 5.1 respectively):

<u>Theorem 5.3</u> : Suppose all above hypotheses are satisfied. We assume that $\lambda < o$. Let ε_o be such that $\lambda_{\varepsilon} < o$, $\forall \varepsilon < \varepsilon_o$.

Then $\gamma_{\varepsilon} \Rightarrow \gamma$, as $\varepsilon \to o$ $(\varepsilon < \varepsilon_o)$.

<u>Proof</u> :Again, the argument of Theorem 3.1 shows that : $\forall t < o$,

$$\int_t^o \Phi_{\varepsilon}(s)^{-1} [F(Y_s^{\varepsilon}) + \frac{1}{\sqrt{\varepsilon}} \Sigma\, G_i\, g_i(Y_s^{\varepsilon})]ds$$

$$\Rightarrow \int_t^o \Phi(s)^{-1} [F\,ds + \sum_{1}^{p} G_i \circ dW_s^i]$$

On the other hand (see [11]):

$$\frac{1}{|t|} \mathrm{Log}\, \| \Phi_{\varepsilon}(t) \|^{-1} \to \lambda_{\varepsilon} \text{ , as } t \to -\infty \text{ and } \exists\, \varepsilon_1 \text{ s.t.}$$

$$\forall \varepsilon \in]o, \varepsilon_1], \; \lambda_{\varepsilon} \leqslant \frac{\lambda}{2} < o \; .$$

It is then not hard to show that $\forall \alpha > 0$, $\exists T$ s.t. $\forall t \leq T, \forall \varepsilon \leq \varepsilon_1$,

$$P(|\int_{-\infty}^{t} \Phi_\varepsilon(s)^{-1}[F(Y_s^\varepsilon) + \frac{1}{\sqrt{\varepsilon}} \sum_1^p G_i\, g_i(Y_s)]\, ds| \geq \alpha) \leq \alpha$$

The result then follows.

□

Remark 5.4 : It is possible to extend Theorems 4.1 and 5.3 to the case where the coefficients are periodic functions of t, using the results in [10],[11].

□

REFERENCES

[1] L. ARNOLD, W. KLIEMANN, E. OELJEKLAUS.- Lyapounov exponents of a linear stochastic system. to appear in Proc. Bremen Workshop on Lyapounov exponents, L. Arnold et V. Wihstutz Eds.

[2] L. ARNOLD, E. OELJEKLAUS, E. PARDOUX.- Almost sure and moment stability for linear Ito equations, to appear in Proc. Bremen Workshop on Lyapounov exponents.

[3] G. BLANKENSHIP, G. PAPANICOLAOU.- Stability and control of stochastic systems with wide-band noise disturbances I, SIAM J. Appl. Math. 34, 437-476 (1978).

[4] R. BOUC, E. PARDOUX.- Asymptotic analysis of PDEs with wide-band noise disturbances, and expansion of the moments, Stoch. Anal. and Applic. 2, 369-422 (1984).

[5] T. KURTZ .- Semigroups of conditional shifts and approximations of Markov processes, Ann. Prob.4, 618-642 (1975).

[6] H. KUSHNER.- Approximation and Weak convergence methods for random processes, M.I.T. Press (1984).

[7] H. KUSHNER.- A cautionary note on the use of singular perturbation methods for "small noise" models. Stochastics 6, 117-120 (1982).

[8] G. PAPANICOLAOU, D. STROOCK, S.R.S. VARADHAN.- Martingale approach to some limit theorems, in Proc. of the 1976 Duke Univ. Conf. on Turbulence.

[9] G. PAPANICOLAOU.- Asymptotic analysis of stochastic equations in, Studies in Probability Theory, M. Rosenblatt Ed, 111-179, MAA Studies in Mathematics 18, M.A.A. (1978).

[10] E. PARDOUX, M. PIGNOL.- Etude de la stabilité de la solution d'une EDS bilinéaire à coefficients périodiques. Application au mouvement des pales d'hélicoptère. in Analysis and Optimization of Systems, Part 2; A. Bensoussan et J.L. Lions Eds, 92-103, Lecture Notes in Control and Info. Sci. 63, Springer-Verlag (1984).

[11] E. PARDOUX, M. PIGNOL.- Stability of periodic bilinear stochastic differential equations with correlated noise inputs, to appear in Ann. Scient. Univ. A.I.Cuza.

[12] E. PARDOUX.- Asymptotic analysis of a semi-linear PDE with wide-band noise disturbances, in Proc. Workshop on "Stochastic space-time models and limit theorems", L. Arnold and P. Kotelenez Eds., D. Reidel (1985).

FILTERING WITH OBSERVATIONS
ON A RIEMANNIAN SYMMETRIC SPACE

Monique PONTIER
UNIVERSITE D'ORLEANS
Dépt Mathématiques et Informatique
45046 ORLEANS CEDEX

Jacques SZPIRGLAS
I. N. T.
9, rue Charles Fourier
91011 EVRY CEDEX

I - INTRODUCTION

In this paper, a filtering model is studied, the multiplicative formulation of which appears to be interesting and non standard. Observation Y of a Markovian signal X with generator $(A, \underline{\underline{D}}(A))$ takes its values in a symmetric space M. Briefly speaking, M is a Riemannian manifold such that there exists a Lie subgroup G of the isometric transformation group I (M) of M, which acts transitively on M. Process Y is assumed to be :

(1.1) $\quad Y_t = g_t(X) \,.\, W_t$

Where W is a Brownian motion on M and $g_t(X)$ is a functional of signal X taking values in G, such that $t \rightarrow g_t(X)$ is absolutely continuous on $[0,T]$:

(1.2) $\quad dg_t = \zeta(t, X_s\, ;\, s \leqslant t)(g_t)\,dt \;;\; g_o = e \;;\; t \in [0,T],$

Where ζ is a bounded semimartingale functional of t and the sample path of X, taking values in the set of the left invariant vector fields of G.

This short lecture is devoted to the construction of the filtering model and the computation of the stochastic differential equation which is satisfied by the conditional law of signal X, given its observation Y, i. e. filter π of X given Y. We construct this model by the reference probability method ([2], [12], [13]) using an important result by S H I G E K A W A [10] about equivalence of the laws of g. W and W. It can be seen that the problem is finally reduced to a non trivial multivariate problem with a non bounded signal. The results of this paper generalize the following example on $\mathbb{R}^d$, considered as a symmetric space ; observation process Y is then such that :

(1.3) $Y_t = \Sigma_t(X)\, W_t + H_t(X)$

Where W_t is a d-dimensional Brownian motion independent of X, $\Sigma_t(X)$ is a matrix of $SO(d)$ and $H_t(X)$ is a vector of $\mathbb{R}^d$; thus Y_t is the transformation of W_t by a space deplacement (Σ, H) : rotation Σ and translation H being functional of X. We suppose that :

(1.4) $\dot{\Sigma}_t = \Sigma_t \sigma_t$, $\Sigma_o = I_d$; $\dot{H}_t = \Sigma_t h_t$, $H_o = 0$,

where σ is a bounded skew matrix and h is a bounded d-dimensional vector. Then we get the filtering equation for a bounded function f in domain $\underline{\underline{D}}(A)$ of X :

(1.5) $\pi_t(f) = \pi_o(f) + \int_o^t \pi_s(Af)ds + \int_o^t (\pi_s(f\varphi) - \pi_s(f)\pi_s(\varphi))(dY_s - \pi_s(\varphi)ds)$

where π is the filter of X given Y ; and :

(1.6) $\varphi_t(X,Y) = \Sigma_t(X)\sigma_t(X)\Sigma_t^{-1}(X)(Y_t - H_t(X)) + \Sigma_t(X)h_t(X)$

If Σ equals to I_d (and σ to 0) we obtain the standard filtering equation.

Our results are similar the ones by NG and CAINES [8] for the general Riemannian case. But our restriction to a symmetric space allows a more general formulation of the filtering problem. DUNCAN [2] was also concerned with filtering on symmetric spaces, but he used this last structure to ensure the non explosion of the Brownian motion and studied the same problem as NG and CAINES.

In the second part, a few necessary results about symmetric spaces and Brownian motion on symmetric space are given. The third one is devoted to the construction of the filtering model and to the filtering equation. We conclude with an example on the sphere.

NOTATIONS : All filtrations in this paper are assumed to be complete and right continuous. For any process Z defined on a probability space $(\Omega, \underline{\underline{A}}, \mathbb{P})$, $\underline{\underline{G}}^z$ will denote the natural filtration generated by Z, completed with respect to probability $\mathbb{P}$, and right continuous.

Morecver, Stratonovitch integral is denoted by o : $\int Y_o\, dB$, and Itô one by . : $\int Y.\, dB$

II - STOCHASTIC CALCULUS ON SYMMETRIC SPACES

In this part, some definitions about symmetric spaces are recalled ; Brownian motion on symmetric space is defined and a result about its natural filtration is given : last paragraph is devoted to the stochastic calculus for semimartingales on Lie groups [10]

II.1 - Symmetric Spaces. A d– dimensional symmetric space M is a C^∞ d– dimensional Riemannian manifold such that, for each p belonging to M, the geodesic symmetry s_p with respect to p is an isometry and p is an isolated fixed point of s_p. The following properties [4] will be useful in the sequel. Let I (M) be the isometric transformation group of M and G be a subgroup of I (M) connected component containing identity e ; it is well known that the Lie group G acts on M on the left transitively. Take a point 0 of M and let K be the isotropy subgroup of G at 0 . K is a compact subgroup of G ; furthermore G/K and M are diffeomorphic. Let $\mathfrak{G}$ and $\mathfrak{K}$ denote recpectively Lie algebras of G and K. It can be proved that there exists a d – dimensional subspace $\mathfrak{M}$ of $\mathfrak{G}$ such that $\mathfrak{G} = \mathfrak{M} \oplus \mathfrak{K}$; and if we define mapping i : G → M by i (g) = g. 0, its differential at point e, i_*, is an isomorphism between $\mathfrak{M}$ and T_o (M), the tangent space at point 0. As in SHIGEKAWA's paper, for a better understanding, we use indices I, J, K... for vectors $x = (x_I \; ; \; I = 1 \ldots n)$ in algebra $\mathfrak{G}$, indices i, j, ... for vectors $x = (x_i \; ; \; i = 1 \ldots d)$ in $\mathfrak{M}$, indices α, β, ... for vectors $x = (x_\alpha \; ; \; \alpha = 1 \ldots p = n - d)$ in $\mathfrak{K}$.

Now take a point $(0, u_o)$ in 0 (M), the bundle of orthonormal frames. We extend mapping i to 0 (M) as follows :

$$(2.1) \quad i(g) = g.\, u_o = \{(g_*)_e \, u_o^i \; ; \; i = 1 \ldots d\}$$

Where the u_o^i are the basic vectors of frame u_o. To any vector C of $\mathfrak{G}$ is associated a vector field C* on M :

$$(2.2) \quad C^* f(x) = \frac{d}{dt} f(\text{expt}\, C \,.\, x)\Big]_{t=o} \quad , \; x \in M$$

Then we define an isomorphism $L : \mathfrak{M} \to T_o(M)$ by $L(C) = C^*(o)$ and another one from $\mathbb{R}^d$ to $\mathfrak{M}$:

$$(2.3) \quad \xi \to L^{-1}(u_o(\xi)) = A(\xi)$$

Hence from a standard basis (e_i) of $\mathbb{R}^d$ a basis of $\mathfrak{M}$ can be constructed $(A_i = A(e_i) \; ; \; i = 1 \ldots d)$. Take any basis $(A_{d+1} \ldots A_{d+p})$ of $\mathfrak{K}$, and $(A_1 \ldots A_n)$ forms a basis of $\mathfrak{G}$.

Let us recall that adjoint representation Ad g denotes the differential at e of the automorphism of G : $h \longrightarrow g h g^{-1}$; mapping Ad is an homomorphism of group from G to GL ($\mathfrak{g}$), the linear group of $\mathfrak{g}$; $\mathcal{M}$ and $\mathcal{K}$ are invariant under Ad (K) ; moreover, $\mathrm{Ad}\, k/_{\mathcal{M}}$ belongs to O (d) for any k of K. For all C in $\mathfrak{g}$ with decomposition $C_{\mathcal{K}} + C_{\mathcal{M}}$, the decomposition of Ad k (C) is in the same way $\mathrm{Ad}\, k (C_{\mathcal{K}}) + \mathrm{Ad}\, k (C_{\mathcal{M}})$.

We are concerned in the sequel only with left invariant vector fields over $\mathfrak{g}$ (deduced from vectors of $\mathfrak{g}$) and we generally omit the point of G where they are taken.

II.2 - Brownian motion on Lie groups and symmetric spaces. Let B be a standard d– dimensional Brownian motion, defined on its canonical space $(\Omega^B, \underline{\underline{G}}^B, \mathbb{P}^B)$. There are two standard ways to construct a Brownian motion Y taking values in a symmetric space M : SHIGEKAWA's construction [10] and IKEDA - WATANABE's one [5]. In [5], process Y is defined as the canonical projection from bundle of orthonormal frames O (M) to M.

(2.4) $Y_t = \tau (U_t)$

Process U = (Y, u) is solution of the stochastic differential equation :

(2.5) $dU_t = B(e_i)(U_t) \circ dB_t^i \quad ; \; U_0 = (0, u_0)$

Where $B(e_i)$ are the canonical horizontal vector fields of O (M), associated to the Riemannian connexion of M [4]. It is easy to prove that the natural filtrations of Y and U coincides.

(2.6) $\underline{\underline{G}}^Y = \underline{\underline{G}}^U$

SHIGEKAWA defines Brownian motion Y as the image by mapping i of a "Brownian motion" h taking values in G, which is easier to deal with because h is a diffusion on G and stochastic calculus on Lie group is well developped ([1] [10]) :

(2.7) $Y_t = i(h_t) = h_t \cdot 0$.

Process h is solution of the stochastic differential equation :

(2.8) $dh_t = A_j(h_t) \circ dB_t^j \quad ; \; h_0 = e,$

Where the A_j's are vector fields defined in part II.1. It can easily be proved that the two definitions coincide and moreover that $h_t \cdot u_0 = u(t)$ in $O(M)$.

Lemma 2.1 The natural filtrations of Y, h, B, U coincide.
Proof : Because of (2.6), (2.7), (2.8), we have :

(2.9) $\underline{\underline{G}}^Y = \underline{\underline{G}}^U \subset \underline{\underline{G}}^h \subset \underline{\underline{G}}^B$

In [11] SHIGEKAWA expresses B with respect to U, by means of multivariate 1 - form ω defined as :

(2.10) $\omega_u(\xi) = u^{-1}(\tau_*)_u(\xi)$; $u \in O(M)$; $\xi \in T_u M$;

Then B satisfies :

(2.11) $B_t = \int_0^t \omega_{U_s} \circ dU_s$

This implies that B is $\underline{\underline{G}}^U$– adapted and concludes the proof.

II.3 - Stochastic Calculus on Lie Groups. To enlighten the next part we give some stochastic rules for semimartingales on Lie groups [10]. Let k and l be solution of the following stochastic differential equations on G :

(2.12) $dk_t = \xi_{i,t} \circ dB_t^i + \xi_{0,t}\, dt$; $k_0 = k$

(2.13) $dl_t = \eta_{i,t} \circ dB_t^i + \eta_{0,t}\, dt$; $l_0 = l$

where $(\xi_i, \eta_i$; $i = 1 \ldots d)$ are $\mathfrak{G}$ - valued continuous semimartingales and ξ_0, η_0 are $\mathfrak{G}$ - valued locally integrable processes. From [10] we have :

Proposition 2.2 Let us denote m_t the product $k_t \cdot l_t$; then process m satisfies the following stochastic differential equation on G :

(2.14) $dm_t = (\mathrm{Ad}\, l_t^{-1}(\xi_{i,t}) + \eta_{i,t}) \circ dB_t^i + (\mathrm{Ad}\, l_t^{-1}(\xi_{0,t}) + \eta_{0,t})\, dt$;

$m_0 = kl$

Let $(\mathrm{Ad}\, l_t^{-1})_J^I$ denote the elements of matrix $\mathrm{Ad}\, l_t^{-1}$:

(2.15) $d(\mathrm{Ad}\, l_t^{-1})_J^I = C_{KJ}^I (\mathrm{Ad}\, l_t^{-1})_L^K (\eta_{i,t}^L \circ dB_t^i + \eta_{0,t}^L\, dt)$;

$(\mathrm{Ad}\, l_0^{-1})_J^I = (\mathrm{Ad}\, l^{-1})_J^I$,

Where C_{KJ}^I's are the structure constants of Lie group G :

(2.16) $[A_K, A_L] = C_{KL}^I A_I$.

Moreover, process l^{-1} satisfies :

(2.17) $d\, l_t^{-1} = -\, Ad\, l_t\, (\xi_{i,t}) \circ d B_t^i - Ad\, l_t\, (\xi_{o,t})\, dt \; ; \; l_o^{-1} = l^{-1}$

III - CONSTRUCTION OF THE MODEL. FILTERING EQUATIONS.

1. In this part we first construct the filtering model by the reference probability method ([2], [12], [13]) and we use some SHIGEKAWA's results. Let signal X be a Markov process taking values in a Lusin space E with generator $(A, \underline{\underline{D}}(A))$ and initial law μ. We consider X on probability space $(\Omega^X, \underline{\underline{G}}^X, \mathbb{P}_\mu^X)$. Let observation Y be a Brownian motion as defined in (2.4) or (2.7) by means of a standard d_ dimensional Brownian motion B, independent of X, on probability space $(\Omega^B, \underline{\underline{G}}^B, \mathbb{P}^B)$. Recall that $\underline{\underline{G}}^B = \underline{\underline{G}}^Y$ (lemma 2.1).

Let us define the probability space product :

$$\Omega = (\Omega^X \times \Omega^B) \; ; \; \underline{\underline{G}} = \underline{\underline{G}}^X \otimes \underline{\underline{G}}^Y \; ; \; \mathbb{P}_\mu = \mathbb{P}_\mu^X \otimes \mathbb{P}^B .$$

Processes X, B, Y can be extended trivially to $(\Omega, \underline{\underline{G}}, \mathbb{P}_\mu)$. Let us notice that B and Y are still $\underline{\underline{G}}$_ Brownian motions because of the $\mathbb{P}_\mu$-independence of $\underline{\underline{G}}^X$ and $\underline{\underline{G}}^Y$. The probability space modelizing the system signal - observation (1.1) - (1.2) will be constructed by a change of equivalent probability measures.

Let a transformation g_t be given on G with the following assumptions, stronger than SHIGEKAWA's ones, because of application to filtering :

(3.1) $dg_t = \zeta(t, X_s \; ; \; s \leqslant t)(g_t)\, dt \, , \; t \in [0,T] \; ; \; g_o = e,$

Where ζ is a $\underline{\underline{G}}^X$ continuous semimartingale (hence $\underline{\underline{G}}$- semimartingale), uniformly bounded on $[0,T] \times \Omega^X$, taking values in the space of left invariant vector fields.

We have the following proposition, using SHIGEKAWA's results :
Proposition 3.1 with assumption (3.1), a process L can be defined :

(3.2) $L_t = \exp \int_0^t (\varphi_s^i \, d B_s^i - \frac{1}{2} \| \varphi_s \|^2 ds)$,

Where φ_s is the projection on $\mathcal{M}$ of $Adh_s^{-1} \, . \, Adg_s \, . \, \zeta_s$. Moreover, process L is a strictly positive uniformly integrable $(\Omega, \underline{\underline{G}}, \mathbb{P}_u)$ - Martingale which is $\underline{\underline{G}}^Y$ - locally square integrable.
Therefore, let us define $Q_\mu = L_T \, . \, \mathbb{P}_\mu$; Q_μ and $\mathbb{P}_\mu$ are equivalent probability measures. Furthermore, there exists an $(\Omega, \underline{\underline{G}}, Q_\mu)$ -

Brownian motion W on M such that :

(3.3) $Y_t = g_t(X) \cdot W_t$

hint of the proof : Because of independence of X and Y, process X can be considered as a parameter and thus we use the whole SHIGEKAWA's proof working on $(\Omega^Y, \underline{\underline{G}}^Y, \mathbb{P}^Y)$. The idea of his proof is to find a decomposition of h as :

(3.4) $h_t = g_t \cdot h'_t \cdot p_t$

where p is a semimartingale on K ; h' is a "Brownian motion" constructed on G as (2.8), with Z a d-dimensional Q_μ- Brownian motion :

(3.5) $dh'_t = A_i(h'_t) \circ dZ^i_t$

Then we get $W_t = h'_t \cdot 0 = h'_t\, p_t \cdot 0$ and $Y_t = g_t \cdot W_t$. Process p, solution of $dp_t = -(\mathrm{Ad}\, h_t^{-1}\, \mathrm{Ad}\, g_t\, \zeta_t)_{\mathcal{K}}\, dt$, is convenient. We can prove as [10] that :

(3.6) $dZ^i_t = (\mathrm{Ad}\, p_t)^i_j\, (dB^j_t - \varphi^j_t\, dt)$.

Matrix Ad p is in O(d) ; so the laws of B and Z are equivalent, as a consequence of fine results in [10] related to the particular stochastic differential equation satisfied by processus Ad p, Ad h^{-1}, Ad g.

Then, the only thing to prove is that L is $\underline{\underline{G}}^Y$ - locally square integrable : Ad g is bounded on [0,T] because of (3.1) ; Ad h^{-1} is a continuous $\underline{\underline{G}}^Y$ - adapted process ; hence φ is $\underline{\underline{G}}^Y$ - locally bounded. So L is well defined and satisfies the conclusions of Proposition 3.1.

2. Now, we devine filtering equations with respect to observation process. This implies Stratonovich calculus and the extension of some projections theorems to Stratonovich integrals. But before, we define the filter π and the unnormalized filter $\tilde{\pi}$: filter π of X given Y is defined as the unique measure valued process such that, for all bounded function f on E, $\pi_t(f)$ is the Q_u- optional projection with respect to filtration $\underline{\underline{G}}^Y$ of process $f(X_t)$. Briefly speaking, $\pi_t(f)$ is a "smooth version" of the conditional mean $E_{Q_\mu}(f(X_t) / \underline{\underline{G}}^Y_t)$. The reference probability method allows simple computations, thanks to the existence of probability $\mathbb{P}_\mu$ equivalent to Q_μ such that, under $\mathbb{P}_\mu$, $\underline{\underline{G}}_t$ and $\underline{\underline{G}}^Y_\infty$ are independent conditionally on $\underline{\underline{G}}^Y_t$, i.e :

(3.7) if $Z_t \in \underline{\underline{G}}_t$, $E_{\mathbb{P}_\mu}(Z_t / \underline{\underline{G}}^Y_t) = E_{\mathbb{P}_\mu}(Z_t / \underline{\underline{G}}^Y_\infty)$.

Here, this property is implied by the $\mathbb{P}_\mu$ - independence of X and Y.

This allows us to express a complex operator of optional projection by means of simpler operators of conditional means :

$$(3.8)\quad E_{Q_\mu}(Z_t / \underline{\underline{G}}_t^Y) = \frac{E_{IP_\mu}(L_t Z_t / \underline{\underline{G}}_\infty^Y)}{E_{IP_\mu}(L_t / \underline{\underline{G}}_\infty^Y)}$$

This induces the definition of a new measure valued process : the unnormalized filter $\tilde{\pi}_t$, such that, for all bounded function f on E , $\tilde{\pi}_t$ (f) is the conditional mean of process Lf (X) with respect to $\underline{\underline{G}}_\infty^Y$:

$$(3.9)\quad \pi_t(f) = \tilde{\pi}_t(f) / \tilde{\pi}_t(1) .$$

We get classically the filtering equations by projecting with respect to $\underline{\underline{G}}_t^Y$ the product $L_t\, f(X_t)$ and its expression from Itô formula ; this expression contains multivariate Brownian motion B ; although B is observable, we prefer to compute only with respect to observation Y. So we need some Stratonovitch calculus. Let us define the integral of 1 - form ω_C (associated to vector field C on M) along the path of Brownian motion Y (Cf [11]).

$$(3.10)\quad \int_0^t \omega_{C_s} \circ dY_s = \int_0^t c_s^i \circ dy_s^i = \int_0^t c_s^i\, u_j^i(s) \circ dB_s^j$$

where c^i and y^i are the local coordinates of C and Y ; and the y^i's are satisfying equation (2.5) expressed in local coordinates. We define now the following processes Ψ, φ , $\mathcal{X}$, such that φ and ψ take values in $\mathfrak{G}$ and χ is a real process :

$$(3.11)\quad \psi_t = Ad\, g_t \cdot \zeta_t$$

$$(3.12)\quad \varphi_t = Ad\, h_t^{-1} \cdot Ad\, g_t \cdot \zeta_t = Ad\, h_t^{-1} \cdot \psi_t$$

$$(3.13)\quad \chi_t = C_{Ki}^i\, (Ad\, h_t^{-1})_I^K\, (Ad\, g_t \cdot \zeta_t)^I = C_{Ki}^i\, \varphi_t^K$$

Then we get filtering equations depending only on the observation.

Proposition 3.2 Let us consider the model constructed in part III.1 ; for all f in $\underline{\underline{D}}$ (A), we have :

$$(3.14)\quad \tilde{\pi}_t(f) = p(f) + \int_0^t \tilde{\pi}_s(Af)ds + \int_0^t \omega_{\tilde{\pi}_s(f\Psi^*)} \circ dY_s - \frac{1}{2}\int_0^t \tilde{\pi}_s(f(\mathcal{X}_s + \|\Psi_{Y_s}^*\|^2))ds$$

$$(3.15)\quad \pi_t(f) = \mu(f) + \int_0^t \pi_t(Af)ds + \int_0^t \omega_{\pi_s}(\bar{f}\Psi^*) \circ dY_s - \frac{1}{2}\int_0^t \pi_s(\bar{f}(\mathcal{X}_s + \|\Psi_{Y_s}^*\|^2))ds$$

where $\overline{f}_t = f(X_t) - \pi_t\ (f)$, $\mathcal{X}$ and Ψ are defined above and vector field Ψ^* in (2.2)

Proof : The first step of the proof is to express process L with respect to Y. Because $h_t u_o = u(t)$, as in [10] we get :

$$(3.16) \quad \varphi^i_t = u^j_i(t)\,(\Psi^*(Y_t))^j$$

Hence using the definition of Stratonovitch integral and definition 3.10, we get :

$$(3.17) \quad \int_o^t \varphi^i \,.\, dB^i = \int_o^t \varphi^i \circ dB^i - \frac{1}{2} <\varphi^i, B^i>_t = \int_o^t \omega_{\Psi^*_s} \circ dY_s - \frac{1}{2} <\varphi^i, B^i>_t$$

From (2.15), (3.12) and the independence of X and B, it follows :

$$(3.18) \quad d <\varphi^i, B^i>_t = C^i_{Ki}\ (Ad\, h_t^{-1})^K_l\ \Psi^l_t\ dt = \mathcal{X}_t\, dt$$

At last, from (3.16) and because u(t) is an orthonormal frame :

$$(3.19) \quad \|(\varphi_t)_{\mathcal{M}}\|^2 = \|\Psi^*(Y_t)\|^2$$

and with (3.17) - (3.19), process L can be written :

$$(3.20) \quad L_t = \exp \int_o^t \left(\omega_{\Psi^*_s} \circ dY_s - \frac{1}{2}\,(\mathcal{X}_s + \|\Psi^*(Y_s)\|^2)\, ds\right)$$

We now compute product $L_t\ f(X_t)$ with the formula of integration by parts :

$$(3.21) \quad L_t\, f(X_t) = f(X_o) + \int_o^t L_s\, Af(X_s)\, ds + \int_o^t L_s \circ dM_s f + \int_o^t \omega_{L\, f \Psi^*_s} \circ dY_s$$

$$- \frac{1}{2} \int_o^t L_s\, f(\mathcal{X}_s)\,(\mathcal{X}_s + \|\Psi^*(Y_s)\|^2)\, ds$$

where Mf is the martingale part of $\underline{\underline{G}}^X$- semimartingale f(X). Let us notice that the bracket of processes L and M(f) vanishes ; then process $\int L \circ dMf$ is a $\underline{\underline{G}}^X$ - martingale the projection with respect to $(\underline{\underline{G}}^Y, \mathbb{P}^\mu)$ of which classically vanishes. The other terms, except the integral of 1 - form along the path of Y, can be easily projected.

Lemma 3.3 Let ω_C be the 1 - form associated to vector field C as defined in (3.10), the local coordinates of which are continuous $\underline{\underline{G}}$ - semimartingales, $\underline{\underline{G}}^Y$ - locally square integrable. Then :

$$(3.22) \quad E_{\mathbb{P}_\mu}\left(\int_o^t \omega_{C_s} \circ dY_s \,/\, \underline{\underline{G}}^Y_\infty\right) = \int_o^t \omega_{\tilde{C}_s} \circ dY_s$$

where $\tilde{C}_s$ is the vector field on M defined by $E_{IP_\mu}(C_s / \underline{\underline{G}}^Y_\infty)$

Proof : A sequence of $\underline{\underline{G}}^Y$ - stopping times T_n is defined recursively as follows : T_n is the last exit time of chart (U^{n-1}, φ^{n-1}), neighbourhood of $Y_{T_{n-1}}$ in M. So the semimartingale to be projected can be written with $c^{n,i}$ and $y^{n,i}$, local coordinates of C and Y in (U^n, φ^n).

(3.23) $$N_t = \int_0^t \omega_{C_s} \circ dY_s = \sum_{n,i} \int_{T_n \wedge t}^{T_{n+1} \wedge t} c_s^{n,i} \circ dy_s^{n,i}$$

Thus, we shall work only on real Stratonovitch integrals, which can be written in the Itô form :

(3.24) $$N_t = \int_0^t c_s \circ dy_s = \int_0^t c_s \cdot dy_s + \frac{1}{2} <c , y>_t$$

(indices n and i are omitted, and E denotes E_{IP_μ} , for simplicity).

Because local coordinates c are $\underline{\underline{G}}^Y$ - locally square integrable, the Itô integral can be projected as follows :

(3.25) $$E \left(\int_0^t c_s \cdot dy_s / \underline{\underline{G}}^Y_\infty \right) = \int_0^t E (c_s / \underline{\underline{G}}^Y_\infty) \cdot dy_s$$

Let Z be a $\underline{\underline{G}}^Y$ (hence a $\underline{\underline{G}}$) bounded martingale. Then, computing the mean of product $Z_t \cdot <c , y>_t$ with results of [6], using $\underline{\underline{G}}$ - martingale part of c_t and $E(c_t / \underline{\underline{G}}^Y_\infty)$, it follows :

(3.26) $$E \left(Z_t \cdot <c, y>_t\right) = E \left(Z_t <E (c / \underline{\underline{G}}^Y_\infty) , y>_t\right)$$

and this achieves the proof of lemma 3.3 .

As L is $\underline{\underline{G}}^Y$ - locally square integrable, this lemma can be applied to vector field $C_t = L_t f(X_t) \Psi^*$ and we get the unnormalized filtering equation (3.14).

At last, equation (3.15) is obtained by computing ratio $\pi_t (f) = \tilde{\pi}_t (f) / \tilde{\pi}_t (1)$ as a Stratonovitch integral :

(3.27) $$u_t / v_t = u_0 / v_0 + \int_0^t \frac{1}{v_s} \circ du_s - \int_0^t \frac{u_s}{(v_s)^2} \circ dv_s ,$$

and this concludes the proof of Proposition 3.2 .

Remark : We get ordinary filtering equations if process Ψ is a bounded function of X_t taking values in $\mathfrak{g}$.

(3.28) $$\Psi_t = Ad\, g_t \cdot \zeta_t = \Psi (X_t)$$

This situation is quite frequent. Indeed, let us consider the case of a matrix group G ; then $\mathfrak{g}$ is also a set of matrices and in this case :

$$(3.29) \quad Ad\, g_t \cdot \zeta_t = g_t \cdot \zeta_t \cdot g_t^{-1}$$

Then process g is the solution of the following matrix equation :

$$(3.30) \quad \frac{dg_t}{dt} = \Psi(X_t) \cdot g_t \ , \ g_o = I$$

This solution is classically called [3] the "matrizing" of Ψ and is expressed by :

$$(3.31) \quad g_t = \Omega_o^t(\Psi(X.)) = I + \int_o^t \Psi(X_s)ds + \int_o^t \Psi(X_{s_1}) \int_o^{s_1} \Psi(X_{s_2})ds_2\, ds_1 + \ldots + \int_o^t \Psi(X_{s_1}) \int_o^{s_1} \Psi(X_{s_2}) \int_o^{s_2} \ldots \int_o^{s_{n-1}} \Psi(X_{s_n})ds_n \ldots ds_1 + \ldots$$

Hence process ζ is given :

$$(3.32) \quad \zeta_t = [\Omega_o^t(\Psi(X.))]^{-1}\, \Psi(X_t)\, \Omega_o^t(\Psi(X.))$$

and satisfies hypothesis (1.2).

IV - EXAMPLE OF THE SPHERE

An other application of our results (the multivariate case was described in part I) is the case of the sphere S_2 considered as a 2-dimensional symmetric space. Sphere S_2 is, embedded in R^3, defined as the set $\{x = (x_1, x_2, x_3) / \sum_i x_i^2 = 1\}$. As a manifold, S_2 is endowed with an open covering of two charts U and U' defined by spherical coordinates around polar axis, (either third or second axis in $\mathbb{R}^3$). Fix a point $0 = (1,0,0)$ in S_2 and an orthonormal frame u_o in $O(S_2)$ at point 0, defined by tangent vectors (e_1, e_2) : $e_1 = (0,1,0)$ and $e_2 = (0,0,1)$. As a symmetric space, S_2 is endowed, for any p in S_2, with the involutive symmetry s_p which associates to any x its symmetric point on the great circle containing p and x. Let us consider Lie group G = SO (3), the isometric transformation group of S_2 ; G naturally acts on S_2 : let be given g in SO (3) and $x = (x_1, x_2, x_3)$ in S_2, the action g . x is defined by the product of matrices g (3,3) and x (3,1). The isotropy subgroup of G at 0, K, is the group of rotations around first axis. The Lie algebra $\mathfrak{SO}(3)$ of SO (3) is the set of (3,3) skew matrices ; the sub - Lie algebra $\mathcal{K}$ of K is generated by the following matrix :

$$(4.1) \quad A_3 = \begin{pmatrix} 0 & 0 & 0 \\ 0 & 0 & -1 \\ 0 & 1 & 0 \end{pmatrix}$$

Let A_1 and A_2 be the following matrices of $\mathcal{SO}(3)$:

$$(4.2)\quad A_1 = \begin{pmatrix} 0 & -1 & 0 \\ 1 & 0 & 0 \\ 0 & 0 & 0 \end{pmatrix} \qquad A_2 = \begin{pmatrix} 0 & 0 & -1 \\ 0 & 0 & 0 \\ 1 & 0 & 0 \end{pmatrix} \;;$$

it is easy to show that (A_1, A_2) is a basis of $\mathcal{M}$, vector space defined in part II.1. The structure constants are now simple to compute :

$$(4.3)\quad C^{I}_{KI} = 0 \;;\; I = 1, 2, 3 \quad \text{and} \quad C^{2}_{13} = -C^{3}_{12} = -C^{1}_{23} = 1$$

The filtering model is now constructed in the spherical case : let $g_t(X)$ be defined as $\Omega^{\cdot t}_{0}(\Psi(X.))$ (3.31) where $\Psi(X_t)$ is a matrix in $\mathcal{SO}(3)$:

$$(4.4)\quad \Psi(X_t) = \Psi^{I}(X_t)\, A_I$$

and $\Psi^I(X.)$ are bounded continuous real semimartingales. We define on $(\Omega, \underline{\underline{G}}, \mathbb{P}_\mu)$ a standard 2-dimensional Brownian motion B and a Brownian motion Y taking values in S_2 :

$$(4.5)\quad Y_t = h_t \,.\, 0 \;;\; dh_t = h_t\, A_1 \circ dB^1_t + h_t\, A_2 \circ dB^2_t \;,\; h_o = e \,.$$

To apply proposition (3.2) we need vector field $(\Psi(X_t))^*$; because map $C \to C^*$ is a Lie algebra homomorphism [7], we get :

$$(4.6)\quad (\Psi(x))^*(y) = \Psi^I(x)\, A^*_I(y)$$

and we have to compute $A^*(y)$. Coming back to definition (2.2) we easily compute local coordinates of A^*_I, for instance in chart U :

$$(4.7)\quad A^*_1(\theta,\varphi) = \frac{\partial}{\partial\theta} \;;\; A^*_2(\theta,\varphi) = \operatorname{tg}\varphi \sin\theta \frac{\partial}{\partial\theta} + \cos\theta \frac{\partial}{\partial\theta}$$

$$A^*_3(\theta,\varphi) = -\operatorname{tg}\varphi \cos\theta \frac{\partial}{\partial\theta} + \sin\theta \frac{\partial}{\partial\varphi}$$

Then it only remains to apply proposition 3.2 to this spherical filtering model. In this case, for all f in $\underline{\underline{D}}(A)$ we get :

$$(4.8)\quad \pi_t(f) = \mu(f) + \int_0^t \pi_s(Af)ds + \int_0^t \pi_s(\overline{f}\,\Psi^I)\omega_{A^*_I} \circ dY_s - \frac{1}{2}\int_0^t \pi_s(\overline{f}\, \|\Psi^*_{Y_s}\|^2)ds$$

CONCLUSION - We studied a non standard filtering model. Observation process Y takes values in a symmetric space M. This particular assumption allows to consider process Y under a multiplicative form and then generalizes the previous cases [8] [9] More precisely, process Y depends on signal X by means of a stochastic isometric transformation : $Y_t = g_t(X) \cdot W_t$ where W_t is a Brownian motion taking values in M. We get an intrinsical filtering equation under Stratonovitch form.

REFERENCES

[1] **J.M. BISMUT** : "Mécanique Aléatoire", L.N. in Mathematics N° 866 SPRINGER VERLAG (1981)

[2] **T.E. DUNCAN** : "Stochastic Filtering in Manifolds" Proc. I F A C World Congress. PERGAMON (1981)

[3] **F.R. GANTMACHER** : "Théorie des Matrices" Tome 2. DUNOD Paris (1966)

[4] **S. HELGASON** : "Differential Geometry and Symmetric Spaces" ACADEMIC PRESS. New York (1962)

[5] **N. IKEDA - S. WATANABE** : "Stochastic Differential Equations and Diffusion Processes". NORTH - HOLLAND. Amsterdam (1981)

[6] **J. JACOD** : "Calcul Stochastique et problème de martingales" L.N. in Math N° 715. SPRINGER VERLAG (1979)

[7] **S. KOBAYASHI - K. NOMIZU** : "Fundations of Differential Geometry" I-II. New York. INTERSCIENCE (1963-1969)

[8] **S.K. NG-P.E. CAINES** : "Non Linear Filtering in Riemannian Manifolds" Preprint

[9] **M. PONTIER - J. SZPIRGLAS** : "Filtrage Non Linéaire avec observation sur une variété". Stochastics (1985) to appear

[10] **I. SHIGEKAWA** : "Transformations of the Brownian Motion on a Riemannian Symmetric Space". Zeit Für Wahrsch. (65) p. 493 - 522 (1984)

[11] **I. SHIGEKAWA** : "On Stochastic Horizontal Lifts". Zeit für Wahrsch. 59 (2) p. 211 - 222 (1982)

[12] **J. SZPIRGLAS - G. MAZZIOTTO** : "Modèle Général de Filtrage Non Linéaire et Equations Différentielles Stochastiques Associées" Ann. I.H.P. XV N° 2 p. 147-173 (1979)

[13] **M. ZAKAI** : "On the Optimal Filtering of Diffusion Processes" Zeit Für Wahrsch. 11 p. 230 - 249 (1969)

TO THE THEORY OF THE GENERALIZED DIFFUSION

N.I.Portenko

Institute of Math.Academy of Science
of the Ukrainian SSR
Kiev, USSR

1. The aim of the present report is to construct a solution of the stocahstic differential equation

$$dx(t) = a(x(t))dt + dw(t), \tag{1}$$

where $w(t)$ is the one-dimensional Wiener process, $a(x)$ is a generalized function of the type of a derivative of an arbitrary finite measure on $\mathcal{R}^1$ with atoms whose weights are less than unit.In other words it will be proved that there exist the (Markov)solution of the equation (1) for every function $a(x)$ such that there exist a finite measure $A(x)$ on the Borel subsets of $\mathcal{R}^1$, possessing the properties:

a) $A(\{x\}) < 1$ for each $x \in \mathcal{R}^1$ (denote by $\{x\}$ a one-point set);

b) $$\int_{\mathcal{R}^1} a(x)\varphi(x)\,dx = \int_{\mathcal{R}^1} \varphi(x)\,A(dx),$$

whatever continuous finite function $\varphi(x)$, $x \in \mathcal{R}^1$ is.Further we denote by $C_0(\mathcal{R}^1)$ the family of all continious finite functions on $\mathcal{R}^1$.

Previously a solution of equation (1) turned out well to be constructed for $a \in L_1(\mathcal{R}^1)$ or in the case when and only when the number of atoms of $\mathcal{R}^1$ was finite ([3] , [4]).

One will see later that the condition a) is a essential one in this case.The method of proof is based on passage to the limit in equations of the type (1), so we obtain some limit theorems in the process of proof.

2. Denote by $\mathfrak{P}$ the space of all continuous functions defined on $[0, \infty)$ with the values in $\mathcal{R}^1$ and by $\mathcal{M}_t^s$ - σ-algebra of subsets of $\mathfrak{P}$ generated by the class of sets of the form $\{x(\cdot) \in \mathfrak{P}: (x(\tau) \in \Gamma\}$ at $\tau \in [s, t]$ and the Borel sets $\Gamma \in \mathcal{R}^1$. Let a sequence of bounded continuous functions $(a_n(x))_{n \geq 1}$ is given on $\mathcal{R}^1$. Then there exists a unique measure $P_x^{(n)}$ on σ-algebra $\mathcal{M}_\infty^0 \ (= \bigvee_t \mathcal{M}_t^0)$ for every $n=1,2,\ldots$ and $x \in \mathcal{R}^1$ so that $P_x^{(n)}\{x(0)=x\}=1$ and the process $x(t)-$ $-x(0)-\int_0^t a_n(x(\tau))d\tau$ is a square integrable martingal with characteristic t. In other words the measure $P_x^{(n)}$ corresponds to the solution of the stochastic differential equation

$$d x(t) = a_n(x(t))dt + dw(t), \quad x(0) = x. \tag{2}$$

In this connection $(x(t), \mathcal{M}_t^0, P_x^{(n)})$ is a homogeneous Markov process.

Our aim is to describe all possible limit measures of the set of measures $(P_x^{(n)})_{n \geq 1}$. With respect to local behavior of functions $a_n(x)$ the next result is apparent to be the most general condition of compactness.

<u>Lemma 1</u>. Let $(A_n(x))_{n \geq 1}$ be a sequence of functions on $\mathcal{R}^1$ such that

1) $A_n(x)$ is absolutely continuous and its derivative coinsides with $a_n(x)$;

2) $\sup_{x,n} |A_n(x)| < \infty.$

Then at every fixed $x \in R^1$ the class of measures $(P_x^{(n)})_{n \geq 1}$ is a compact one in the next sense: a subsequence $n_{k_j} \to \infty$ can be extracted from every sequence $n_k \to \infty$ that the restrictions of measures $P_x^{(n_{k_j})}$ to σ-algebras $\mathcal{M}_T^0$ for any $T < \infty$ converges weakly.

One can find proof of this lemma in [4].

The next estmate will play an important part in limit passage from equation (2) to equation (1). It is a generalization of corresponding result in [4].

Lemma 2. Let the sequence $(a_n(x))_{n \geq 1}$ be the same as the one in lemma 1. Then the inequality

$$M_x^{(n)} \int_s^t \varphi(x(\tau))d\tau \leq K(t-s)^{1/2} \|\varphi\|_1 \tag{3}$$

uniformly with respect to $x \in R^1$ and $n = 1,2,\ldots$ for nonnegative function $\varphi \in L_1(R^1)$.

Here $\|\varphi\|_1$ is the L_1-norm of function φ, K is an absolute constant, $M_x^{(n)}$ is a symbol of a mathematical expectation with respect to measure $P_x^{(n)}$.

Proof. Denote by $\Phi^{(n)}(x)$ a general solution of equation

$$\frac{1}{2}\frac{d^2\Phi^{(n)}(x)}{dx^2} + a_n(x)\frac{d\Phi^{(n)}(x)}{dx} = \varphi(x). \tag{4}$$

Since $\varphi \in L_1(R^1)$ then the equality is realized almost everywhere. $\Phi^{(n)}(x)$ can be written in the form

$$\Phi^{(n)}(x) = 2\int_{x_0}^{x} e^{-2A_n(x)} \left[\int_{x_1}^{y} e^{2A_n(x)}\varphi(z)dz\right]dy + C' f_n(x) + C'',$$

where $f_n(x)=\int_{x_0}^{x} e^{-2A_n(y)}dy$, x_0, x_1, c', c'' are real numbers. According to Ito formula at $s<t$

$$M_x^{(n)}\left[\Phi^{(n)}(x(t))-\Phi^{(n)}(x(s))\right]=M_x^{(n)}\int_s^t \varphi(x(\tau))d\tau \tag{5}$$

for every $x\in R^1$, $n=1,2,\ldots$ It is obvious that

$$\Phi^{(n)}(x)-\Phi^{(n)}(y)=$$

$$=c'(f_n(x)-f_n(y))+2\int_y^x e^{-2A_n(z)}\left[\int_{x_1}^{z} e^{2A_n(u)}\varphi(u)du\right]dz.$$

Hence

$$M_x^{(n)}\left|\Phi^{(n)}(x(t))-\Phi^{(n)}(x(s))\right|\le K'\|\varphi\|_1\sqrt{M_x^{(n)}(f_n(x(t))-f_n(x(s)))^2}.$$

Here K' is a constant depending on c' and on the value of $\sup_{n,x} e^{2A_n(x)}$ only. Since $f_n(x(t))$ is a square integrable martingale and $\int_0^t e^{-4A_n(x(s))}ds$ is its characteristic, then

$$M(f_n(x(t))-f_n(x(s)))^2\le\sup_{n,z} e^{-4A_n(z)}(t-s).$$

Substituting this inequality into previous one the inequality

$$M_x^{(n)}\left|\Phi^{(n)}(x(t))-\Phi^{(n)}(x(s))\right|\le K\|\varphi\|_1(t-s)^{1/2}$$

can be obthained. Here $K=K'\sup_{n,z} e^{-4A_n(z)}$. (3) follows from (5) and the last inequality. Thus the lemma is proved.

3. Let the sequence of function $(a_n(x))_{n\geq 1}$, satisfying the conditions of lemma 1, be given. We hold some $\bar{x}\in R^1$, $c', c'', x_0, x_1\in R^1$ fixed and choose $n_k\longrightarrow\infty$ so that:

(i) the restriction of measure $P_{\bar{x}}^{(n_k)}$ to σ-algebra $\mathcal{M}_T^0$

for any $T < \infty$ weakly converges to the restriction of some measure $P_{\bar{x}}$, defined on $\mathcal{M}^{0}_{\infty}$, to the same σ-algebras;

(ii) $\Phi^{(n_k)}(x) \longrightarrow \Phi(x)$ uniformly with respect to $x_0, x_1, c'' \in R^1$ and locally uniformly with respect to $x \in R^1$, $c' \in R^1$, $\varphi \in L_1(R^1)$;

(iii) $f_{n_k}(x) \longrightarrow f(x)$ locally uniformly with respect to $x \in R^1$;

(iv) there is a function $g \in L_\infty(R^1)$ such that a sequence of function $e^{2A_n(x)}$ converges to the function $g(x)$ $*$-weakly;

(v) there is a function $h \in L_\infty(R^1)$ such that a sequence of functions $e^{-2A_{n_k}(x)}$ converges to the function $h(x)$ $*$-weakly.

Realisation of (i) is by lemma 1 guaranted. Properties (ii) and (iii) follows from obvious inequalities

$$| \Phi^{(n)}(x) - \Phi^{(n)}(y) | \leq K (c' + \|\varphi\|_1) | x-y |, \tag{6}$$

$$| f_n(x) - f_n(y) | \leq K | x-y | \tag{7}$$

for every $x,y \in R^1$, $c' \in R^1$, $\varphi \in L_1(R^1)$, $n=1,2,\ldots$ Here K is a constant depending on $\sup_{n,x} | A_n(x) |$. Properties (iv) and (v) follow from the fact that the sequences of functions $(e^{2A_n(x)})_{n \geq 1}$ and $(e^{-2A_n(x)})_{n \geq 1}$ belong to some spheve in the space $L_1(R^1)$. Hence we can conclude that

$$\int_{R^1} e^{2A_{n_k}(z)} \varphi(z)\,dz \longrightarrow \int_{R^1} \varphi(z) g(z)\,dz,$$

$$\int_{R^1} e^{-2A_{n_k}(z)} \varphi(z)\,dz \longrightarrow \int_{R^1} \varphi(z) h(z)\,dz,$$

as $n_k \longrightarrow \infty$ for every $\varphi \in L_1(R^1)$. Moreover since $\inf_{n,z} e^{2A_n(z)} > 0$ and $\inf_{n,z} e^{-2A_n(z)} > 0$ one can choose functions $h(x)$ and $g(x)$ to be strictly positive for any $x \in R^1$. The function $f(x)$ is obvious to be absolutely continuous and its derivative exists almost everywhere and coinsides with $h(x)$ for almost all x. So one can consider $h(x) = f'(x)$. Similarly the function $\Phi(x)$ is absolutely continuous and can be written in a form:

$$\Phi(x) = \int_{x_0}^{x} \left[2 \int_{x_1}^{y} \varphi(z)\, dG(z) + C' \right] df(y) + C'' , \tag{8}$$

where $\varphi \in L_1(R^1)$, $df(y) = f'(y)dy$, $dG(z) = g(z)dz$.

It follows from this that at almost all $x \in R^1$ $\frac{1}{2} D_G D_f \Phi(x) = \varphi(x)$ where D_G and D_f are differential operators with respect to monotone functions $G(x)$ and $f(x)$ respectively. Moreover if φ is a continuous function then the last equality holds at all $x \in R^1$.

Lemma 3. $P_{\bar{x}}\{x(0) = \bar{x}\} = 1$ and at all $\varphi \in L_1(R^1), C', C'', x_0, x_1 \in R^1$ the process

$$\Phi(x(t)) - \Phi(x(0)) - \int_0^t \varphi(x(\tau))d\tau , \quad t \geq 0 \tag{9}$$

is a square integrable martingal with respect to $(\mathcal{M}_t^0, P_{\bar{x}})$ and

$$\int_0^t D_f \Phi(x(\tau)) D_G \Phi(x(\tau))d\tau = \int_0^t \frac{[\Phi'(x(\tau))]^2}{f'(x(\tau))\, g(x(\tau))} d\tau \tag{10}$$

is its characteristic.

Proof. Let ξ be a continuous bounded $\mathcal{M}_s^0$ -measurable functional and $\varphi \in C_0(R^1)$. The relation

$$M_x^{(n)} \xi \int_0^t \varphi(x(\tau))d\tau = M_x^{(n)} \left[\Phi^{(n)}(x(t)) - \Phi^{(n)}(x(s)) \right]$$

takes place at all $x \in R^1$, $s < t, \varphi \in L_1(R^1)$, $n=1,2,\dots$ and it is a corollary of Ito formula and of equation (4). Let us set $x = \bar{x}$, $\varphi \in C_0(R^1)$ and find the limit as the sequence $n_k \to \infty$ in this relation. The equality

$$M_{\bar{x}} \eta \int_s^t \varphi(x(\tau))d\tau = M_{\bar{x}} \eta \left[\Phi(x(t)) - \Phi(x(s)) \right] \tag{11}$$

will be recived. It is easy to spread it on all $\mathcal{M}_s^0$-measurable variable η and all $\varphi \in L_1(R^1)$ using lemma 2. Equality (11) means that the process (9) is a $(\mathcal{M}_t^0, P_{\bar{x}})$-martingal for all $\varphi \in L_1(R^1)$. All moments of the martingal are evidently exist. Further (11) and the equality $\frac{1}{2} D_G D_f \Phi^2(x) = D_G \Phi D_f \Phi + 2\varphi \Phi$ imply

$$M_{\bar{x}} \eta \left[\Phi^2(x(t)) - \Phi^2(x(s)) \right] =$$

$$= M_{\bar{x}} \Phi \int_s^t D_G \Phi(x(\tau)) D_f \Phi(x(\tau))d\tau + 2M_{\bar{x}} \eta \int_s^t \varphi(x(\tau)) \Phi(x(\tau)) d\tau .$$

Thus, for every $\mathcal{M}_s^0$-measurable variable η and $\varphi \in L_1(R^1)$ the equality

$$M_{\bar{x}} \eta \left[\Phi(x(t)) - \Phi(x(s)) - \int_s^t \varphi(x(\tau)) d\tau \right] =$$

$$= M_{\bar{x}} \eta \int_s^t D_G \Phi(x(\tau)) D_f \Phi(x(\tau))d\tau$$

holds. Hence, (10) is the characteristic of martingal (9). The lemma is proved.

<u>Remark</u>. The measure $P_{\bar{x}}$ is determined uniquely by conditions (9), (10). In fact, choosing $\varphi(x) \equiv 0$ one recieves that the process $f(x(t)) - f(\bar{x})$ with respect to the measure $P_{\bar{x}}$ is a square integrable martingal with the characteristic

$$\int_0^t \frac{f'(x(\tau))}{q(x(\tau))} d\tau .$$

Setting $y(t)=f(x(t))$ we define the measure $P_{\bar{x}}$ on the space of continuous functions $y(t)=f(x(t))$ with the properties: $P_{\bar{x}}\{y(0)=f(\bar{x})\}=1$ and $y(t)$ is a $(\mathcal{M}^0_t, P_{\bar{x}})$-square integrable martingal with characteristic

$$\int_0^t \frac{f'(f^{-1}y(\tau))}{g(f^{-1}y(\tau))}d\tau.$$

According to the uniqueness theorem from [1] this measure is unique.

Now we can prove that for every $x\in R^1$ the measures $P_x^{(n_k)}$ converge in the sence pointed out above to the measure P_x characterized in the same way as the measure P_x (see(9),(10)). Really, the family of measures $(P_x^{(n_k)})$ (for fixed x) can not have more than one limit measure,and the compactness of this family implies our statement.We obtained the theorem.

Theorem 1. There exists a family of probability measures $(P_x)_{x\in R^1}$ on the σ-algebra $\mathcal{M}^0_\infty$ such that for every $x\in R^1$ the measure P_x is uniquely determined by the properties: $P_x\{x(0)=x\}=1$ and the process (9) is a $(\mathcal{M}^0_t, P_x)$-square integrable martingale with the characteristic (10).The process $(x(t), \mathcal{M}^0_t, P_x)_{x\in R^1}$ is a Markov one.

4. Theorem 1 characterized the family of measures $(P_x)_{x\in R^1}$ which are the limits for the family of measures $(P_x^{(n)})_{x\in R^1}$ with respect to a subsequance $n_k\longrightarrow\infty$. Putting $\varphi(x)\equiv 0$ we obtain a characterisation of the family of measures $(P_x)_{x\in R^1}$ given in [3]. To characterise the process $x(t)$ itself one has to take $\varphi(x)$, x_0, x_1, c', c'', satisfying $\Phi(x)\equiv x$, i.e. $\varphi(x)=\frac{1}{2}D_G D_f x$. Even envolving the theory of generalized functions,

we don't always succeed in putting the proper meaning of the expression $D_G D_f x$. In fact, since almost everywhere $D_f x = (f'(x))^{-1}$ the function $g(x)$ has the form

$$\varphi(x) = \frac{1}{2}\cdot\frac{1}{g(x)}\cdot\frac{d}{dx}\left(\frac{1}{f'(x)}\right). \qquad (12)$$

Even in the case if $\frac{d}{dx}\left(\frac{1}{f'(x)}\right)$ is a generalized function,the last expression can have no meaning because the last function must be multiplyed by the function $\frac{1}{g(x)}$ which is nonregular in general.Let us consider some particular cases when one can put the proper meaning for expression (12).

A) Let $f'(x)$ be absolutely continuous.Setting for almost all $x \in R^1$

$$a(x) = \frac{1}{2} D_G\left(\frac{1}{f'(x)}\right) = -\frac{1}{2}\cdot\frac{f''(x)}{g(x)(f'(x))^2},$$

one reaeives that the process $x(t) - x(0) - \int_0^t a(x(s))ds$ is a $\mathcal{M}_t^0$-square integrable martingale with respect to measure P_x and $\int_0^t \frac{d\tau}{f'(x(\tau))\, g(x(\tau))}$ is its characteristic.In other words the measure P_x corresponds to the solution of the stochastic differential equation $dx(t) = -\frac{1}{2}\frac{f''(x(t))dt}{g(x(t))(f'(x(t)))^2} + \left[f'(x(t))g(x(t))\right]^{-\frac{1}{2}}dw(t)$.

If in particular $f'(x)g(x) \equiv 1$ almost everywhere then the limit equation has such form $dx(t) = -\frac{1}{2}\frac{f''(x(t))}{f'(x(t))}dt + dw(t)$.

The next example [2] shows that $f'(x)g(x) \equiv 1$ not ever.

<u>Example</u>. Let $a_n(x) = n\cos nx$. Putting $A_n(x) = \sin nx$ it is easy to find $f'(x) = g(x) = k = \sum_{j=0}^{\infty} (j!)^{-2}$. Besides $f''(x) \equiv 0$ and so $a(x) \equiv 0$. Thus the limit measure P_x corresponds to the process $x(t) = x + \frac{1}{k} w(t)$, where $w(t)$ - is the Wiener process, $w(0) = 0$. In this case the measure $P_x^{(n)}$ corresponds to the solution of the equation $x(t) = x + \int_0^t n\cos(nx(s))ds + w(t)$.

Therefore when passing to the limit as $n \to \infty$ the drift vanishes and the diffusion decreases ($K > 1$).

B) Let an arbitrally monotone non-decreasing left-continuous function $A(x)$ be given and its variation is bounded. Let the sequence $A(x)$ is such that $A_n(x) \to A(x)$ as $n \to \infty$ at all point where $A(x)$ is continuous. In this case $f'(x) = e^{-2A(x)}$; $g(x) = e^{2A(x)}$, so $f'(x)g(x) \equiv 1$. Set $\Delta_A(x) = A(x+) - A(x)$, $S_A = \{y \in R^1 : \Delta_A(y) > 0\}$, $A_d(x) = \sum_{y<x} 1_{S_A}(y)\Delta_A(y)$, $A_c(x) = A(x) - A_d(x)$ for $x \in R^1$. Here $1_\Gamma(x)$ is the indicator function of set Γ. Define the function

$$\tilde{A}(x) = A_c(x) + \sum_{y<x} 1_{S_A}(y) \,\mathrm{th}\, \Delta_A(y), \tag{13}$$

where $\mathrm{th}\, z = (e^z - e^{-z})/(e^z + e^{-z})$. Let $a(x) = \frac{d\tilde{A}(x)}{dx}$ in the sense of generalized function, i.e. $\int_{R^1} \varphi(x)a(x)dx = \int_{R^1} \varphi(x)dA_c(x) + \sum_{x \in S_A} \varphi(x) \,\mathrm{th}\, \Delta_A(x)$ for every continuous bounded function $\varphi(x)$. A generalized function $a(x)$ can be applied to function without second order discontinuities if considering $\int_{R^1} \varphi(x)dA_c(x) + \sum_{x \in S_A} \frac{1}{2}[\varphi(x) + \varphi(x-)] \,\mathrm{th}\, \Delta_A(x) = \int_{R^1} \varphi(x)a(x)dx$ in this case. Put $\Phi(x) \equiv x$ in (8). Affer differentiating with respect to x we get: the equation for $\varphi(x)$

$$e^{2A(x)} = c' + 2\int_{x_1}^{x} e^{2A(z)}\varphi(z)dz \tag{14}$$

holds for almost all $x \in R^1$. Let us show that the function $a(x)$, defined above, satisfies this equation. In other words, if we substitute the generalized function $a(z)$ instead of $\varphi(z)$ in (14), this equation will be correct for all points where the function $A(x)$ is continuous. In fact, if $x_1 = -\infty$, $c' = 1$, we have for these x

$$2\int_{-\infty}^{x} e^{2A(z)} dA_c(x) = \int_{-\infty}^{x} e^{2A_d(z)} de^{2A_c(z)} = e^{2A(x)} - 1 - \int_{-\infty}^{x} e^{2A_c(z)} de^{2A_d(z)} =$$

$$= e^{2A(x)} - 1 - \sum_{y<x} 1_{S_A}(y)\, e^{2A(y)} \left[e^{2\Delta_A(y)} - 1 \right].$$

Denoting $N(x)=\sum_{y<x} 1_{S_A}(y)\,\mathrm{th}\,\Delta_A(y)$ and as before considering x as the point where $A(x)$ is continuous, we find $2\int_{-\infty}^{x} e^{2A(z)}\,dN(z) =$
$=\sum_{y<x} 1_{S_A}(y)(\mathrm{tg}\,\Delta_A(y))\left[e^{2A(y+)}+e^{2A(y)}\right]=\sum_{y<x} 1_{S_A}(y)e^{2A(y)}\left[e^{2\Delta_A(y)}+\right.$
$\left.+1\right]\mathrm{th}\,\Delta_A(y)=\sum_{y<x} 1_{S_A}(y)\,e^{2A(x)}-1.$
Hence $2\int_{-\infty}^{x} e^{2A(z)}a(z)\,dz = e^{2A(x)}-1$ for all points of the function $A(x)$, i.e. $a(x)$ satisfies the equation(14). It means that if the generalized function $a(x)$ is substituted instead of $\varphi(x)$, we have $\Phi(x)\equiv x(C''=x_0)$. It can be easy proved that in the case under consideration the additive continuous homogeneous non-negative functional $\xi(t)=\int_0^t a(x(s))\,ds$ exists for Markov process $(x(t), \mathcal{M}_t^0, P_x)$ and it is the limit in the square mean of usual integral functionals, besides the process $x(t)-x(0)-\int_0^t a(x(s))\,ds$ is a continuous square integrable martingal with respect to every measure P_x and the flow of σ-algebras $\mathcal{M}_t^0$, whose characteristic is equal t. In other words the measure P_x defines the generalized diffusuin process $dx(t)=a(x(t))\,dt + dw(t)$, $x(0)=x$, where $a(x)$ is the generalized derivative of the function (13) constructed above. Some special cases of this result were considered in [3], [4].

1. Veretennikov A.U. On the strong solutions of stochastic differential equations.-Theory Prob.Appl., 2(1979), p.348-360. 2. Kulinich G.L. The limit theorems for homogeneous stochastic differential equations under nonregular dependence of coefficients on parameter.-Theory Prob.and Math.Stat 15(1976), p.99-113. 3. Kulinich G.L. On necessary and sufficient conditions of convergence of the solutions of one-dimensional stochastic diffusion equations under nonregular dependence of coefficients on parameter.-The theory of prob.and appl., N 4

(1982), p. 795-801. 4. Portenko N.I. Generalized diffusion process. Lecture Notes in Mathematics, 550 (1976), p.500-523.

THE LINEAR OPERATOR-VALUED STOCHASTIC EQUATIONS

A.V.Skorokhod

Institute of Mathematics of the
Academy of Sciences UkrSSR
Kiev, Repin str.,3

1. We consider the objects of two kinds, namely 1) the solutions of equation

$$dX_t = X_t \, dY_t \, , \qquad (1)$$

where Y_t is an operator-valued process with independent increments and X_t - an operator-valued process; 2) stochastic semigroups X_t^s $(0 \le s \le t)$; each of them being a family of random operators, for which a) there exists a family of σ-algebras $\mathcal{F}_t^s$ such that a1) $\mathcal{F}_t^s \subset \mathcal{F}_v^u$ if $[s,t] \subset [u,v]$; a2) $\mathcal{F}_t^s$ and $\mathcal{F}_v^u$ are independent, if $s < t \le u < v$; a3) X_t^s is $\mathcal{F}_t^s$-measurable; b) for $s < t < u$

$$X_t^s X_u^t = X_u^s \qquad (2)$$

(the semigroup property); c) $X_t^s \longrightarrow I$, if $s \to u$, $t \to u$ for all u (I is the identity operator).

Suppose the equation (1) has a unique solution. Denote by X_t^s the solution of (1) for $t > s$ and $X_s^s = I$. Then X_t^s is a stochastic semigroup. We are interested in the conditions, under which there exists a process with independent increments, Y_t, such that semigroup X_t^s satisfies equation (1). If such a process Y_t does exist, we can investigate the nonlinear object, that is a semigroup, with the help of the linear one, that is the process with independent increments.

My report is devoted to the investigation of existence and uniqueness of the solution of the equation (1) and possibility of construction of stochastic semigroups as the solution of such equations.

2. <u>Examples.</u>

I. Let ξ_t be a solution of the linear stochastic equation in R^d

$$d\xi_t = A\xi_t + \sum_{k=1}^{m} B_k \xi_t \, dw_k(t), \tag{3}$$

A, B_k are linear operators in R^d, $w_k(t)$ - independent Wiener processes in R. If $\xi_{x,s}(t)$ is a solution of (3) for which $\xi_{x,s}(s) = x$, Then $\xi_{x,s}(t) = (\mathcal{U}_t^s)^* x$, where $\mathcal{U}_t^s$ is a stochastic semigroup in $L(R^d)$,

$$d_t \mathcal{U}_t^s = \mathcal{U}_t^s A^* dt + \sum_{k=1}^{m} \mathcal{U}_t^s B_k^* \, dw_k(t) \tag{4}$$

(S^* is the operator conjugate to S). This is an equation of type (1) with $Y(t) = A^* t + \sum_{k=1}^{m} B_k(t) w_k(t)$.

II. Let ξ_t be a solution of stochastic differential equation in R^1

$$d\xi_t = a(\xi_t)dt + B(t, \xi_t) \, dw(t), \tag{5}$$

where $a(t,x): R_+ \times R^d \to R^d$; $B(t,x): R_+ \times R^d \to L(R^d)$, $w(t)$ is the Wiener process in R^d.

Denote by $\xi_{x,s}(t)$ the solution of (5), for which $\xi_{x,s}(s) = x$, and suppose that the solution of (5) exists and is unique. Then for $s < t < u$

$$\xi_{x,s}(u) = \xi_{\xi_{x,s}(t),t}(u).$$

We consider the linear space $C(R^d)$ of continuous bounded

functions $\varphi: R^d \longrightarrow R$ and random operators

$$X_t^s \varphi(x) = \varphi(\xi_{x,s}(t)).$$

Then X_t^s is a stochastic semigroup and

$$dX_t^s \varphi(x) = X_t^s A\varphi(x)dt + \sum_{k=1}^{d} X_t^s B_k \varphi(x) d(w(t), e_k), \quad (6)$$

where

$$A\varphi(x) = (\varphi'(x), a(x)) + \frac{1}{2} \operatorname{tr} B(x)\varphi''(x),$$

$$B_k \varphi(x) = (B^*(x)\varphi'(x), e_k),$$

$\{e_k\}$ is an orthonormal basis in R^d, $\varphi'(x) \in R^d$, $\varphi''(x) \in L(R^d)$, $(\cdot,\cdot)$ is the scalar product.

The equation (6) is of the form (1) with

$$Y(t) = tA + \sum_{k=1}^{d} (w(t), e_k) B_k,$$

the latter being the process with the unbounded operator values.

III. Now we consider the diffusion with a generalized random drift. Usually the equation of diffusion is of the form

$$\frac{du}{dt} = c\Delta u + (a, \nabla u).$$

Suppose that a is a Gaussian "white noise" with independent components. Then the equation of such "diffusion" takes the form:

$$du(t,x) = c\Delta u(t,x)dt + \sum_{k=1}^{d} \frac{\partial u(t,x)}{\partial x_k} dw_k(t).$$

This equation can be written down in the following way:

$$d\tilde{X}_t = d\tilde{Y}_t \tilde{X}_t,$$

where $\tilde{Y}_t$ is a process with independent increments and unbounded operator values. The latter equation turns into (1) under conjugation.

3. <u>Hilbert-Schmidt's stochastic semigroups.</u>

Let H be a separable Hilbert space, $|\cdot|$ and $(\cdot,\cdot)$ be the norm and scalar product in H. Denote by $L_2(H)$ the space of Hilbert-Schmidt operators on H ; $L_2(H)$ is the Hilbert space with the scalar product

$$(A,B) = \operatorname{tr} AB^*, \qquad A, B \in L_2(H).$$

We consider the semigroup of operators $\{I + A,\ A \subset L_2(H)\}$, denote it by $GL_2(H)$. We introduce the distance in $GL_2(H)$ like this:

$$\| C - \mathcal{D} \|_2 = r(C, \mathcal{D}) = (C - \mathcal{D}, C - \mathcal{D})^{1/2}, \quad C, \mathcal{D} \in GL_2(H).$$

Hilbert-Schmidt's stochastic semigroup is the semigroup X_t^s with values in $GL_2(H)$, for which $r(X_t^s, I) \to 0$ in probability of $s \to u$, $t \to u$ for all $u \in R_+$.

There exists such a modification of X_t^s, which is left and right continuous in s and t, $s \leq t$. For all $\varepsilon > 0$ there exists the sequence of stopping times $\tau_1, \tau_2, \dots$ such that $\| X_{\tau_k^+}^{\tau_k^-} - I \|_2 > \varepsilon$ and $\| X_{t+}^{t-} - I \| \leq \varepsilon$, if $t \neq \tau_k$, $k = 1, 2, \dots$

I. Let S be a Borel subset in $L_2(H)$, such that the distance between S and 0 is positive. Then

$$\nu_t(S) = \sum_{s \leq t} I_{\{X_{s+}^{s-} - I \in S\}} \tag{7}$$

is finite (I_C is the indicator of C). $\nu_t(S)$ is a Poisson random measure with independent values on the Borel subsets in $L_2(H)$.

II. Let τ_k be such as above. For $\tau_{k-1} < s \leq \tau_k < \dots < \tau_\ell \leq t < \tau_{\ell+1}$ define

$$X_t^s(\varepsilon) = X_{\tau_k^-}^s \, X_{\tau_{k+1}^-}^{\tau_k^+} \dots X_t^{\tau_\ell^+},$$

we put $X_{\tau_k^-}^{\tau_k} = I$ and $X_{\tau_\ell^-}^{\tau_\ell^+} = I$ therein. $X_t^s(\varepsilon)$ are the

Hilbert-Schmidt's stochastic semigroups, for which

$$\sup_{s} \| X^{s-}_{s+}(\varepsilon) - I \|_2 \le \varepsilon .$$

If $\varepsilon < 1$, then the operators $X^{s-}_{t}(\varepsilon)$ are reversible and $\| (X^{s}_{t}(\varepsilon))^{-1} - I \|$ is locally bounded with probability 1.

III. There exist all moments $E \| X^{s}_{t}(\varepsilon) - I \|^{k}$, $k \geq 0$. If

$$M^{s}_{t}(\varepsilon) = E X^{s}_{t}(\varepsilon)$$

(this means that for all $x, y \in H$ $E(X^{s}_{t}(\varepsilon)x, y) = (M^{s}_{t}(\varepsilon)x, y)$), then

$$M^{0}_{s}(\varepsilon) X^{s}_{t}(\varepsilon) \left(M^{0}_{t}(\varepsilon) \right)^{-1} = \tilde{X}^{s}_{t}(\varepsilon)$$

is also a stochastic semigroup, it is a semimartingale in t, for which $E \tilde{X}^{s}_{t}(\varepsilon) = I$.

IV. There exists the stochastic integral

$$\tilde{Y}_{t}(\varepsilon) = \int_{0}^{t} (\tilde{X}^{0}_{s}(\varepsilon))^{-1} d \tilde{X}^{0}_{s}(\varepsilon), \tag{8}$$

$\tilde{Y}_{t}(\varepsilon)$ is a process with independent increments in $L_2(H)$, it is stochastically continuous square integrable martingale.

V. Let $M^{0}_{s}(\varepsilon)$ be a function of bounded variation in $GL_2(H)$. Then $X^{0}_{s}(\varepsilon)$ is a semimartingale and

$$Y_{t}(\varepsilon) = \int_{0}^{t} (X^{0}_{s}(\varepsilon))^{-1} d X^{0}_{s}(\varepsilon)$$

is also a process with independdnt increments in

$$Y_{t}(\varepsilon) = \int_{0}^{t} (M^{0}_{s}(\varepsilon))^{-1} d M^{0}_{s}(\varepsilon) + \int_{0}^{t} (M^{0}_{s}(\varepsilon))^{-1} d \tilde{Y}_{s}(\varepsilon) (M^{0}_{s}(\varepsilon)). \tag{9}$$

<u>Theorem 1</u>. The semigroup X^{s}_{t} is a solution of equation (1), where Y_{t} is a process with independent increments in $L_2(H)$, iff X^{s}_{t} is a semimartingale in t. In this

case M_t^0 is a function of bounded variation and

$$Y_t = Y_t(\varepsilon) + \int_{L_2(H),\, \|\mathcal{U}\|_2 > \varepsilon} \mathcal{U} \nu_t(d\mathcal{U}). \tag{10}$$

4. <u>The square integrable strong stochastic semigroups.</u>

Let $H(\Omega)$ be the space of H-valued random variables which are defined on $\{\Omega, \mathcal{F}, P\}$ with topology of convergence in probability. $H(\Omega)$ is a linear space. The linear transformation of H into $H(\Omega)$ is called a strong random linear operator on H, if it is bounded in probability on a sphere in H of radius 1. The space of all strong linear operators on H is denoted by $L_s(H, \Omega)$. If $\mathcal{U} \in L_s(H, \Omega)$ and for all $x, y \in H$ $E|(\mathcal{U}x, y)| < \infty$, then there exists an operator $A \in L(H)$ such that

$$E(\mathcal{U}x, y) = (Ax, y).$$

We put $A = E\mathcal{U}$. If for all $x \in H$ $E|\mathcal{U}x|^2 < \infty$, then there exists $E\mathcal{U}^*\mathcal{U} \in L(H)$, it is a nonnegative symmetric operator.

<u>Lemma</u>. If $\mathcal{U}, \mathcal{V} \in L_s(H, \Omega)$, $\mathcal{U}, \mathcal{V}$ are independent as random variables, then

$$\mathcal{U}\mathcal{V} \in L_s(H, \Omega).$$

Put $\|E\mathcal{U}^*\mathcal{U}\|^{1/2} = \|\mathcal{U}\|_s$. The set $L_s^2(H, \Omega)$ of such operators $\mathcal{U}$, for which $\|\mathcal{U}\|_s < \infty$ is a Banach space. For independent operators $\mathcal{U}, \mathcal{V} \in L_s^2(H, \Omega)$ we have

$$\|\mathcal{U}\mathcal{V}\|_s \leq \|\mathcal{U}\|_s \cdot \|\mathcal{V}\|_s .$$

In this section we consider the semigroups of strong random operators $X_t^s \in L_s^2(H,\mathfrak{P})$ for which $\| X_t^s - I \|_s \to 0$, if $s \to u$, $t \to u$ for all u.

Let $M_t^s = E X_t^s$. M_t^s is a semigroup of bounded operators on H : for $s < t < u$ $M_t^s M_u^t = M_u^s$. $\| M_t^s - I \| \to 0$, if $t \le t_0$, $t - s \to 0$ for all t_0. This implies that M_t^s is a continuous function in s and t in $L(H)$, $(M_t^s)^{-1}$ exists and is also a continuous function. We define

$$\tilde{X}_t^s = M_s^0 X_t^s (M_t^0)^{-1} .$$

$\tilde{X}_t^s$ is also a semigroup of strong random operators in $L_s^2(H,\mathfrak{P})$ with the same properties of continuity, $E\tilde{X}_t^s = I$ and $\tilde{X}_t^s$ is a martingale.

<u>Theorem 2</u>. There exists the limit in norm $\|\cdot\|_s$

$$\tilde{Y}_t = \overline{\lim_{\max \Delta t_n \to 0}} \sum_{k=0}^{n-1} \left(\tilde{X}_{t_{k+1}}^{t_k} - I \right), \quad 0 = t_0 < \dots < t_n = t, \tag{11}$$

$\tilde{Y}_t$ is a process with independent increments in $L_s^2(H,\mathfrak{P})$.
$\tilde{Y}_t$ is a martingale and

$$\tilde{X}_t^s = I + \int_s^t \tilde{X}_u^s \, d\tilde{Y}_u . \tag{12}$$

The semigroup X_t^s satisfies an equation of the form (1) with Y_t being the $L_s^2(H,\mathfrak{P})$ -valued process with independent increments, iff $\tilde{X}_t^s$ is a semimartingale in $L_s^2(H,\mathfrak{P})$. This means that M_t^s has a bounded variation as a function of t in $L(H)$. In this case

$$Y_t = \int_0^t (M_s^0)^{-1} dM_s^0 + \int_0^t (M_s^0)^{-1} d\tilde{Y}_s M_s^0 . \tag{13}$$

The stochastic integrals in (12) and (13) are defined in the obvious way.

5. Stochastic equations with constant coefficients.

In this section we consider the equation

$$dX_t = X_t A\,dt + \sum_{k=1}^{\infty} X_t B_k\,dw_k(t). \tag{14}$$

The operators A, B_k may be unbounded, but all of them are defined on some linear set $\mathcal{D}$, which is dense in H. The equation (14) holds, if for all $x \in \mathcal{D}$

$$d(X_t x) = X_t(Ax)\,dt + \sum_{k=1}^{\infty} X_t(B_k x)\,dw_k(t).$$

We shall be interested in solutions X_t, which are the $L_S^2(H,\Omega)$-valued processes and $E|X_t x - x|^2 \longrightarrow 0$ for all $x \in H$.

If equation (14) has a unique solution, then X_t^s is the $L_S^2(H,\Omega)$-valued stochastic semigroup, for which $E|X_t^s x - x|^2 \longrightarrow 0$ as $t-s \to 0$.

Let for $C \in L(H)$

$$V_t(C) = E(X_t^0)^* C X_t^0 .$$

Then $V_t(C)$ is a semigroup on $L(H)$:

$$V_{t+s}(C) = V_t(V_s(C)).$$

For all $x, y \in \mathcal{D}$, $C \in L(H)$

$$\lim_{t\downarrow 0} \frac{1}{t}\left[(V_t(C)x,y) - (Cx,y)\right] = (Q(C)x,y),$$

where

$$Q(C) = A^* C + CA + \sum_{k=1}^{\infty} B_k^* C B_k . \tag{15}$$

Theorem 3. The equation (14) has a unique solution, for which

$$\lim_{t \to 0} E|X_t x - x|^2 \longrightarrow 0 \qquad \text{for all } x \in H$$

only in the case when $Q(C)$ is a generator of semigroup in $L(H)$. If there exists an $\alpha > 0$, for which

$$2(Ax,x) + \sum_{k=1}^{\infty} |B_k x|^2 \leq \alpha |x|^2 , \tag{16}$$

then the solution of (14) exists and

$$\| E X_t^* X_t \| \leq \| X_0 \|^2 e^{\alpha t} .$$

The condition (16) is sufficient for the existence and uniqueness of solution of (14).

6. Continuous homogeneous semigroups in $L_s^2(H,\Omega)$.

We consider the stochastic semigroup X_t^s in $L_s^2(H,\Omega)$, such that the distribution of X_t^s depends only on $t - s$ (homogeneous case), for all $x \in H$ $X_t^0 x$ is a continuous function in H with probability 1 and $E|X_t^0 x - x|^2 \to 0$ as $t \to 0$. There exists $EX_t^0 = T_t$ and T_t is a homogeneous semigroup of bounded linear operators on H. This semigroup is strongly continuous: $T_t x \longrightarrow x$ for all $x \in H$. Let A be a generator of semigroup T_t, A is defined on a dense set $\mathfrak{D} \subset H$.

I. For all $x \in \mathfrak{D}$ the process

$$\xi(t,x) = X_t^0 x - \int_0^t X_s^0 A x \, ds$$

is a continuous martingale in H.

II. Let

$$V_t(C) = E(X_t^0)^* C X_t^0 .$$

$V_t(C)$ is a homogeneous semigroup of bounded linear operators on $L(H)$. If $Q(C)$ is a weak generator of $V_t(C)$, that is

$$(Q(C)x,y)=\lim_{t\downarrow 0}\frac{1}{t}\left[(V_t(C)x,y)-(Cx,y)\right],\quad x,y\in H$$

and $Q(C)$ is defined if this limit exists, then the set $\mathcal{D}_1$ of such C is dense in weak sense in $L(H)$.

III. For $x,y\in\mathcal{D}$, $C\in\mathcal{D}_1$

$$(C\xi(t,x),\xi(t,y))-\int_0^t(Q_1((X_s^0)^*CX_s^0)x,y)ds$$

is a continuous martingale, where

$$Q_1(C)=Q(C)-A^*C-CA.$$

There exists the sequence of operators B_k, such that for all $C\in\mathcal{D}_1$

$$Q_1(C)=\sum_k B_k^*CB_k.$$

<u>Theorem 4.</u> There exist Wiener processes $w_k(t)$ in R such that for all $x\in\mathcal{D}$ and C for which $C^*C\in\mathcal{D}_1$

$$dCX_t^0x=X_t^0Axdt+\sum_{k=1}^{\infty}CX_t^0B_kx\,dw_k(t). \tag{17}$$

References

1. A.V.Skorokhod. Martingales and stochastic semigroups.Theory of random processes,4,Kiev,1976,86-94.
2. E.P.Butsan. Stochastic semigroups.Kiev,Naukova Dumka,1977.
3. A.V.Skorokhod. Random linear operators. D.Reidel Publ.Comp. 1983.
4. A.V.Skorokhod. The operator stochastic differential equations and stochastic semigroups.Uspehi Matematiczeskih Nauk,37:6,1982, 157-183.

STOCHASTIC CALCULUS OF VARIATIONS REVISITED

Kunio Yasue
Notre Dame Seishin University
Notre Dame Hall - Box 151
2-16-9 Ifuku-cho, Okayama 700
Japan

Apology

I have been worrying over my thoughtlessness as to have coined the notion of stochastic calculus of variations. When I did so four years ago in Geneva, Switzerland, I was young and completely ignorant of the long and profound history of calculus of variations. I knew neither the classical contributions of Weierstrass and of Hilbert nor the modern treatment of Morse. What remained in my mind with the name of calculus of variations was merely a formal sense used in classical mechanics as principles of least action.

What I wished was indeed to show the validity of least action principles in quantum mechanics. For this aim I needed a new idea that extends calculus of variations in classical mechanics to that in quantum mechanics. This, of course, demands the explicit use of spatial trajectories of particles in quantum mechanics. Such a formulation was already known as stochastic mechanics introduced by Edward Nelson. [1] There, particle trajectories are given by stochastic processes, and then I looked for the minimal extension of calculus of variations in classical mechanics to the corresponding calculus in stochastic mechanics. I called it stochastic calculus of variations, though it was, I think now, only a beginning of the future theory. [2-5] However, I have been proud of my stochastic calculus of variations regardless of its formality, because of ideas and concepts which guide future generations of official stochastic calculus of variations.

I left Geneva looking forward to future generations, and

hibernated for two years in Japan. During my hibernation, Jean-Claude Zambrini succeeded me in both mathematical and physical approaches. [6-10] A concrete mathematical approach from the martingale theory was presented by Zheng and Meyer. [11] It is time, in June 1985, to declare the future generation of stochastic calculus of variations to be respectable. Zambrini will provide us with what we look for in the near future, I believe. [12]

I will talk in this instance about ideas, methods and concepts of stochastic calculus of variations looking forward to seeing the future generation.

Wavepackets and Quantum Mechanics

Let us see what quantum mechanics is. The symbolic term of quantum mechanics may be the notion of non-commuting dynamical variable or operator. There, dynamical variables are self-adjoint operators acting on wavefunctions or state vectors belonging to an infinite dimensional Hilbert space. A naive image of particle trajectory has been lost in quantum mechanics, though Schrödinger himself had it in mind. Extracted from his heart full of realistic view of quantum dynamics in physical space were wavepackets subject to the famous Schrödinger equation.

Although a wavepacket well represents quantum dynamics of a particle, it is no longer the particle itself. A particle is thought to be something like a point mass, called quantum, being somewhere in the wavepacket drawn by a complex valued function ψ. The position of the particle is subject to quantum fluctuation, and it happens to be a random variable with probability distribution density $p = |\psi|^2$. A wavepacket ψ is then written as $\psi = \sqrt{p}\cdot\exp(iW)$ for a certain phase function W.

Stochastic Mechanics

Nelson's idea of stochastic mechanics is to see the particle kinematics behind the wavepacket as a stochastic process. There, a particle goes zigzag as a stochastic process, while a wavepacket spreads out smoothly according to the Schrödinger equation.

$$i\hbar\frac{\partial \psi}{\partial t} = -\frac{\hbar^2}{2m}\Delta\psi + V\psi \tag{1}$$

Here, m denotes the mass of a particle and interactions are included in the expression of the potential energy V . A universal constant $\hbar$ is the Planck constant divided by 2π.

The stochastic process $X = \{X_t | -\infty<t<\infty\}$ in question is a conservative diffusion subject to a stochastic differential equation

$$dX_t = b(X_t,t)dt + \sqrt{\frac{\hbar}{2m}}\,dW_t\ . \tag{2}$$

The local drift b is given by the wavefunction ψ through a relation

$$b(x,t) = \frac{\hbar}{m}\,(\,\mathrm{Re} + \mathrm{Im}\,)\,\frac{\nabla\psi(x,t)}{\psi(x,t)}\ . \tag{3}$$

Effect of quantum fluctuation in position is given by a standard Wiener process $W = \{W_t | -\infty<t<\infty\}$ times the magnitude of fluctuation $\sqrt{\frac{\hbar}{2m}}$. [1]

Stochastic mechanics claims the revival of the notion of particle trajectory in quantum theory. It seems closer to Schrödinger's vision of realistic quantum dynamics than the wavepacket. Stochastic processes stand for spatial trajectories of quanta much more finely than wavepackets, I think.

Stochastic Processes and Stochastic Variations

Since the notion of particle trajectory arose from stochastic mechanics, we are in a good position to see if least action principles remain valid in quantum mechanics. This is the very starting point of stochastic calculus of variations.

Let us forget stochastic mechanics for the moment and see stochastic processes as merely stochastic processes. There are many ways to see stochastic processes. Among them, the most convenient one for developing the idea of stochastic calculus of variations may be the L^2-theory. There, a stochastic process is nothing else but a single curve in a real Hilbert space $H = L^2(\Omega, \mathbb{R}^f)$ composed of random variables with finite variances. The nature of a stochastic process can be understood as that of a corresponding curve in the Hilbert space H . On the other hand, a stochastic process is an assembly of possible spatial trajectories, called sample paths, and the nature of a stochastic process can be understood as that of almost all sample paths, too.

We consider a class of stochastic processes X's indexed by a time interval $[0,\infty)$ which draw continuous curves in the Hilbert space H and continuous sample paths in the configuration space $\mathbb{R}^f$. For simplicity, all the stochastic processes in this class are assumed to have the same initial value Z in H . This class will be denoted by C . Associated with this class is another class of stochastic processes which draw continuous curves in H and continuous sample paths in $\mathbb{R}^f$ passing through the origin at the initial time $t = 0$. This class will be denoted by C_0 , and stochastic processes therein are said to be stochastic variations.

Let us see the randomness of stochastic processes through two reference frames of filtrations $\mathcal{P}$ and $\mathcal{F}$. The former is an increasing family of sigma-algebra of measurable events and the latter is a decreasing one. We will fix these two reference frames, since we are mainly interested in stochastic processes of the same origin of fluctuation. Now, remember that we are seeing possible spatial trajectories of a particle in quantum mechanics. The common origin of randomness in spatial trajectories is, of course, the quantum fluctuation. The two reference frames $\mathcal{P}$ and $\mathcal{F}$ thus reflect the universal uncertainty of quantum phenomena. The two classes of stochastic processes C and C_0 are from now on assumed to contain only those adapted to $\mathcal{P}$ and $\mathcal{F}$.

Principles of Least Action

Among many stochastic processes in the class C , those appearing in Nature as particle trajectories are all subject to a dynamical law. Nature seems to demand the dynamical law to be simple and universal.

So, we look for a universal dynamical law that keeps its simple form unchanged between quantum mechanics and classical mechanics. It will be Hamilton's principle of least action in terms of stochastic calculus of variations.

The fundamental physical quantity is the action integral along a spatial trajectory of a particle. The action integral can be defined on stochastic processes X's in the class C as follows.

$$J(X) = E[S(X_t,t)+\int_0^t \mathcal{L}(X_s,DX_s,D_*X_s,s)ds] \tag{4}$$

Here, S is a smooth function of position and time to be specified, and $\mathcal{L}$ is a given function of position, two velocities and time,

$$\mathcal{L} = \frac{1}{2}(\frac{m}{2}|DX_t|^2+\frac{m}{2}|D_*X_t|^2) - V(X_t,t) , \tag{5}$$

called Lagrangian. Of course, the action integral J(X) remains finite only for a restricted class of stochastic processes X's depending on the forms of S and $\mathcal{L}$ in the right-hand side.

Two velocities DX_t and D_*X_t are, respectively, the mean forward derivative and the mean backward one of the stochastic process X given by

$$DX_t = \lim_{h\to 0+} E[\frac{X_{t+h} - X_t}{h} | \mathcal{P}_t] , \tag{6}$$

$$D_*X_t = \lim_{h\to 0+} E[\frac{X_t - X_{t-h}}{h} | \mathcal{F}_t] . \tag{7}$$

Those are assumed for each time t to be random variables in the Hilbert space H . Then, finiteness of the action integral J(X) is essentially a matter of the forms of two functions S and V .

Now let us see the stochastic variations which are nothing else but stochastic processes in the class C_0 . For any stochastic variation Y in C_0 and any stochastic process X in C , the sum X + Y will be another stochastic process in the class C . The action integral along X + Y is J(X+Y) . Hamilton's principle of least action would claim that a dynamically admissible spatial trajectory of a particle in quantum mechanics is a stochastic process X in C along which the action integral keeps its minimal. Namely,

$$J(X+Y) \geqq J(X) \tag{8}$$

locally. This physical assertion may be better formulated in the context of functional differential analysis.

Stochastic Calculus of Variations

The action integral $J(X)$ along each X in the class C forms a functional J on C. Hamilton's principle of least action, then, can be restated by saying that the functional J takes its minimal value on a stochastic process X which represents a dynamically admissible trajectory of a quantum.

What is needed now is the differential analysis of the functional J defined on the class C of stochastic processes. To do so, a way that measures the vicinity of stochastic processes is needed. It will be given by a norm $|||\cdot|||$. For any stochastic processes X's in the classes C and C_0,

$$|||X||| = \sup_s [\, \|X_s\|, \|DX_s\|, \|D_*X_s\| \,] , \tag{9}$$

where $\|\cdot\|$ is the norm of random variables R's in the Hilbert space H, that is, $\|R\| = (E[|R|^2])^{1/2}$. For example, a stochastic variation Y is said to be smaller than the other Y' if the norm $|||Y|||$ is smaller than $|||Y'|||$.

A functional J defined on the class of stochastic processes C is said to be differentiable on a stochastic process X if for any stochastic variation Y in C_0,

$$J(X+Y) = J(X) + dJ(X,Y) + o(|||Y|||) . \tag{10}$$

Here, the functional $dJ(X,Y)$ depends on Y linearly. It stands for the first order change of value of the functional J in the vicinity of X. This quantity $dJ(X,Y)$ is a differential of the functional J on the stochastic process X along the stochastic variation Y. It is often said to be a variation of the functional J corresponding to the stochastic variation Y. Local behaviors of the functional such as the maximal property are seen by the differential. If the differential is known to exist, it can be calculated in a weak sense

$$dJ(X,Y) = \frac{d}{dr} J(X+rY)\Big|_{r=0} . \tag{11}$$

From the practical point of view, the maximal condition in Hamilton's principle of least action can be relaxed to a stationary condition. The functional J is said to be stationary on a stochastic process X if the differential $dJ(X,Y)$ vanishes identically for any stochastic variations Y's in C_0, that is, $dJ(X,Y)=0$, for any Y in C_0. Clearly, maximal property implies stationary property. It is worthwhile to notice that only the stationary property is required for describing the fundamental law of motion by means of the least action principle. Let us see this.

First, we calculate the differential of the functional J.

$$\begin{aligned} J(X+Y) &= E[S(X_t+Y_t,t)+\int_0^t \mathcal{L}(X_s+Y_s, DX_s+DY_s, D_*X_s+D_*Y_s, s)ds] \\ &= E[S(X_t,t)+\int_0^t \mathcal{L}(X_s, DX_s, D_*X_s, s)ds \\ &\quad + \nabla S(X_t,t)\cdot Y_t + \int_0^t (\tfrac{m}{2}DX_s\cdot DY_s + \tfrac{m}{2}D_*X_s\cdot D_*Y_s - V(X_s,s)\cdot Y_s)ds] \\ &\quad + o(|||Y|||) \\ &= J(X) + dJ(X,Y) + o(|||Y|||) \end{aligned} \tag{12}$$

By means of the integration by parts formula for stochastic processes in the classes C and C_0, [1]

$$E[\int_0^t DX_s\cdot X'_s ds] = -E[\int_0^t D_*X'_s\cdot X_s ds] + E[X_s\cdot X'_s\Big|_0^t] , \tag{13}$$

$$E[\int_0^t D_*X_s\cdot X'_s ds] = -E[\int_0^t DX'_s\cdot X_s ds] + E[X_s\cdot X'_s\Big|_0^t] , \tag{14}$$

this expression becomes

$$\begin{aligned} dJ(X,Y) = E[(\nabla S(X_t,t) + m\tfrac{1}{2}(DX_t+D_*X_t))\cdot Y_t \\ - \int_0^t (m\tfrac{1}{2}(DD_*X_s+D_*DX_s)+V(X_s,s))\cdot Y_s ds] . \end{aligned} \tag{15}$$

Here, the condition $Y_0=0$ is used for stochastic variations Y's. In terms of the Lagrangian $\mathcal{L}$, this may be written

$$dJ(X,Y) = E[(\nabla S(X_t,t)+(\frac{\partial\mathcal{L}}{\partial DX_t} + \frac{\partial\mathcal{L}}{\partial D_*X_t}))\cdot Y_t + \int_0^t (\frac{\partial\mathcal{L}}{\partial X_s} - D\,\frac{\partial\mathcal{L}}{\partial D_*X_s} - D_*\,\frac{\partial\mathcal{L}}{\partial DX_s})\cdot Y_s ds] \;. \tag{16}$$

Hamilton's principle of least action asserts that the right-hand side of this expression vanishes identically for all stochastic variations Y's. Under certain regularity assumptions on the stochastic process X, Lagrangian $\mathcal{L}$, and the function S, this implies

$$(\frac{\partial\mathcal{L}}{\partial DX_t} + \frac{\partial\mathcal{L}}{\partial D_*X_t}) + \nabla S(X_t,t) = 0 \;, \tag{17}$$

$$\frac{\partial\mathcal{L}}{\partial X_t} - D\,\frac{\partial\mathcal{L}}{\partial D_*X_t} - D_*\,\frac{\partial\mathcal{L}}{\partial DX_t} = 0 \;, \tag{18}$$

for all t in the time interval $[0,\infty)$. The regularity assumptions needed above are, for example, that the two stochastic processes X' and X'' appearing in the left-hand sides of Eqs. (17) and (18) fall into the class C_0. These assumptions may be reduced to those on the local drift b if the stochastic process X is a conservative diffusion subject to the stochastic differential equation (2). For such a conservative diffusion X, we have

$$DX_t = b(X_t,t) \;, \tag{19}$$

$$D_*X_t = b_*(X_t,t) \;. \tag{20}$$

Here, b_* is a backward local drift given by the local drift b and the probability distribution density p through a formula

$$b_* = b - (\frac{\hbar}{m})\frac{\nabla p}{p} \;. \tag{21}$$

The two stochastic processes X' and X'' become

$$\begin{aligned} X'_t &= -\nabla V(X_t,t) - \frac{m}{2}Db_*(X_t,t) - \frac{m}{2}D_*b(X_t,t) \\ &= -\nabla V(X_t,t) - \frac{m}{2}(\frac{\partial}{\partial t} + b(X_t,t)\cdot\nabla + \frac{\hbar}{2m}\Delta)b_*(X_t,t) \\ &\quad - \frac{m}{2}(\frac{\partial}{\partial t} + b_*(X_t,t)\cdot\nabla - \frac{\hbar}{2m}\Delta)b(X_t,t) \\ &= f'(X_t,t) \;, \end{aligned} \tag{22}$$

$$X''_t = m\frac{1}{2}(b(X_t,t)+b_*(X_t,t)) + \nabla S(X_t,t) = f''(X_t,t) \ , \tag{23}$$

and they fall into the class C_0 provided that the functions f' and f'' are of class $C^{2,1}$ and vanish identically for $t=0$. We have no other choice than assuming them, though it is not trivial at all.

Thus, Hamilton's principle of least action claims that the stochastic process X representing a dynamically admissible spatial trajectory of a quantum satisfies the Euler-Nelson equation

$$m\frac{1}{2}(DD_*X_t+D_*DX_t) = -\nabla V(X_t,t) \tag{24}$$

as well as the transversality condition

$$m\frac{1}{2}(DX_t+D_*X_t) = -\nabla S(X_t,t) \ . \tag{25}$$

In terms of local drifts b and b_* and the probability distribution density p , the differential of the functional J becomes

$$dJ(X,Y)$$
$$= \int f''(x,t)g(x,t)p(x,t)d^f x + \int_0^t \int f'(x,s)g(x,s)p(x,s)d^f x ds \ . \tag{26}$$

Here, Y is any stochastic variation of the type $Y_t=g(X_t,t)$, with g any bounded function of class C^2 such that $g(\cdot,0)=0$. Then, the condition of vanishing differential for any such function g turns out to be both $f'(x,t)=0$ and $f''(x,t)=0$ for all (x,t) such that $p(x,t)\neq 0$. Namely, we have

$$\frac{m}{2}\left(\frac{\partial b_*}{\partial t} + b\cdot\nabla b_* + \frac{\hbar}{2m}\Delta b_* + \frac{\partial b}{\partial t} + b_*\cdot\nabla b - \frac{\hbar}{2m}\Delta b\right) = -\nabla V \tag{27}$$

and

$$\frac{m}{2}(b+b_*) = -\nabla S \ . \tag{28}$$

The stochastic differential equation (2) implies the Fokker-Planck equation for the probability distribution density p ,

$$\frac{\partial p}{\partial t} = -\nabla\cdot(\ b\ p\ ) + \frac{\hbar}{2m}\Delta p \ , \tag{29}$$

which yields the continuity equation

$$\frac{\partial p}{\partial t} = -\nabla\cdot\left(\frac{1}{2}(b+b_*)p\right) . \tag{30}$$

The Euler-Nelson equation (27), the transversality condition (28) and the continuity equation (30) are all now converted into a single equation for a complex valued function

$$\psi = \sqrt{p}\,\exp(-\frac{i}{\hbar}S) . \tag{31}$$

This very equation is the Schrödinger equation (1), and the validity of Hamilton's principle of least action in quantum mechanics is shown.

References

[1] E. Nelson, "Quantum Fluctuation," (Princeton University Press, Princeton, 1985).
[2] K. Yasue, J. Funct. Anal. 41 (1981), 327.
[3] K. Yasue, J. Math. Phys. 22 (1981), 1010.
[4] K. Yasue, J. Math. Phys. 23 (1982), 1577.
[5] K. Yasue, J. Funct. Anal. 51 (1983), 133.
[6] J.-C. Zambrini, Doctoral thesis, University of Geneva, 1982.
[7] J.-C. Zambrini and K. Yasue, Ann. Phys. (N.Y.) 143 (1982), 54.
[8] J.-C. Zambrini, J. Math. Phys. 25 (1984), 1314.
[9] J.-C. Zambrini and K. Yasue, Phys. Rev. Lett. 52 (1984), 2107.
[10] J.-C. Zambrini, Intern. J. Theor. Phys., to be published.
[11] W. A. Zheng and P. A. Meyer, Lecture Notes in Mathematics, Séminaire de Probabilités XVIII (1984), 223.
[12] J.-C. Zambrini, private communication.

STABILITY UNDER SMALL PERTURBATIONS

J. Zabczyk
Institute of Mathematics
Polish Academy of Sciences
Warsaw, Poland
and
Department of Mathematics
Heriot-Watt University
Edinburgh, Great Britain

I. Introduction

Let f be a continuously differentiable, Lipschitz function from R^n into R^n and let the origin of R^n be asymptotically stable equilibrium for the dynamical system

(1) $$\dot{z} = f(z)$$

Stochastically perturbed version of (1) is usually given in the form of a stochastic Ito equation

(2) $$dX = f(X)dt + \varepsilon g(X)dW_t$$

where (W_t) is an m-dimensional Wiener process defined on a probability space (Ω, F, P), ε is a real number and g is a continuously differentiable matrix valued function defined on R^n. Solutions to the equation (2) starting from x will be denoted in the sequel as $X^{x,\varepsilon}$. Small disturbances correspond to small parameters ε. If D is a fixed reference domain containing $0 \in R^n$, then in many cases of interest the process $X^{x,\varepsilon}$ hits the boundary ∂D with the probability one. Let

$$\tau^{x,\varepsilon} = \inf\{t \geq 0;\ X^{x,\varepsilon}(t) \in \partial D\} .$$

Exit theorems, see [3], are concerned with the limit behaviour of the expected exit time $\mathbb{E}(\tau^{x,\varepsilon})$ and the exit place $X^{x,\varepsilon}(\tau^{x,\varepsilon})$ as $\varepsilon \downarrow 0$.

II. Finite dimensional case

In addition to systems (1) and (2) it is convenient to introduce a control system

(3) $$\dot{y} = f(y) + g(y)u$$

where $u(\cdot)$ is a function with values in R^m.
The following function V - so called quasipotential - plays an important role in the discussed theory. If $a,b \in R^n$ then

$$V(a,b) = \frac{1}{2} \inf \int_0^T |u(s)|^2 ds$$

where the infimum is taken with respect to all control functions $u(\cdot)$ such that the corresponding $y(\cdot)$ from (3) satisfies $y(0) = a$, $y(T) = b$, with $T > 0$ being an arbitrary number.

We impose the following conditions

(4) There exist open neighbourhoods D_1 and D_2 of 0 and ∂D respectively such that V is continuous on $D_1 \times D_2$.

(5) Rank $[B, AB, \ldots, A^{n-1}B] = n$
where $A = f_x(0)$ and $B = g(0)$.

(6) $\partial D = \partial \bar{D}^c$ and there exists an open set $D_3 \supset \bar{D}$ such that for arbitrary $\delta > 0$ there exists $T > 0$ such that

$$|z(t,x)| < \delta \qquad \text{for all} \quad x \in D_3, \quad t \geq T,$$

where $z(\cdot,x)$ is the unique solution of (1) starting from x.

Let finally

$$D_0 = \{x : z(t,x) \in D \quad \text{for all} \quad t \geq 0\}.$$

The following theorem holds

Theorem 1. If conditions (4) - (6) hold then

(i) For all $x \in D_0$

(9) $$2 \langle QF^{*}x,x \rangle + |\sigma^{*}x|^{2} = 0 , \qquad x \in D(F^{*})$$

has a unique, non-negative definite solution Q being a trace type operator, see [4]. Let for arbitrary $\gamma>0$, Γ_{γ} denote the following, infinite dimensional ellipsoid

$$\Gamma_{\gamma} = \{ x \in H; \quad \langle Q^{-1}x,x \rangle \leq 2\gamma \}$$

In a similar way as Theorem 1 the following result can be proved

Theorem 2. If the condition (8) holds then

$$\overline{\lim_{\varepsilon \downarrow 0}} \; \varepsilon^{2} \ln \mathbb{E}(\tau^{x,\varepsilon}) \leq \sup \{ \gamma ; \Gamma_{\gamma} \subset \overline{D} \} = \overline{r}$$

$$\underline{\lim_{\varepsilon \downarrow 0}} \; \varepsilon^{2} \ln \mathbb{E}(\tau^{x,\varepsilon}) \geq \sup \{ \gamma : \Gamma_{\gamma} \subset D \} = \underline{r} .$$

Moreover for arbitrary $\delta > 0$

$$\lim_{\varepsilon \downarrow 0} \mathbb{P}(\rho(X^{x,\varepsilon}(\tau^{x,\varepsilon}),\overline{E}) < \delta) = 1$$

where $\overline{E} = \partial D \cap \Gamma_{\overline{r}}$.

IV. Some applications

Example 1.

Assumptions of Theorem 1 are satisfied for linear finite dimensional systems (7) when the controllability condition

$$\text{Rank} [G,FG,\dots,F^{n-1}G] = n \qquad \text{holds.}$$

In particular the theorem is applicable for n-th order systems:

$$z^{(n)} + a_{1}z^{(n-1)} + \dots + a_{n}z = \varepsilon \dot{W}_{t}$$

provided that the deterministic part is stable.

Assume that $n = 2$, $a_1 > 0$, $a_2 > 0$ and $D = \{ \binom{z}{\dot z};\ z^2 + \dot z^2 < 1 \}$ then exit sets $\bar{E}$ are of the form:

$$\left\{ \binom{1}{0}, \binom{0}{-1} \right\} \quad \text{or} \quad \left\{ \binom{0}{1}, \binom{-1}{0} \right\} \quad \text{if} \quad a_1 \neq a_2$$

and the unit circle if $a_1 = a_2$.

If $n = 3$, $a_1 > 0$, $a_2 > 0$, $a_3 > 0$ and $a_1 a_2 - a_3 > 0$ then the exit set consists of two points only

$$\bar{E} = \left\{ \begin{pmatrix} \alpha \\ 0 \\ (1-\alpha^2)^{1/2} \end{pmatrix}, \begin{pmatrix} -\alpha \\ 0 \\ -(1-\alpha^2)^{1/2} \end{pmatrix} \right\}, \quad 0 < \alpha < 1.$$

See [5].

Example 2.

In the infinite dimensional case if (8) holds and $D = \{ x \in H,\ |x| < 1 \}$, then

$$\lim_{\varepsilon \downarrow 0} \varepsilon^2 \ln \mathbb{E}(x, \ ) = \frac{1}{2\Lambda},$$

where Λ is the maximal eigenvalue of Q, see (9). Moreover the exit set $\bar{E}$ consists of all eigenvectors of Q corresponding to Λ. Theorem 2 is applicable to a Hilbert space version of the following system

$$\ddot z(t) + \lambda(0)\dot z(t) + \int_0^{+\infty} \lambda'(s) z(t-s)\,ds = \varepsilon \dot W_t, \quad t \geq 0,$$

which describes oscillations of a particle suspended by a tight, linearly viscoelastic string. Here $\lambda(\cdot)$ is a relaxation modulus and H has to be defined as a Hilbert space of triplets (α, β, v), where α, β are real numbers and v is a real function defined on $[0, +\infty)$.

Example 3.

Let us consider a finite dimensional controlled, stochastic system

$$\lim_{\varepsilon \downarrow 0} \varepsilon^2 \ln \mathbb{E}(\tau^{x,\varepsilon}) = \inf\{V(0,b);\ b \in \partial D\} = r$$

(ii) For arbitrary $\delta > 0$ and $x \in D_0$

$$\mathbb{P}(\rho(X^{x,\varepsilon}(\tau^{x,\varepsilon}),E) < \delta) \longrightarrow 1, \quad \text{as} \quad \varepsilon \downarrow 0$$

where $E = \{b \in \partial D;\ V(0,b) = r\}$.

In contrast to the exit theorems of Freidlin and Wentzell [3] we do not assume that the matrix $a(x) = g(x)g^*(x)$ is invertible, that D is an invariant set for (1) and that the exit set E defined in Theorem 1 consists of one point only. All these conditions limited the number of applications considerably. If the matrix valued function $a(\cdot)$ is invertible then the crucial condition (4) is automatically satisfied as in this case V is continuous on $R^n \times R^n$, see [3]. In general however V is not continuous on $R^n \times R^n$ even if it is a finite valued function, see [5]. An earlier version of Theorem 1 is contained in [5]. The present version uses exponential estimates due to Azencott [1].

III. Infinite dimensional case

Let us consider a linear version of (2) :

(7) $$dX = FXdt + \varepsilon G dW_t$$

on a Hilbert space H. We assume that F is the infinitesimal generator of an exponentially stable C_0-semigroups (S_t) and that $X^{x,\varepsilon}$ is the mild solution of (7) starting from $x \in H$. As before (W_t) is a standard Wiener process on R^m. In addition we assume that

(8) Linear operator G transforms R^m into $D(F)$, the domain of F.

Assumption (8) implies in particular that trajectories of $X^{x,\varepsilon}$ are continuous. Exponential stability of (S_t) implies that the following Liapunov's equation

$$(10) \qquad dX = FXdt + Budt + \varepsilon GdW_t$$

where u denotes a control parameter from R^1 and the pair (F,B) is controllable: Rank $[B, FB, \dots, F^{n-1}B] = n$.
There are many $1 \times n$ matrices K such that A + BK is a stable matrix. Let $X^{x,\varepsilon,K}$ denote the solution of (10) with u replaced by KX and let

$$\tau^{x,\varepsilon,K} = \inf \{ t \geq 0; \ X^{x,\varepsilon,K}(t) \in \partial D \}$$

The following theorem, which proof follows from Theorem 2 and Theorem 3.2.5 of M. Ehrhard's Ph D Theses 2 , gives a condition under which the exit rate can be made arbitrarily large.

Theorem 3. For arbitrary $\Delta > 0$ there exists K such that

$$\lim_{\varepsilon \downarrow 0} \varepsilon^2 \ln E(\tau^{x,\varepsilon,K}) > \Delta \quad , \text{ for all } x \in D_0 ,$$

if and only if Image $G \subset$ Image B .

References

[1] R.G.Azencott, Sur les grand deviations, Lecture Notes in Math. 774, (1978)

[2] M.Ehrhardt, Zur kontrollierbarkeit linearer stochastischen Systeme; Ph.D. Dissertation, Universität Bremen, 1983

[3] M.I.Freidlin and A.Wentzell, Random Perturbations of Dynamical Systems, Springer Verlag 1984

[4] J.Zabczyk, Structural properties and limit behaviour of linear stochastic systems in Hilbert spaces; Banach Center Publications, vol.14, 1985

[5] J.Zabczyk, Exit problem and control theory, to appear in Systems and Control Letters, 1985

S. Albeverio
Mathematisches Institut
Universitaet Bochum
Universitaetsstr. 150 NA
4630 Bochum 1
W. Germany

G. Alsmeyer
Mathematisches Seminar
Universitaet Kiel
Olshausenstr. 40-60
Haus 12a
2300 Kiel 1
W. Germany

A.V. Balakrishnan
Dept. Electrical Engineering
School Eng. & Appl. Sciences
University of California
Los Angeles Calif. 90024
U. S. A.

J.S. Baras
Dept. Electrical Engineering
University of Maryland
College Park MD 20742
U. S. A.

V.E. Benes
Bell Laboratories
Murray Hill
New Jersey 07974
U. S. A.

A. Bensoussan
INRIA
Domaine de Voluceau-
Rocquencourt
Boite Postale 105
78150 Le Chesnay
France

R. Boel
Lab. theoret. Elektriciteit
Rijksuniversiteit Gent
Grote Steenweg Noord 12
9710 Gent (Zwijnaarde)
Belgien

A. Calzolari
Dip. di Matematica
Universita di Roma
Via Orazio Raimondo
00173 Roma (La Romanina)
Italy

N. Christopeit
Inst. f. Oekonometrie &
Operations Research
Adenauerallee 24-42
5300 Bonn 1
W. Germany

R. Cohen
C. N. E. T.
PAA. TIM. MTI.
38-40 Ave du General Leclerc
92131 Issy-les-Moulineaux
France

M.H.A. Davis
Dept. Electrical Engineering
Imperial College
London SW7 2BT
Great Britain

R.J. Elliott
Dept. of Pure Mathematics
University of Hull
Hull HU5 2DW
England

H. Foellmer
- Mathematik -
ETH Zuerich
Raemischstr. 101
8092 Zuerich
Switzerland

G. Gorni
Scuola Normale Superiore
56100 Pisa
Italy

G. Del Grosso
Dip. di Matematica
Universita di Roma
Citta Universitaria
Piazzale Aldo Moro 5
00185 Roma
Italy

F. Guerra
Dip. di Matematica
Universita di Roma
Piazzale Aldo Moro 5
00185 Roma
Italy

I. Gyoengy
Fachbereich Mathematik
Universitaet Frankfurt
Robert-Mayer-Str. 6-10
6000 Frankfurt 1
W. Germany

Z. Haba
Inst. of Theoretical Physics
University of Wroclaw
50 - 205 Wroclaw
Cybulskiego 36
Poland

U. Haussmann
Dept. of Mathematics
University British Columbia
Vancouver, B.C. V6T 1W5
Canada

K. Helmes
Inst. f. Angew. Mathematik
Universitaet Bonn
Wegelerstr. 6
5300 Bonn 1
W. Germany

O. Hijab
Dept of Mathematics
Temple University
Philadelphia, PA 19122
U. S. A.

R. Hoepfner
Inst. f. Math. Stochastik
Universitaet Freiburg
Hebelstr. 27
7800 Freiburg i. Br.
W. Germany

N. Ikeda
Dept. of Mathematics
Osaka University
Tyonaka
Osaka 560
Japan

M. Jerschow
FB 6 Mathematik
Gesamthochschule Essen
Universitaetsstr. 2
4300 Essen
W. Germany

I. Karatzas
Dept. of Statistics
Columbia University
New York N.Y. 10027
U. S. A.

D. Koehnlein
Inst. f. Angewandte Mathematik
Universitaet Bonn
Wegelerstr. 6
5300 Bonn 1
W. Germany

M. Kohlmann
Fak. f. Wirtschaftswissen-
scaften & Statistik
Postfach 5560
7750 Konstanz 1
W. Germany

F. Konecny
Inst fuer Mathematik &
Angewandte Statistik
Universitaet f. Bodenkultur
Gregor Mendel-Strasse 33
1180 Wien
Austria

H. Korezlioglu
Ecole Nationale Superieure
des Telecommunications
46 rue Barrault
75634 Paris Cedex 13
France

K. Kubilius
Inst. of Mathematics &
Cybernetics
Lithuanian Academy Sciences
Vilnius
U. S. S. R.

H.J. Kushner
Div. of Applied Mathematics
Lefschetz Center for
Dynamical Systems
Brown University
Providence R.I. 02912
U. S. A.

R. Kwong
Dept. Electrical Engineering
University of Toronto
Toronto M5S 1A4
Canada

G. Leha
Mathematisches Institut
Universitaet Erlangen
Bismarckstr. 1 1/2
8520 Erlangen
W. Germany

W. Leithead
Dept. of Mathematics & Comp.
Paisley College of Techn.
High Street
Paisley, Renfrewshire
Scotland PA1 2BE
Great Britain

P. Mandl
Dept. of Probability &
Mathematical Statistics
Charles University
Sokolovska 83
186 Prague 8
Czechoslovakia

E. Mann
Inst. f. Angew. Mathematik
Universitaet Bonn
Wegelerstr. 6
5300 Bonn 1
W. Germany

P.S. Maybeck
Dept. of the Air Force
Inst. of Technology
Wright-Patterson Air Force
Base
Ohio OH 45433
U. S. A.

R. Mazumdar
Dept. Electrical Engineering
Columbia University
186 Prague 8
Czechoslovakia

G. Mazziotto
Centre National d'Etudes
des Telecommunications
38-40 Ave du General Leclerc
92131 Issy-les-Moulineaux
France

S.K. Mitter
Dept. Electrical Engineering
& Computer Science
Mass. Inst. of Technology
Cambridge MA 02139

C. Moura
Dept Engenharia Electrotecnia
Faculdade de Engenharia
Universidade do Porto
Rua dos Bragas
4000 Porto Codex
Portugal

S. Mueller
Inst. f. Gesellschafts- &
Wirtschaftswissenschaften
Universitaet Bonn
Adenauerallee 24-42
5300 Bonn 1
W. Germany

E. Niehl
Trippstadterstr. 121
App. 103
6750 Kaiserslautern
W. Germany

D. Ocone
Dept. of Mathematics
Hill Center for the Math-
ematical Sciences
Rutgers University
New Brunswick N.J. 08903
U. S. A.

E. Pardoux
U.E.R. de Mathematiques
Universite de Provence
3, place Victor-Hugo 3
13331 Marseille Cedex 3
France

D. Plachky
Inst. f. Math. Statistik
Universitaet Muenster
Einstein-Str. 62
4400 Muenster
W. Germany

H.J. Plum
Inst. f. Angew. Mathematik
Universitaet Bonn
Wegelerstr. 6
5300 Bonn 1
W. Germany

M. Pontier
Universite d'Orleans
Dept. de Mathematiques et
d'Informatique
UER de Sciences Fond.
45046 Orleans Cedex
France

N. Portenko
Institute of Mathematics
Ukraine Academy of Sciences
Kiev
U. S. S. R.

R. Rishel
Dept. of Mathematics
University of Kentucky
Lexington KY 40506
U. S. A.

W. Runggaldier
Universita di Padova
Seminario Matematico
Via Belzoni 7
35131 Padova
Italy

M. Schael
Inst. f. Angew. Mathematik
Universitaet Bonn
Wegelerstr. 6
5300 Bonn 1
W. Germany

K. Schnepper
Inst. f. Dyn. Flugsysteme
DFVLR e.V.
Oberpfaffenhofen
8031 Wessling Obb.
W. Germany

J.H. van Schuppen
Mathematical Centre
P.O. Box 4079
1009 AB Amsterdam
Netherlands

S.E. Shreve
Dept of Mathematics
Carnegie-Mellon University
Pittsburgh, Penn. 15213
U. S. A.

A. Skorokhod
Inst. of Mathematics
Ul. Repina 3
Kiev, 252601
U. S. S. R.

F. Spizzichino
Dip. di Matematica
Universita di Roma
Piazzale Aldo Moro 5
00185 Roma
Italy

P. Spreij
Mathematical Centre
P.O. Box 4079
1009 AB Amsterdam
Netherlands

J. Szpirglas
PAA/ATR/MTI
C. N. E. T.
38-40 Ave du General Leclerc
92131 Issy-les-Moulineaux
France

S. Taniguchi
Dept. of Applied Sciences
Kyushu University 36
Hakozaki
Fukuoka 812
Japan

R. Theodorescu
Universite Laval
Faculte Sciences et Genie
Cite Universitaire
Quebec G1K 7P4
Canada

H. Witting
Inst. f. Math. Stochastik
A.-Ludwigs-Universitaet
Hebelstr. 27
7800 Freiburg i. Br.
W. Germany

G. Wurm-Schoenert
Inst. f. Angew. Mathematik
Universitaet Bonn
Wegelerstr. 6
5300 Bonn 1
W. Germany

K. Yasue
Computer Center
Notre Dame Seishin
University
2-16-9 Ifuku-Cho
Okayama 700
Japan

J. Zabczyk
Institute of Mathematics
Polish Academy of Sciences
Sniadeckich 8
00-950 Warszawa
Poland

M. Zakai
Technion Inst. of Technology
Dept. of Elect. Engineering
Haifa 32 000
Israel

Lecture Notes in Control and Information Sciences

Edited by A. V. Balakrishnan and M. Thoma

Vol. 43: Stochastic Differential Systems
Proceedings of the 2nd Bad Honnef Conference
of the SFB 72 of the DFG at the University of Bonn
June 28 – July 2, 1982
Edited by M. Kohlmann and N. Christopeit
XII, 377 pages. 1982.

Vol. 44: Analysis and Optimization of Systems
Proceedings of the Fifth International
Conference on Analysis and Optimization of Systems
Versailles. December 14–17, 1982
Edited by A. Bensoussan and J. L. Lions
XV, 987 pages, 1982

Vol. 45: M. Arató
Linear Stochastic Systems
with Constant Coefficients
A Statistical Approach
IX, 309 pages. 1982

Vol. 46: Time-Scale Modeling of Dynamic Networks
with Applications to Power Systems
Edited by J. H. Chow
X, 218 pages. 1982

Vol. 47: P. A. Ioannou, P. V. Kokotovic
Adaptive Systems with Reduced Models
V, 162 pages. 1983

Vol. 48: Yaakov Yavin
Feedback Strategies for Partially
Observable Stochastic Systems
VI, 233 pages, 1983

Vol. 49: Theory and Application of Random Fields
Proceedings of the IFIP-WG 7/1
Working Conference
held under the joint auspices of the
Indian Statistical Institute
Bangalore, India, January 1982
Edited by G. Kallianpur
VI. 290 pages. 1983

Vol. 50: M. Papageorgiou
Applications of Automatic Control Concepts
to Traffic Flow Modeling and Control
IX, 186 pages. 1983

Vol. 51: Z. Nahorski, H.F. Ravn, R.V.V. Vidal
Optimization of Discrete Time Systems
The Upper Boundary Approach
V, 137 pages 1983

Vol. 52: A. L. Dontchev
Perturbations, Approximations and Sensitivity Analysis
of Optimal Control Systems
IV, 158 pages. 1983

Vol. 53: Liu Chen Hui
General Decoupling Theory of Multivariable
Process Control Systems
XI, 474 pages. 1983

Vol. 54: Control Theory for Distributed
Parameter Systems and Applications
Edited by F. Kappel, K. Kunisch,
W. Schappacher
VII, 245 pages. 1983.

Vol. 55: Ganti Prasada Rao
Piecewise Constant Orthogonal Functions
and Their Application to Systems and Control
VII, 254 pages. 1983.

Vol. 56: Dines Chandra Saha, Ganti Prasada Rao
Identification of Continuous
Dynamical Systems
The Poisson Moment Functional
(PMF) Approach
IX, 158 pages. 1983.

Vol. 57: T. Söderström, P. G. Stoica
Instrumental Variable Methods
for System Identification
VII, 243 pages. 1983.

Vol. 58: Mathematical Theory of
Networks and Systems
Proceedings of the MTNS-83 International
Symposium
Beer Sheva, Israel, June 20–24, 1983
Edited by P. A. Fuhrmann
X, 906 pages. 1984

Vol. 59: System Modelling and Optimization
Proceedings of the 11th IFIP Conference
Copenhagen, Denmark, July 25-29, 1983
Edited by P. Thoft-Christensen
IX, 892 pages. 1984

Vol. 60: Modelling and Performance
Evaluation Methodology
Proceedings of the International Seminar
Paris, France, January 24–26, 1983
Edited by F. Bacelli and G. Fayolle
VII, 655 pages. 1984

Vol. 61: Filtering and Control of Random
Processes
Proceedings of the E.N.S.T.-C.N.E.T. Colloquium
Paris, France, February 23–24, 1983
Edited by H. Korezlioglu, G. Mazziotto, and
J. Szpirglas
V, 325 pages. 1984

Lecture Notes in Control and Information Sciences

Edited by M. Thoma and A. Wyner

Vol. 62: Analysis and Optimization of Systems
Proceedings of the Sixth International Conference on Analysis and Optimization of Systems
Nice, June 19–22, 1984
Edited by A. Bensoussan, J. L. Lions
XIX, 591 pages. 1984.

Vol. 63: Analysis and Optimization of Systems
Proceedings of the Sixth International Conference on Analysis and Optimization of Systems
Nice, June 19–22, 1984
Edited by A. Bensoussan, J. L. Lions
XIX, 700 pages. 1984.

Vol. 64: Arunabha Bagchi
Stackelberg Differential Games in Economic Models
VIII, 203 pages, 1984

Vol. 65: Yaakov Yavin
Numerical Studies in Nonlinear Filtering
VIII, 273 pages, 1985.

Vol. 66: Systems and Optimization
Proceedings of the Twente Workshop
Enschede, The Netherlands, April 16–18, 1984
Edited by A. Bagchi, H. Th. Jongen
X, 206 pages, 1985.

Vol. 67: Real Time Control of Large Scale Systems
Proceedings of the First European Workshop
University of Patras, Greece, Juli 9–12, 1984
Edited by G. Schmidt, M. Singh, A. Titli, S. Tzafestas
XI, 650 pages, 1985.

Vol. 68: T. Kaczorek
Two-Dimensional Linear Systems
IX, 397 pages, 1985.

Vol. 69: Stochastic Differential Systems – Filtering and Control
Proceedings of the IFIP-WG 7/1 Working Conference
Marseille-Luminy, France, March 12-17, 1984
Edited by M. Metivier, E. Pardoux
X, 310 pages, 1985.

Vol. 70: Uncertainty and Control
Proceedings of a DFVLR International Colloquium
Bonn, Germany, March, 1985
Edited by J. Ackermann
IV, 236 pages, 1985.

Vol. 71: N. Baba
New Topics in Learning Automata Theory and Applications
VII, 231 pages, 1985.

Vol. 72: A. Isidori
Nonlinear Control Systems: An Introduction
VI, 297 pages, 1985.

Vol. 73: J. Zarzycki
Nonlinear Prediction Ladder-Filters for Higher-Order Stochastic Sequences
V, 132 pages, 1985.

Vol. 74: K. Ichikawa
Control System Design based on Exact Model Matching Techniques
VII, 129 pages, 1985.

Vol. 75: Distributed Parameter Systems
Proceedings of the 2nd International Conference, Vorau, Austria 1984
Edited by F. Kappel, K. Kunisch, W. Schappacher
VIII, 460 pages, 1985.

Vol. 76: Stochastic Programming
Edited by F. Archetti, G. Di Pillo, M. Lucertini
V, 285 pages, 1986.

Vol. 77: Detection of Abrupt Changes in Signals and Dynamical Systems
Edited by M. Basseville, A. Benveniste
X, 373 pages, 1986.

Vol. 78: Stochastic Differential Systems
Proceedings of the 3rd Bad Honnef Conference, June 3–7, 1985
Edited by N. Christopeit, K. Helmes, M. Kohlmann
V, 372 pages, 1986.